Agricultural Methodologies, Practices and Production

Agricultural Methodologies, Practices and Production

Edited by **Laura Vivian**

R CALLISTO
REFERENCE

New York

Published by Callisto Reference,
106 Park Avenue, Suite 200,
New York, NY 10016, USA
www.callistoreference.com

Agricultural Methodologies, Practices and Production
Edited by Laura Vivian

International Standard Book Number: 978-1-63239-654-9 (Hardback)

Printed in the United States of America.

Contents

Preface

Agriculture is a vast field of study that focuses on cultivating, harvesting and storage of food grains and other plant species of economic importance. This discipline has evolved over time to create a scope for animal farming as well. As the climatic conditions as well as soil health have changed, many subfields of agriculture have come into existence such as organic farming, agronomy, agrochemicals, plant breeding, etc. This book covers a wide range of topics such as soil sealing, sustainability assessment, biological degradation, soil quality, differentially-expressed genes, etc. It strives to provide a fair idea about this discipline and to help develop a better understanding of the latest advances within this area. With state-of-the-art inputs by acclaimed experts of this field, this book targets students and professionals. It will serve as an essential guide for anyone who wants to delve deeper into the methods and practices of agriculture.

This book is a comprehensive compilation of works of different researchers from varied parts of the world. It includes valuable experiences of the researchers with the sole objective of providing the readers (learners) with a proper knowledge of the concerned field. This book will be beneficial in evoking inspiration and enhancing the knowledge of the interested readers.

In the end, I would like to extend my heartiest thanks to the authors who worked with great determination on their chapters. I also appreciate the publisher's support in the course of the book. I would also like to deeply acknowledge my family who stood by me as a source of inspiration during the project.

Editor

Biochars Derived from Gasified Feedstocks Increase the Growth and Improve Nutrient Acquisition of *Triticum aestivum* (L.) Grown in Agricultural Alfisols

Kristin M. Trippe *, Stephen M. Griffith, Gary M. Banowetz and Gerald W. Whitaker

US Department of Agriculture Agricultural Research Service National Forage Seed and Production Research Center, Corvallis, OR 97331, USA; E-Mails: smgriffith@me.com (S.M.G.); banowetg@gmail.com (G.M.B.); Jerry.Whittaker@ars.usda.gov (G.W.W.)

* Author to whom correspondence should be addressed; E-Mail: Kristin.Trippe@ARS.USDA.GOV

Academic Editor: Bin Gao

Abstract: Biochars are produced by low-oxygen gasification or pyrolysis of organic waste products, and can be co-produced with energy, achieving waste diversion and delivering a soil amendment that can improve agricultural yields. Although many studies have reported the agronomic benefits of biochars produced from pyrolysis, few have interrogated the ability of gasified biochars to improve crop productivity. An earlier study described the ability of a biochar that was derived from gasified Kentucky bluegrass (KB) seed screenings to impact the chemistry of acidic agricultural soils. However, that study did not measure the effects of the biochar amendment on plant growth or on nutrient acquisition. To quantify these effects we conducted a greenhouse study that evaluated wheat grown in agricultural soils amended with either the KB-based biochar or a biochar derived from a blend of woody mixed-waste. Our studies indicated that biochar amended soils promoted the growth of wheat in these agricultural alfisols. Our elemental analysis indicated that an attenuation of metal toxicity was likely responsible for the increased plant growth. The results of our study are placed in the context of our previous studies that characterized KB-sourced biochar and its effects on soil chemistry.

Keywords: *Triticum aestivum* L.; gasification; biochar; seed screenings; wheat; nutrition; acid soil; Kentucky bluegrass; aluminum toxicity

1. Introduction

The thermo conversion of crop residues has the potential to generate value-added agricultural income from the production of on-farm energy, the capture of process heat, and biochar production [1,2]. Biochar, a highly-persistent form of organic matter, has demonstrated agronomic benefits [3]. However, the agricultural benefits of biochar are difficult to predict because highly variable production methods often result in radically different physiochemical properties, which in turn impact the ability of a biochar to amend soil conditions [4]. Although this complexity has led to a proliferation of biochar characterization studies, generalizable principles and agronomic recommendations have been slow to emerge.

Thermo conversion of crop residues occurs either by pyrolysis, an anerobic combustion process that occurs at temperatures below 700 °C, or gasification, a process that occurs in the presence of controlled oxygen concentrations at higher temperatures [5]. The vast majority of studies that interrogate the physiochemical characteristics and utility of biochars are focused on those produced by pyrolysis, and most studies that characterize biochar generated by gasification are focused on woody source material. Studies that use gasified herbaceous feedstocks have looked at the effects of biochar on soil fertility [6], plant germination [7], and plant growth [8]. However, because studies regarding the effect of gasified herbaceous source materials on plants and soils have only occasionally been reported, and because gasified biochar has a strong effect on soils [4], we were interested in determining the effect of gasified herbaceous source material on plant nutritional status.

Recently our research has been focused on the ability of a farm-scale gasifier to produce bioenergy [9], and biochar [10] from the seed screenings and residual straw [1] from perennial grasses. We previously described the physiochemical properties of the biochar produced from the gasified seed screenings of Kentucky bluegrass (KB) [10]. The KB biochar possessed similar physical and chemical characteristics to biochars sourced from alfalfa stems [11] and other gasified herbaceous feedstocks, including a high ash content, moderate alkalinity, and the absence of detectible metals or PAHs. The impact of KB biochar on the chemistry of agricultural soils was recently described [12]. However, that study did not measure the effects of the biochar amendment on plant growth or on nutrient acquisition. Because the KB biochar was similar to that produced from alfalfa, we hypothesized that the application of this biochar to agricultural soils might evoke a similar plant response. The objectives of this study were to evaluate the effects of two types of biochar amendments on the growth and nutritional status of wheat and to compare different application rates of biochar.

To achieve these objectives, we conducted a replicated greenhouse study that used two agricultural soils, each amended with either the KB-based biochar or a biochar derived from a blend of woody mixed-waste. Our studies indicated that the addition of either herbaceous or woody biochar significantly increased the ability of wheat to grow in these highly acidic and weathered soils. Our analyses indicated that the mechanism for increased plant growth was most likely an attenuation of metal toxicity and an increase in soil fertility provided by increased pH, absorbtion of soil metals onto biochar particles, and increased soil nutrient content.

2. Results

2.1. Plant Growth

The above-ground and below-ground biomass of wheat plants increased when either soil series was amended with either biochar (Figure 1). However, this increase was only sometimes significant (Table 1). In the shoots, the increase in the dry weight was proportional to the biochar amendment rate for each biochar ($p \leq 0.05$). However, wood-sourced biochar amendments out performed KB-biochars in their ability to increase shoot mass. In the roots, wood-sourced char also out performed KB-biochars in their ability to increase biomass. However, this effect was not as pronounced as it was in the above-ground biomass, and was more dramatic in the Bernhill soil series, where root growth was initially limited. Biochar sourced from KB seed screenings did not impact root growth in either soil series. Our collective results indicate that KB-sourced biochar specifically impacted the growth of shoots while wood-sourced biochar had broader impacts on the growth, increasing the growth of both the shoot and the root.

The elemental composition of roots and shoots was altered when soils were amended with biochar (Figures 2–8), however these results were not consistent across soil series or biochar sources. For example, in the shoots we observed that the concentration of Al (Figure 2A) decreased as biochar concentrations increased, however this result was not statistically significant. In the roots, the accumulation of Al decreased when either soil series was amended with either biochar, however, the impact of KB-sourced biochar (Figure 2B) on Al accumulation in the root was more pronounced and statistically relevant.

In general, the effect of biochar source on elemental composition of plant roots was more specific than in plant shoots. In plant roots, both KB- and wood-sourced biochars reduced the accumulation of Na (Figure 3B), but increased the accumulation of K (Figure 4B) and Mg (Figure 5B). The roots of plants grown in soils amended with KB biochars accumulated less Fe (Supplementary Table S1), Mn (Supplementary Table S1), and Ni (Figure 6B), but accumulated more P (Figure 7B). Wood-sourced biochar amendments reduced the accumulation of S and Cu (Supplementary Table S1), but facilitated root accumulation of Ca (Figure 8B).

The effects of the biochar amendments in plant shoots were not as sensitive to the biochar source material as the roots. Plants grown in soils amended with either biochar accumulated more K (Figure 4A), Mg (Figure 5A), and P (Figure 7A) but accumulated less Ni (Figure 6A) and Ca (Figure 8A) in their shoots. Wood-sourced biochar amendments reduced shoot accumulation of S (Supplementary Table S2).

The elemental concentration of roots was not necessarily a reflection of soil concentrations. For example, Ca concentrations increased in soils, but decreased in shoots. Likewise, Na (Figure 3B), Mg (Figure 5b), Fe (Supplementary Table S3), and Ni (Figure 6B) concentrations in the soil remained constant, but concentrations in the roots or shoots decreased. However, the concentration of P and K in amended soils correlated nicely with the increase in roots and shoots.

Table 1. The elemental compositions of the root and shoot (mg/kg) were statistically analyzed using the General Linear Model (A+B+C+D+BC+BD+CD+BCD) that included planned comparisons testing whether the linear and quadratic components were significant. The asterisk indicates estimates that are statistically significant ($p \leq 0.05$).

u	Biochar Type		Soil Type		KB Biochar Concentration		Wood Biochar Concentration		Biochar Type × Soil Source		Soil Type × KB Biochar Concentration		Soil Type × Wood Biochar Concentration		Adj R^2
	Estimate	S. E.	Estimate	S. E.	Estimate	S. E.	Estimate	S. E.	Estimate	S. E.	Estimate	S. E.	Estimate	S. E.	
Root mass	-0.037	0.049	0.361 *	0.049	1.213	0.84	5.02 *	0.84	-0.137	0.069	-0.955	1.188	0.689	1.188	0.7
Shoot mass	0.00	0.074	0.436 *	0.074	4.80 *	1.28	7.21 *	1.28	-0.053	0.105	1.07	1.81	10.8 *	1.81	0.85
Root Al	-430 *	131	141	131	-9411 *	2248	-3950	2248	136	185	-3554	3179	-5337	3179	0.44
Root Ca	288 *	67	50.4	67.1	1272	1150	6495 *	1150	-84.7	94.9	-651	1627	-5532 *	1627	0.6
Root Fe	-315 *	104	258 *	104	-4319 *	1790	56.5	1790	169	148	-4177	2531	-6798 *	2531	0.36
Root K	-2013 *	672	-340	672	39564 *	11528	23358 *	11528	287	950	-12726	16303	-22913	16303	0.44
Root Mg	87.1 *	36	-36.4	36	3445 *	617	4302 *	617	-50.3	50.8	-2226 *	873	-3610 *	873	0.64
Root Na	106	111	-540 *	111	-11172 *	1904	-11498 *	1904	121	157	7717 *	2693	5137	2693	0.53
Root Ni	4.211 *	-0.973	3.628 *	-0.973	-33.575 *	-16.683	-130 *	-16.683	-4.732*	-1.376	-28.125 *	-23.6	30.515	-23.593	0.589
Root P	-405	225	-1162 *	225	22,119 *	3854	501	3854	577	318	-200	5451	-8946	5451	0.69
Root S	-134	92.9	-308 *	92.9	963	1593	-5498 *	1593	-73.7	131	-4510 *	2253	-5413 *	2253	0.72
Root Zn	22.3 *	6.76	-45.7 *	6.76	-199	116	-750 *	116	-16.9	9.56	179	164	457 *	164	0.7
Shoot Al	20.1	28.5	101 *	28.5	-535	488	-396	488	-25.8	40.3	417	690	-604	690	0.3
Shoot Ca	-574 *	240	-636 *	240	-30,750 *	4110	-22,400 *	4110	-51.5	339	14,040 *	5813	9901	5813	0.56
Shoot Fe	14.2	23.3	74.3 *	23.3	-285	400	-15	400	2.76	32.9	366	565	-726	565	0.28
Shoot K	4180 *	1010	584	1010	98,648 *	17,322	74,325 *	17,322	-1967	1428	-38,238	24498	-76,180 *	24,498	0.5
Shoot Mg	158	86.8	-349 *	86.8	4311*	1488	6845 *	1488	-41.6	123	1159	2104	-992	2104	0.6
Shoot Na	-2.06	22.9	-27.4	22.9	-148	393	-21.2	393	-13.3	32.4	201	556	80.5	556	0.03
Shoot Ni	-0.8 *	-0.137	-0.522 *	-0.137	-16.733 *	-2.355	-4.064	-2.355	0.632 *	-0.194	11.908 *	-3.33	-0.98	-3.33	0.44
Shoot P	628 *	223	-622 *	223	11664 *	3825	14426 *	3825	-387	315	9446	5410	-8150	5410	0.56
Shoot S	849 *	162	-788 *	162	3639	2786	-14137 *	2786	-735 *	230	-7475	3940	5278	3940	0.73
Shoot Zn	3.14	3.24	-10.8 *	3.25	18.2	55.7	44.6	55.7	2.58	4.59	120	78.8	-20.3	78.8	0.23

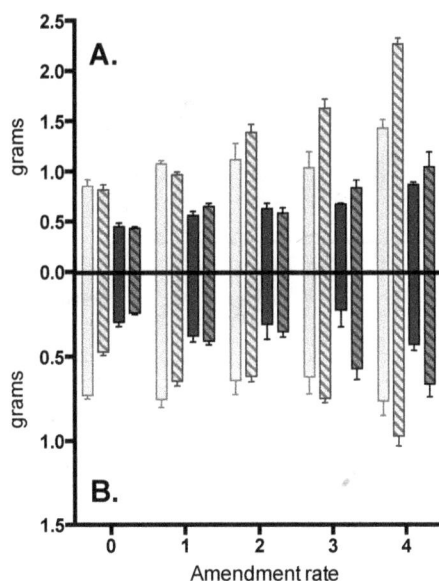

Figure 1. The dry mass of shoots (**A**) and roots (**B**) grown in Freeman (light grey) or Bernhill (dark grey) soils amended with KB- (solid) or wood- (cross-hatched) sourced biochars. Amendment rates are represented as 0–4, which correspond to rates of 0%, 3%, 6%, 16%, and 25% by volume for both biochars, 0%, 0.7%, 1.7%, 3.7%, and 8.6% by mass for wood biochar, and 0%, 0.4%, 1.2%, 2.5%, and 5.8% by mass for KB-sourced biochar.

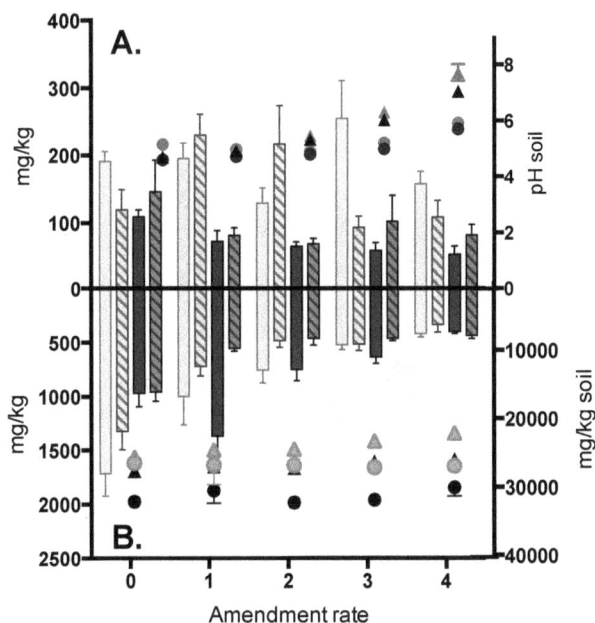

Figure 2. The aluminum concentration of shoots (**A**) and roots (**B**) grown in Freeman (light grey) or Bernhill (dark grey) soils amended with KB- (solid) or wood- (cross-hatched) sourced biochars. The pH of the soils (right axis, (**A**)) and the soil aluminum concentration (right axis, (**B**)) is also plotted for Freeman (light grey) or Bernhill (dark grey) soils amended with KB- (circle) or wood- (triangle) sourced biochars (Trippe 2015). Amendment rates are represented as 0–4, which correspond to rates of 0%, 3%, 6%, 16%, and 25% by volume for both biochars, 0%, 0.7%, 1.7%, 3.7%, and 8.6% by mass for wood biochar, and 0%, 0.4%, 1.2%, 2.5%, and 5.8% by mass for KB-sourced biochar.

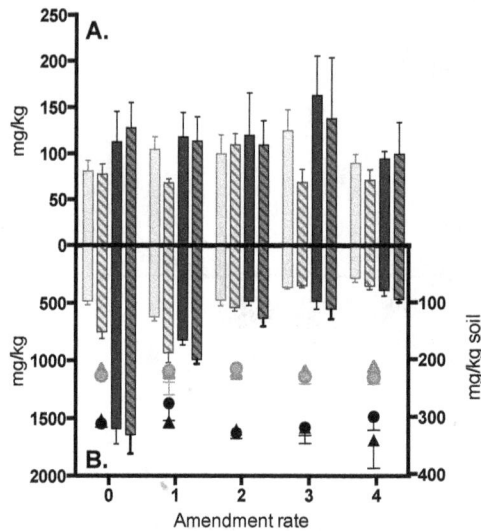

Figure 3. The sodium concentration of shoots (**A**) and roots (**B**) grown in Freeman (light grey) or Bernhill (dark grey) soils amended with KB- (solid) or wood- (cross-hatched) sourced biochars. The sodium of the soils (right axis, (**B**)) is also plotted for Freeman (light grey) or Bernhill (dark grey) soils amended with KB- (circle) or wood- (triangle) sourced biochars. Amendment rates are represented as 0–4, which correspond to rates of 0%, 3%, 6%, 16%, and 25% by volume for both biochars, 0%, 0.7%, 1.7%, 3.7%, and 8.6% by mass for wood biochar, and 0%, 0.4%, 1.2%, 2.5%, and 5.8% by mass for KB-sourced biochar.

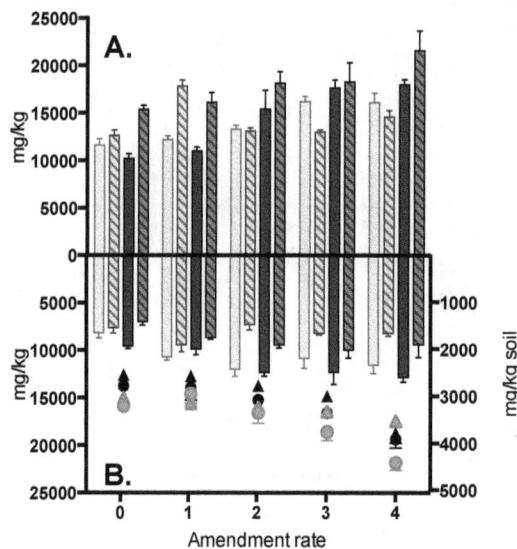

Figure 4. The potassium concentration of shoots (**A**) and roots (**B**) grown in Freeman (light grey) or Bernhill (dark grey) soils amended with KB- (solid) or wood- (cross-hatched) sourced biochars. The potassium of the soils (right axis, (**B**)) is also plotted for Freeman (light grey) or Bernhill (dark grey) soils amended with KB- (circle) or wood- (triangle) sourced biochars (Trippe 2015). Amendment rates are represented as 0–4, which correspond to rates of 0%, 3%, 6%, 16%, and 25% by volume for both biochars, 0%, 0.7%, 1.7%, 3.7%, and 8.6% by mass for wood biochar, and 0%, 0.4%, 1.2%, 2.5%, and 5.8% by mass for KB-sourced biochar.

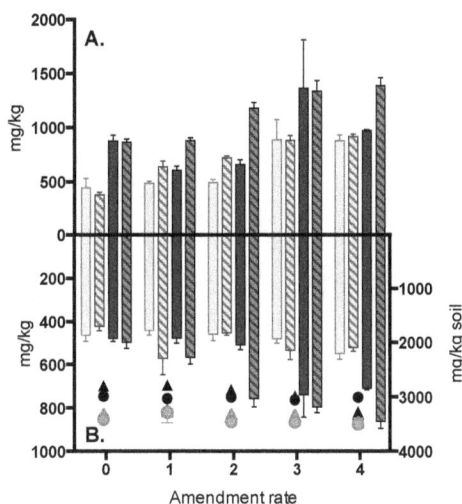

Figure 5. The magnesium concentration of shoots (**A**) and roots (**B**) grown in Freeman (light grey) or Bernhill (dark grey) soils amended with KB- (solid) or wood- (cross-hatched) sourced biochars. The aluminum of the soils (right axis, (**B**)) is also plotted for Freeman (light grey) or Bernhill (dark grey) soils amended with KB- (circle) or wood- (triangle) sourced biochars (Trippe 2015). Amendment rates are represented as 0–4, which correspond to rates of 0%, 3%, 6%, 16%, and 25% by volume for both biochars, 0%, 0.7%, 1.7%, 3.7%, and 8.6% by mass for wood biochar, and 0%, 0.4%, 1.2%, 2.5%, and 5.8% by mass for KB-sourced biochar.

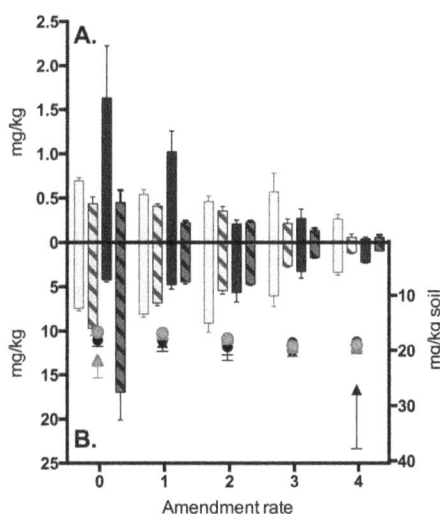

Figure 6. The nickel concentration of shoots (**A**) and roots (**B**) grown in Freeman (light grey) or Bernhill (dark grey) soils amended with KB- (solid) or wood- (cross-hatched) sourced biochars. The nickel of the soils (right axis, (**B**)) is also plotted for Freeman (light grey) or Bernhill (dark grey) soils amended with KB- (circle) or wood- (triangle) sourced biochars. Amendment rates are represented as 0–4, which correspond to rates of 0, 3, 6, 16, and 25% by volume for both biochars, 0%, 0.7%, 1.7%, 3.7%, and 8.6% by mass for wood biochar, and 0%, 0.4%, 1.2%, 2.5%, and 5.8% by mass for KB-sourced biochar.

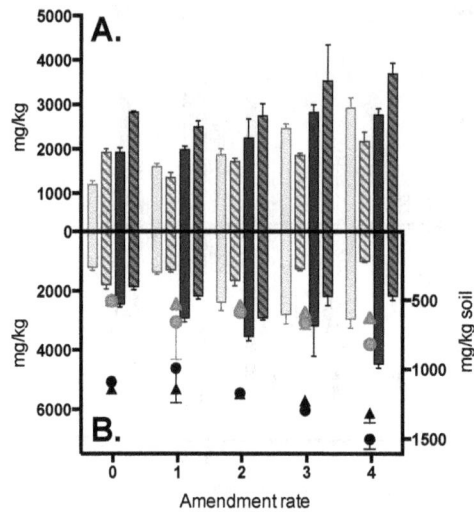

Figure 7. The phosphorus concentration of shoots (**A**) and roots (**B**) grown in Freeman (light grey) or Bernhill (dark grey) soils amended with KB- (solid) or wood- (cross-hatched) sourced biochars. The phosphorus of the soils (right axis, (**B**)) is also plotted for Freeman (light grey) or Bernhill (dark grey) soils amended with KB- (circle) or wood- (triangle) sourced biochars (Trippe 2015). Amendment rates are represented as 0–4, which correspond to rates of 0%, 3%, 6%, 16%, and 25% by volume for both biochars, 0%, 0.7%, 1.7%, 3.7%, and 8.6% by mass for wood biochar, and 0%, 0.4%, 1.2%, 2.5%, and 5.8% by mass for KB-sourced biochar.

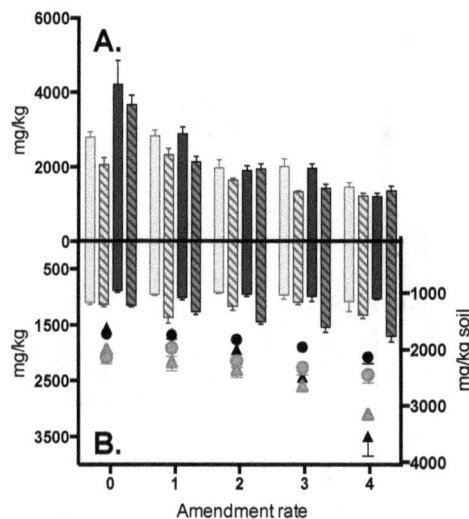

Figure 8. The calcium concentration of shoots (**A**) and roots (**B**) grown in Freeman (light grey) or Bernhill (dark grey) soils amended with KB- (solid) or wood- (cross-hatched) sourced biochars. The calcium of the soils (right axis, (**B**)) is also plotted for Freeman (light grey) or Bernhill (dark grey) soils amended with KB- (circle) or wood- (triangle) sourced biochars (Trippe 2015). Amendment rates are represented as 0–4, which correspond to rates of 0%, 3%, 6%, 16%, and 25% by volume for both biochars, 0%, 0.7%, 1.7%, 3.7%, and 8.6% by mass for wood biochar, and 0%, 0.4%, 1.2%, 2.5%, and 5.8% by mass for KB-sourced biochar.

The general linear statistical model (Table 1) indicated that soil type was correlated with biochar-induced elemental changes in the plants. Elemental concentrations in plants grown in Bernhill soils were more likely to be impacted by biochar amendments. This correlation was especially notable in the elemental changes that occurred in the roots of plants grown in soil amended with wood-sourced biochar. However, the changes that we observed in elemental composition of the shoots did not seem to be strongly correlated to soil series.

3. Discussion

This study compared the effects of two biochars produced either by the gasification of wood or the gasification of KB seed screenings on the growth and nutrient status of wheat. A previous study characterized the effects of these same biochars on soil characteristics, including pH and fertility [12]. However, that study did not report the effects of these biochar amendments on plant growth. The objective of this study was to analyze these effects and to compare them to those observed in the soil study. Within this synthesis, our objective was to draw broader conclusions regarding the ability of the biochar to limit or promote the uptake of toxic metals or growth-promoting nutrients and tease out mechanisms that might lead to increased plant growth.

The previous [12] and the current study were conducted simultaneously in two similar soil series, Freeman and Bernhill. These specific soils were collected from agricultural fields where soil acidity (pH < 5.0) limits crop productivity [13]. Prior to amendment, both the wood- and the KB-sourced biochars had similar pH values (>10), at the highest amendment level, KB-sourced biochar increased the pH of the soil to ~5.7 while the wood-sourced biochar increased the pH even more (pH > 6.4). Overall, our data indicate that the addition of either biochar to either soil increased plant biomass and improved their overall nutrient composition. Although the data indicate that these effects may be due to an increase in soil pH, the addition of biochar also contributed plant nutrients that may have impacted plant growth.

We predicted that the main driver of increased plant growth in this study would be due to a pH-mediated attenuation of Al toxicity in roots. Aluminum toxicity is thought to be a major factor in reduced plant growth in acidic soils [14]. Under experimental conditions, micromolar concentrations of Al severely inhibit root growth and alter the uptake of Ca and P [15]. Although we observed a noticeable reduction of Al concentrations in root tissue (Figure 2), our statistical analysis indicated that only the KB-sourced biochar significantly reduced the accumulation of Al in roots. This observation is interesting because only the highest amendment rate of KB-sourced biochar raised the pH (Figure 2A) beyond the threshold that significantly alters Al bioavailability [16]. This suggests that the KB-sourced biochar reduced Al uptake by another mechanism. Although Al is typically in solution below pH 5.5, once in solution it is able to complex with organic and inorganic compounds. A recent study demonstrated that trivalent Al complexes with rice-straw derived biochars [17], and that the presence of silicon increased surface complexation. Because grasses typically contain more silicon than wood, our observation that KB-sourced biochar is more effective at lowering Al root concentrations than wood-sourced biochar may be attributable to fundamental biochar properties rather than a simple attenuation of pH.

The concentration of Ni in shoots and roots similarly declined. KB-sourced biochar was again more effective at reducing Ni uptake in the roots than wood-derived char, however, both biochars were effective at reducing Ni uptake in the shoots. These results mirrored those observed for Al, again indicating that declining concentrations of Ni are likely due to a combination of increasing pH and other soil or biochar properties. Previous studies have indicated that Ni bioavailability is dependent on soil organic matter [18] and on biochar particle size [19]. Overall, the combinatorial effects of an in increase in pH and intrinsic biochar properties likely led to most of the source-specific elemental changes we observed, including decreased Fe concentrations in roots grown with KB-sourced amendments and increased Zn concentrations in shoots grown in wood-sourced biochar.

In our study, root and shoot concentrations of potassium, phosphorus, and magnesium were enhanced with increasing quantities of biochar, regardless of the source material. These increases were likely due to the concentration of these elements in the biochar itself [12], and not to a change in pH. Concentrations of Ca in shoots dramatically decreased (Figure 8A). This result was surprising because it has long been thought that Al elicits a phytotoxic response by disrupting Ca homeostasis [15]. Therefore, we predicted that alleviating Al toxicity would lead to an increase in root and shoot Ca concentrations. Although Ca levels decreased, even at the highest amendment rate, Ca concentrations were well above the critical deficiency threshold [20].

Sodium is not an essential nutrient for plant growth, and often causes phytotoxicity by disrupting potassium homeostasis [21]. Our results indicated that while Na concentrations stayed relatively constant in our amended soils, the concentration of Na in the roots dramatically decreased, suggesting that each biochar was able to sequester soil Na.

An extensive meta-analysis of 371 biochar studies recently concluded that biochar amendments generally improve plant productivity, soil fertility, and nutrient content [3]. The results of the current study echo their conclusions, and reiterate that interactions between soil type and biochar properties are critical in determining whether biochar will have beneficial effects on plant growth. The final pH of the soil:biochar milieu will likely impact the effects of the amendment on plant growth, but it seems clear that additional soil and biochar characteristics influence whether plant productivity is impacted. The KB-sourced biochar used in this study has been fairly well-characterized in terms of its production [9], chemical characteristics [10], influence on soil fertility and nutrient content [12], and growth-promoting potential. Future studies that focus on the longevity of plant growth promoting properties and the economic impacts of biochar production will ultimately determine the consequences and potential for this char as an agricultural amendment prior to its widespread application.

4. Experimental Section

4.1. Experimental

Single plants of wheat (*Triticum aestivum* L. cv. Madsen, a commonly planted cultivar in the state of Washington, U.S.A.) were grown to a Feekes stage 5 (74 days) in 650 cm^3 black plastic pots, containing either a Freeman or Bernhill soil with varying proportions of KB or wood biochar (0%, 3%, 6%, 16%, and 25% by volume). Due to slight differences in of KB and wood biochar densities, the final mass concentration of KB biochar to soil was slightly different. In the soils amended with KB biochar, the highest amendment rate by mass was 5.8% and 8.6% with wood-based biochar. For simplicity, amendment rates in figures are represented as 0–4, which correspond to rates of 0%, 3%, 6%, 16%, and 25% by volume for both biochars, 0%, 0.7%, 1.7%, 3.7%, and 8.6% by mass for wood biochar, and 0%, 0.4%, 1.2%, 2.5%, and 5.8% by mass for KB-sourced biochar. Plants were grown in a greenhouse at Corvallis, OR, U.S.A. with supplemental lighting using high-pressure sodium lamps, with an average quantum flux density of 150 $\mu E\ m^{-2}s^{-1}$ and extended day length to 16 h. Average daily maximum and minimum temperature was 21.1 and 15.6 °C, respectively. Plants were watered twice a week for 74 d. During the duration of the study, no supplemental nutrients were added to the treatments other than what was present initially in the soil and biochar.

The two biochars used in this study were Class 2 biochars as defined by the International Biochar Initiative [22]. The KB biochar was produced using a small-scale gasification unit located on a farm near Rockford, WA, U.S.A. [9]. The stainless steel reactor of 0.45 m^3 was air-blown at a rate of 3 to 3.5 $m^3\ min^{-1}$ and operated at temperatures ranging from 650 to 750 °C at a feed rate of 60 to 80 kg h^{-1}. The KB source material was obtained during a post-harvest seed cleaning procedure and consisted largely of small Kentucky bluegrass straw components, immature seeds, and remnant seed coats. The moisture content of the KB was 14% and the predominant particles in the feedstock were less than 5 cm in length. An extensive characterization of KB biochar was previously reported by Griffith *et al.* [10]. The second biochar was produced from gasification of chipped residues of a Douglas-fir (*Pseudotsuga menziesii* (Mirb.) Franco), Ponderosa pine (*Pinus ponderosa* C. Lawson), white fir (*Abies concolor* (Gord. & Glend.) Lindl. ex Hildebr.), sugar pine (*Pinus lambertiana* Douglas), and incense cedar (*Calocedrus decurrens* (Torr.) Florin) harvest. The wood biochar was produced using a small-scale down-draft gasification unit that was air-blown to produce gas at about 170 $m^2\ h^{-1}$ and operated at maximum char temperatures ranging from 1100 to 1400 °C at a feed rate of about 70 kg h^{-1}. The moisture content of the woody source material was 9% to 15%. Elemental and proximate analyses of these biochars were previously published [10,12].

The two cultivated farm soils used in this study were obtained near Rockford, WA, U.S.A. from two different fields. The Freeman soil was collected at Lat. 47.506539 N, Long. −117.096256 W and the Bernhill soil at Lat. 47.51686 N, Long. −117.068213 W. The Freeman soil series is an Aquandic Palexeralf [23] with very deep, moderately well drained soil formed in loess with a minor amount of volcanic ash mixed in the surface. Freeman soils are found on undulating to rolling loess hills on slopes of 0% to 30%. The Bernhill series consists of very deep, well-drained soils formed in glacial till with a component of loess and volcanic ash at the surface. Bernhill soils are Vitrandic Haploxeralfs [23] on glaciated foothills and glaciolacustrine influenced ground moraines with slopes of 0% to 65%. The two soils were selected in an area predominated by Kentucky bluegrass seed and wheat production and a farming area with historically acidic soils and other nutrient deficiencies [13].

4.2. Plant Analysis

After 74 d from the time of planting, wheat shoots and roots were separated, dried at 40 °C for 24 h, dry mass recorded, and then ground using a Cyclotec 1093 Sample Mill (Tecator, Hoganas, Sweden) fitted with 1 mm^2 screen. Roots were quickly rinsed to remove all adhering soil prior to drying. Plant material was dried at 40 °C, then ground and passed through a 2-mm sieve. We analyzed the total concentrations in ground shoot and root tissue Ca, Cu, Fe, K, Mg, Mn, Na, P, S, and Zn, essential plant macro- and micronutrients, and Al, As, Cd, Na, and Ni, which are potentially toxic. These analyses were performed as described in Milestone Digestion Application Report #03-001. This procedure involved an initial sample digestion in a Milestone Ethos D microwave (Milestone Inc., Shelton, CT) with HNO_3 and HCl, followed by inductively coupled plasma atomic emission spectrometry (Perkin-Elmer Optima 46300 DV ICP-OES) analysis. Total C and N were quantified using a C/N combustion analyzer (LECO TruSpec, St. Joseph, MI).

4.3. Soil Analysis

The effect of biochar additions on pH and on soil concentrations of Al, Ca, K, Mg, and P were previously published [12], but were collected during the greenhouse study described above. The concentrations of Ni, Fe, Na, and Zn were also measured during that study, however the concentrations of these elements were not reported. The soil concentrations of each of these elements are plotted here to provide context.

4.4. Statistical Analysis

This study used a completely randomized block design of five blocks with treatments randomized within each block and block identities were retained in the greenhouse through the duration of the experiment. Blocks were entered in the model as random effects and soil type, biochar type, and biochar concentration were treated as fixed effects. Data were statistically analyzed using the General Linear Model (A+B+C+D+BC+BD+CD+BCD) function in the R basic statistics package [24] that included planned comparisons testing whether the linear, quadratic, cubic and quartic components were significant. The results were analyzed as linear models with nested fixed effects, where biochar amendment rate was nested within biochar type. The fixed effects were biochar type, soil type, amendment rate within biochar type, biochar type × soil type interaction, and biochar concentration × soil type interaction within biochar type.

5. Conclusion

This study compared the ability of biochars produced by the gasification of readily available local feedstocks to impact the growth and nutrient status of wheat. To quantify these effects we conducted a greenhouse study that evaluated these parameters in wheat grown in highly weathered agricultural soils amended with either KB or wood-sourced biochar. Our studies indicated that biochar amended soils promoted the growth of wheat. Our elemental analysis indicated that this growth was likely attributed to an attenuation of metal toxicity and an increase in pH. However, the ability of the biochars to raise soil pH was not solely responsible for the change in metal accumulation in the roots and the shoots, and that other intrinsic biochar properties also contributed to metal sequestration. Likewise, each of these biochars provided plants with essential plant nutrients that are typically limited in these agricultural alfisols.

Acknowledgments

We thank Jennifer Young for her technical assistance, David Gady and Tom Jobson for the production of the biochars, and Mark Johnson for his insightful comments. Reference to trade names does not imply endorsement by the U.S. Government. USDA is an equal opportunity provider and employer. Mention of trade names or commercial products in this publication is solely for the purpose of providing specific information and does not imply recommendation or endorsement by the U.S. Department of Agriculture.

Conflict of Interest

The authors declare no conflict of interest.

References

1. Banowetz, G.M.; Boateng, A.; Steiner, J.J.; Griffith, S.M.; Sethi, V.; El-Nashaar, H. Assessment of straw biomass feedstock resources in the Pacific Northwest. *Biomass Bioenergy* **2008**, *32*, 629–634. doi:10.1016/j.biombioe.2007.12.014.

2. Mueller-Warrant, G.W.; Banowetz, G.M.; Whittaker, G.W. Geospatial identification of optimal straw-to-energy conversion sites in the Pacific Northwest. *Biofuels Bioprod. Biorefining* **2010**, *4*, 385–407.

3. Biederman, L.A.; Stanley Harpole, W. Biochar and its effects on plant productivity and nutrient cycling: A meta-analysis. *GCB Bioenergy* **2013**, *5*, 202–214, doi:10.1111/gcbb.12037.

4. Novak, J.M.; Busscher, W.J.; Laird, D.L.; Ahmedna, M.; Watts, D.W.; Niandou, M.A.S. Impact of Biochar Amendment on Fertility of a Southeastern Coastal Plain Soil. *Soil Sci.* **2009**, *174*, 105–112, doi:10.1097/SS.0b013e3181981d9a.

5. Lehmann, J.; Joseph, S. *Biochar for Environmental Management: Science and Technology*; Earthscan: Sterling, VA, USA, 2012; pp. 127–146.

6. Maestrini, B.; Herrmann, A.M.; Nannipieri, P.; Schmidt, M.W.I.; Abiven, S. Ryegrass-derived pyrogenic organic matter changes organic carbon and nitrogen mineralization in a temperate forest soil. *Soil Biol. Biochem.* **2014**, *69*, 291–301, doi:10.1016/j.soilbio.2013.11.013.

7. Rogovska, N.; Laird, D.; Cruse, R.M.; Trabue, S.; Heaton, E. Germination tests for assessing biochar quality. *J. Environ. Qual.* **2012**, *41*, 1014–1022.

8. Mozaffari, M.; Russelle, M.P.; Rosen, C.J.; Nater, E.A. Nutrient supply and neutralizing value of alfalfa stem gasification ash. *Soil Sci. Soc. Am. J.* **2002**, *66*, 171–178.

9. Banowetz, G.M.; El-Nashaar, H.; Steiner, J.J.; Gady, D. Non-Densified Biomass Gasification Method and Apparatus. US20110220846 A1, 15 September 2011.

10. Griffith, S.M.; Banowetz, G.M.; Gady, D. Chemical characterization of chars developed from thermochemical treatment of Kentucky bluegrass seed screenings. *Chemosphere* **2013**, *92*, 1275–1279, doi:10.1016/j.chemosphere.2013.02.002.

11. Mozaffari, M.; Rosen, C.J.; Russelle, M.P.; Nater, E.A. Chemical characterization of ash from gasification of alfalfa stems: Implications for ash management. *J. Environ. Qual.* **2000**, *29*, 963–972.

12. Trippe, K.M.; Griffith, S.M.; Banowetz, G.M.; Whitaker, G.W. Changes in Soil Chemistry following Wood and Grass Biochar Amendments to an Acidic Agricultural Production Soil. *Agron. J.* **2015**, doi:10.2134/agronj14.0593.

13. Koenig, R.; Schroeder, K.; Carter, A.; Pumphery, M.; Paulitz, T.; Campbell, K.; Huggins, D. Soil acidity and aluminum toxicity in the Palouse Region of the Pacific Northwest. Available online: http://cru.cahe.wsu.edu/CEPublications/FS050E/FS050E.pdf (accessed on 15 July 2015).

14. Ryan, P.R.; Kochian, L.V. Interaction between aluminum toxicity and calcium uptake at the root apex in near-isogenic lines of wheat (*Triticum aestivum* L.) differing in aluminum tolerance. *Plant Physiol.* **1993**, *102*, 975–982.

15. Kochian, L.V. Cellular mechanisms of aluminum toxicity and resistance in plants. *Annu. Rev. Plant Biol.* **1995**, *46*, 237–260.

16. Sparling, D.W.; Lowe, T.P. *Environmental Hazards of Aluminum to Plants, Invertebrates, Fish, and Wildlife*; Springer: New York, NY, USA, 1996; pp. 1–127.

17. Qian, L.; Chen, B. Interactions of aluminum with biochars and oxidized biochars: Implications for the biochar aging process. *J. Agric. Food Chem.* **2014**, *62*, 373–380, doi:10.1021/jf404624h.

18. Weng, L.P.; Wolthoorn, A.; Lexmond, T.M.; Temminghoff, E.J.M.; van Riemsdijk, W.H. Understanding the Effects of Soil Characteristics on Phytotoxity and Bioavailability of Nickel Using Speciation Models. *Environ. Sci. Technol.* **2004**, *38*, 156–162, doi:10.1021/es030053r.

19. Rees, F.; Simonnot, M.O.; Morel, J.L. Short-term effects of biochar on soil heavy metal mobility are controlled by intra-particle diffusion and soil pH increase. *Eur. J. Soil Sci.* **2014**, *65*, 149–161, doi:10.1111/ejss.12107.

20. Genc, Y.; Tester, M.; McDonald, G.K. Calcium requirement of wheat in saline and non-saline conditions. *Plant Soil* **2010**, *327*, 331–345, doi:10.1007/s11104-009-0057-3.

21. Kronzucker, H.J.; Coskun, D.; Schulze, L.M.; Wong, J.R.; Britto, D.T. Sodium as nutrient and toxicant. *Plant Soil* **2013**, *369*, 1–23, doi:10.1007/s11104-013-1801-2.

22. Initiative, I.B. Standardized product definition and product testing guidelines for biochar that is used in soil. *IBI biochar Stand.* **2012**. Availiable online: http://www.biochar-international.org/sites/default/files/IBI_Biochar_Standards_V2%200_final_2014.pdf (accessed on 14 August 2015).

23. USDA-NRCS. Soil Series of Spokane County. *Official Soil Series Descriptions*; USDA-NRCS, United States Department of Agriculture: Lincoln, NE, USA, 2006. Available online: http://www.nrcs.usda.gov/Internet/FSE_MANUSCRIPTS/washington/spokaneWA1968/spokaneWA1968.pdf. (accessed on 14 August 2015).

24. R Development Core Team. *R: A Language and Environment for Statistical Computing*; R Foundation for Statistical Computing: Vienna, Austria, 2014.

Estimated Fresh Produce Shrink and Food Loss in U.S. Supermarkets

Jean C. Buzby [1,*]**, Jeanine T. Bentley** [1]**, Beth Padera** [2]**, Cara Ammon** [3] **and Jennifer Campuzano** [4]

[1] Economic Research Service, U.S. Department of Agriculture, 1400 Independence Ave., Mail Stop 1800, SW, Washington, DC 20250-1800, USA; E-Mail: jbentley@ers.usda.gov

[2] MobiSave, 712 5th Avenue, 14th Floor, New York, NY 10019, USA; E-Mail: BethPadera@hotmail.com

[3] Beacon Research Solutions, 4556 N. Beacon St. No. 3, Chicago, IL 60640, USA; E-Mail: Cara@beaconresearchsolutions.com

[4] Nielsen Perishables Group Inc., 1700 West Irving Park Road, Suite 310, Chicago, IL 60613, USA; E-Mail: Jennifer.Campuzano@nielsen.com

* Author to whom correspondence should be addressed; E-Mail: jbuzby@ers.usda.gov

Academic Editor: Michael Blanke

Abstract: Data on fresh fruit and vegetable shrink in supermarkets is important to help understand where and how much shrink could potentially be reduced by supermarkets to increase their profitability. This study provides: (1) shrink estimates for 24 fresh fruits and 31 fresh vegetables in U.S. supermarkets in 2011 and 2012; and (2) retail-level food loss. For each covered commodity, supplier shipment data was aggregated from a sample of 2900 stores from one national and four regional supermarket retailers in the United States, and this sum was then compared with aggregated point-of-sale data from the same stores to estimate the amount of shrink by weight and shrink rates. The 2011–2012 average annual shrink rates for individual fresh vegetables varied from 2.2 percent for sweet corn to 62.9 percent for turnip greens and for individual fresh fruit ranged from 4.1 percent for bananas to 43.1 percent for papayas. When these shrink estimates were used in the Loss-Adjusted Food Availability data series, annual food loss for these commodities totaled 5.9 billion pounds of fresh fruit and 6.1 billion pounds of fresh vegetables. This study extends the literature by providing important information on where and how much shrink could potentially be reduced. Precise comparisons across studies are difficult. This information, combined with information on available and cost-effective technologies and practices, may help supermarkets target food loss reduction efforts though food loss will never be zero.

Keywords: food loss; food waste; fruit; retail; shrink; supermarket; vegetables

1. Introduction

Understanding the amount of fresh fruits and vegetables that goes unsold in supermarkets is important to understanding the extent that food loss reductions could be made to increase a supermarket's profitability. This amount could also be considered the upper bound on the potential amount of produce that could be donated to food banks and other hunger-relief organizations. *Shrinkage* is a term that sometimes is used for wholesale and retail losses [1]. *Shrink* is the shorthand version of shrinkage and is defined here as the produce that is delivered into supermarkets for sale but is not sold for any reason. Shrink includes both the edible and inedible portions of food (e.g., peels, pits).

Taking a broader view of all food and non-food items sold in U.S. supermarkets, the Food Marketing Institute (FMI) and the Retail Control Group analyzed the causes or drivers for shrink and estimate that 64 percent of shrink in 2012 was caused by operational breakdowns and 36 percent was caused by theft and misdeeds [2]. These estimates are in terms of retail sales in U.S. dollars. The study size was 64 supermarket companies/chains in the United States [3]. The *operations category* (*i.e.*, operations breakdowns) includes loss due to ordering inefficiencies, production planning, product handling errors, employee/cashier errors, poor rotation, damaged/unsaleable goods, scanning errors by cashier, and accounting errors. The *theft category* includes loss due to shoplifting, cashier theft, general employee theft, and vendor theft [2] but "these are all high ticket items that are also in demand. Items like: razor blades, family planning items, high end creams, tobacco, liquor, *etc.*" [3]. The shrink for produce departments was the second-highest shrink by department (after the meat department) contributing to 16 percent of total store shrink. The average total shrink for produce departments was 4.8 percent of retail sales. Theft is a problem for the meat department and the relatively high prices per pound for meat department items compared to the prices per pound of foods items in other departments (e.g., items in the produce department) may make shrink look relatively worse for meat department items when calculated in terms of retail sales. The shares of these estimates specifically for food loss of produce, in general, and for damaged/unsaleable produce were not specified.

Food loss is a much narrower concept than shrink. The U.S. Department of Agriculture's Economic Research Service (USDA/ERS) defines *food loss* as the edible amount of food, postharvest, that is available for human consumption but is not consumed for any reason. It includes cooking loss and natural moisture loss; loss from mold, pests, or inadequate climate control; and food waste. ERS uses the term food loss since it represents a broader category of components in the data than the term food waste and because food waste is a value-laden term—a term that doesn't reflect the remaining economic value in some products (e.g., value for feeding animals, composting, or to make biofuels). Because retail-level food loss estimates are lacking in the United States for all of the individual commodities in the ERS Loss-Adjusted Food Availability data (LAFA) series, ERS uses supermarket shrink estimates (*i.e.*, rates) for individual fresh fruits, vegetables, meat, poultry, and seafood to represent food loss at the retail level for these foods [4]. The primary purpose of this data series is to provide estimates of food availability as a proxy for consumption in the United States for roughly 215 commodities. The LAFA series is also used to estimate the amount, value, and calories of food loss at the retail and consumer levels in the United States [5–7]. Currently, ERS uses 2005–06 shrink estimates for 24 individual fresh fruits and 31 individual fresh vegetables in U.S. supermarkets in the LAFA data series documented in Buzby *et al.* [8] (Buzby *et al.* (2009) predominately use the term "food loss" because the shrink estimates are used as food loss assumptions in LAFA but here, we use the term *shrink* because it is the more precise term for what the current data collection captures). This study found that annual supermarket shrink in the United States for 2005 and 2006 averaged 11.4 percent for fresh fruit and 9.7 percent for fresh vegetables by weight (*i.e.*, not value of retail sales). The primary purpose of the current study is to help determine which fresh fruits and vegetables have the highest amount of food loss and thus would be potential candidates for loss-reducing strategies (*i.e.*, technologies and practices, such as better inventory management systems). The shrink estimates for these fresh fruits and vegetables may be used in the future by ERS to update the underlying 2005–2006 retail-level loss assumptions in the LAFA data series.

Reliable U.S. food loss data for each fresh fruit and vegetable in LAFA at the retail level is not available and thus supermarket shrink for these commodities is used as a proxy. We recognize that using produce shrink estimates as a proxy for food loss is a limitation to some unquantified extent. Shrink is a broader category than food loss as described above and includes several components that have little relevance to food availability, loss, or waste (e.g., theft). Thus, shrink is an upper bound for food loss. Given the perishable nature of fresh fruits and vegetables, we presume that food loss is a large component of supermarket shrink for these items and that supermarket shrink for fresh produce is an adequate proxy for food loss of fresh produce items at the retail level. Supporting this notion is the finding that the meat department and the produce department were the two departments found to have the highest shrink in the FMI study—and that these and perhaps the dairy, bakery, and deli departments are those departments with the greatest relative perishability of all supermarket or retail departments leading us to assume that food loss is a major component for these departments. In the future, greater information about shrink, such as the ultimate destination of product counted as shrink, would help clarify the appropriateness of shrink as a proxy for food loss. For example, theft of fresh fruit and vegetables may ultimately lead to the consumption of these products so stolen product would not be considered food loss. However, with that said, produce is not a high ticket item, relative to meat, poultry, and seafood and many other non-food products in supermarkets: the implication is that theft is likely not a notable issue for fresh produce.

1.1. Aim and Scope of This Study

This article provides two kinds of data and information on fresh fruit and vegetables in U.S. supermarkets. First, this article provides estimates of the percent or share of *shrink* for 24 individual fresh fruit and 31 individual fresh vegetables in U.S. supermarkets in 2011 and 2012. Second, this article provides ERS estimates of the amount of retail-level food loss for these fresh fruits and vegetables using the new shrink estimates in the LAFA data series.

1.2. Definitions and the Difficulty Comparing Findings across Studies

Definitions of food loss and waste vary among studies worldwide and this complicates the comparison of estimates and the identification of trends [5]. Recognizing the variation in food loss and waste definitions (and sometimes the same term has different meanings) and that there are benefits to a harmonized approach, the United Nation's Food and Agriculture Organization (FAO) Global Initiative on Food Loss and Waste Reduction published a 2014 working paper that provides a draft definitional framework for food loss as a global reference for stakeholders [9]. This 2014 FAO working paper defines *food loss* as the "decrease in quantity or quality of food" where quantitative reductions are decreases in mass/weight (p. 3) and qualitative reductions are decreases in "nutritional value, economic value, food safety, and/or consumer appreciation" [9] (p. 2). This FAO working paper considers food waste an important although not sharply distinguished component of food loss and indicates that *food waste* "refers to the removal from the FSC (food supply chain) of food which is fit for consumption, by choice, or which has been left to spoil or expire as a result of negligence by the actor–predominantly, but not exclusively the final consumer at household level" [9] (p. 4). Previously, FAO distinguished food loss as occurring from the producer through manufacturing/processing and food waste as discards later in the food supply chain (*i.e.*, from retail to consumption (e.g., [10] citing [11]). Another study defines food waste as "any solid or liquid food substance, raw or cooked, which is discarded, or intended or required to be discarded. Food waste includes the organic residues (such as carrot or potato peels) generated by the processing, handling, storage, sale, preparation, cooking, and serving of foods" [12]. The current study focuses solely on the quantitative loss in terms of pounds and percent shrink. It does not provide information on the value of shrink and food loss, the shrink percentages in terms of sales or calories, or the natural and monetary depreciation of fresh produce over time.

In addition to the lack of common definitions, there are other factors which make it difficult to precisely and meaningfully compare the data and information across studies. Not only are there different definitions of the measured variable (e.g., shrink, food loss, and food waste) but studies may also use different reference bases (e.g., volume of sales *vs.* food supply values *vs.* quantities or weight delivered; edible *vs.* non-edible food), and different areas of coverage (e.g., stages in the farm-to-fork chain, such as at the farm, retail, or consumer levels, or the specific fruits, vegetables, and mixtures covered) in the analyses. For example, Kader [13] provides loss estimates for fruits and vegetables "between production and consumption sites" and not specifically at supermarkets.

Additionally, data in other studies may not be sufficiently disaggregated for comparison with the estimates for individual fruits and vegetables provided here. For example, Nahman and De Lange [14] provide estimates of the amount of food waste along the value chain in South Africa but do not provide loss estimates for individual foods, only food groups (e.g., fruits and vegetables combined). Additionally, Nahman and de Lange provide estimates for the 'distribution' stage and not for the retail level specifically. However, some studies delve deeper into individual commodities and provide a wealth of data and technical information, such as for papaya loss by Paull *et al.* [1].

The Waste Resources Action Programme (WRAP) in the United Kingdom analyzes food and drink waste along the food supply chain. Some WRAP studies place food losses in three categories: unavoidable, possibly avoidable, and avoidable waste [15] while other WRAP studies use two categories (*i.e.*, unavoidable and avoidable) [16]. Some other studies use these or similar categories, such as

Beretta *et al.* [17], which use unavoidable, possibly avoidable, and avoidable "losses". Even if it was possible to compare WRAP's unavoidable category with the non-edible portions (e.g., peach pits, bones, and peels) in the LAFA data series, U.S. data and information is lacking to disaggregate the U.S. food loss estimates into possibly avoidable and avoidable loss or waste. Additionally, food loss comparisons across studies can be hindered when one study includes multi-ingredient foods (e.g., lasagna) and another only includes individual commodities (e.g., wheat flour, beef, and eggs). Also, studies may handle the inclusion of the packaging weight differently.

1.3. The Causes of Produce Loss and the Retail Environment

As background, Box 1 provides examples of the causes or drivers for the narrower categories of food loss and waste in developed countries at the retail level (e.g., in supermarkets). Some of these may occur at more than one stage of the farm-to-fork chain (e.g., spoilage can occur at each stage) and many are similar across developed countries (e.g., consumer confusion over date labeling). Whereas, some causes have greater variation, such as the socio-demographic characteristics and cultural traditions manifested through individual behavior [5]. For example, one study found that Spanish retail managers did not perceive food waste as an important problem, whereas managers in the United Kingdom were more conscious and concerned about food waste [18].

Box 1. Causes or drivers for food loss and waste at the retail level in developed countries.

- Consumption or damage by insects, rodents, or microbes (e.g., molds, bacteria) [19],
- Un-purchased holiday foods,
- Damage due to poor or excessive handling (e.g., bruised fruit due to rough handling) and spillages (e.g., apples accidentally dropped to the floor),
- Dented cans and damaged packaging (e.g., crushed containers of blueberries).
- Damage from inappropriate packaging (e.g., abrasion which bruises or otherwise damages produce) [11],
- Produce packaging (*i.e.*, if some fresh apples within a bag are rotten and cannot be removed, the whole bag may be discarded),
- Product-specific issues that decrease the longevity of certain foods, such as exposure to light leading to in-store food waste [20],
- Supply chain inefficiencies, such as poor coordination between manufactures, distributors, wholesalers, and retailers leading to food waste and potentially shifting food waste across the supply chain [20],
- Damage from equipment or technical malfunction (e.g., faulty cold or cool storage) [19],
- Damage from excessive or insufficient temperature (e.g., inadequate cold or cool storage or when the temperature settings are inappropriate) [19],
- Sprouting (e.g., tubers) and biological aging in fruit, [19], vegetables, and other fresh foods.
- Seasonal factors, such as heat damage during harvest, transportation, or display near the retail store doors where fresh produce is more exposed to outside temperatures,
- Out of shelf life product [18] in depot (e.g., imported produce expired in transit) or in store (e.g., expired packaged salads) [21]

- Difficulty predicting number of buyers/customers leading to overstocking or over-preparing,
- Insufficient shelf space available [18],
- Consumer confusion over "use-by" and "best before" dates and other date labeling [20], which can result in the discard of edible, within-date, packaged food. (See [22] for more information on date labeling.)
- Marketing standards [20] and out-grading of blemished, misshapen, or wrong-sized foods by retail stores in an attempt to meet consumer demand for high quality, cosmetically-appealing, and convenient foods,
- Consumer demand for high cosmetic standards leading to produce going unsold,
- Some promotions by retailers can cause food waste in retail stores due to reduced forecasting accuracy [18] or later at the consumer level if more food with short-shelf life is purchased than can be eaten (e.g., buy one get one free promotions),
- Unsold fragile produce at the end of the day.

Source: Constructed by authors.

Shrink to retailers means lost revenue and decreased profitability. Therefore, retailers in developed countries routinely incorporate strategies to reduce shrink, food loss and waste. For example, Saucede *et al.* [23] found that department upkeep (*i.e.*, state of displays experienced by shoppers) and shrinkage control are the two important variables to optimize the performance of retail produce departments in France. Shrinkage control in Saucede *et al.* [23] mostly refers to internal actions by retailers, such as retailers removing produce damaged by shoppers' sorting and handling when making their selections so that the department displays remain attractive to shoppers. The handling of produce and sometimes destructive testing for ripeness at the grading stage or by consumers can accelerate product deterioration and lead to high levels of waste. For example, avocadoes can be damaged when tested for ripeness [24]. Vigneault *et al.* provide greater background on temperature, humidity, and other factors affecting produce quality and systems to preserve quality during transportation, many of which have implications later in the supply chain [25]. Tight inventory control (*i.e.*, stock management) and good coordination between manufacturers, wholesalers, distributors, and retailers can help reduce food waste [20]. Mena *et al.* [18] provide a more complete description of trends in the marketplace, natural causes of food waste, and management-related causes that can be acted upon to reduce spoilage (e.g., training retail staff about good practices for stock rotation and handling).

It is important to understand that some level of food loss may be economically justifiable and it is unrealistic to think that food loss will ever be entirely eliminated [7] and the same could be said more specifically for supermarket shrink in produce. Technical factors (e.g., perishable nature of most foods), spatial factors (e.g., time and distance to food markets or food banks), and economic factors limit how much food supermarkets and other food retailers can realistically reduce or prevent, recover and donate for human consumption, or reuse for an alternative purpose (e.g., composting, energy generation). Some loss is inevitable because spoiled, deteriorated, or potentially contaminated food must be discarded to ensure the safety and wholesomeness of the food supply. For example, at the end of the day, any leftover cut-up fresh fruit offered as samples to supermarket shoppers are appropriately discarded out of health considerations. Also, some meat, poultry, and other foods are recalled when there are health or safety concerns. Additionally, supermarkets may see a decrease in their marginal or incremental returns or benefit of implementing additional food-loss reducing technologies and practices (*i.e.*, diminishing returns). Supermarkets will adopt new food-loss reducing technologies and practices when it makes economic sense to do so.

1.4. Earlier Studies on Food Loss and Waste at the Retail Level

There are few peer-reviewed articles on national food loss and waste for developed countries and the estimates that do exist vary widely [11]. This is particularly true for articles on food loss or waste generated on farms [26] and at the food manufacturer/retail level more broadly [18,27] and for the subset of articles on national fresh produce loss and waste at the retail level. There are also data gaps by individual causes of food loss and waste. The amount of food loss and waste worldwide is substantial. Gustavsson *et al.* estimate that global food loss and waste tally about 1.3 billion tons per year or about one-third of all food produced for human consumption [10]. According to a food waste study for the European Commission, food waste generated in the EU27 totaled around 89 million mt in 2006, using EUROSTAT data and available best estimates by member states [20]. Of this amount, 3.8 million mt were for the wholesale/retail sector and the remainder was for three other sectors (manufacturing, food service/catering, and consumer). (Retail estimates were not provided separately from wholesale.) In another study on the EU27, estimated food loss and waste in the wholesale/retail sector was 3.6 percent of the total losses in the supply chain but data was limited for this sector so the authors emphasize that this estimate should be treated with caution [28]. In the United States, some studies (such as those by ERS) have focused on food loss as defined previously (*i.e.*, which is broader than food waste). For example, a recent ERS study found that in the United States, food loss accounted for 31 percent—or 133 billion pounds—of the 430 billion pounds of the available food supply at the retail and consumer levels in 2010, and of this amount, retail-level losses represented 10 percent (43 billion pounds) [7].

Another study in the United States narrowed the focus to food waste. The Food Waste Reduction Alliance contracted with BSR to conduct an analysis of food waste in the United States among food manufacturers, retailers, and wholesalers (The Food Waste Reduction Alliance (FWRA) is an industry-wide initiative in the United States to address food waste in food manufacturing, retail stores, and restaurants. The Grocery Manufacturers Association (GMA), the Food Marketing Institute (FMI), and the National Restaurant Association (NRA) lead this effort). This study defined food waste as "any solid or liquid food substance, raw or cooked, which is discarded, or intended or required to be discarded. Food waste includes the organic residues (such as carrot or potato peels) generated by the processing, handling, storage, sale, preparation, cooking, and serving of foods" [12] (p. 5). The BSR study estimates that the total amount of food waste in both retail and wholesale sectors in the United States combined was 3.8 billion pounds. Of this amount, 55.6 percent or 2.1 billion pounds was diverted from landfills and incineration to higher uses, *i.e.*, donations (17.9 percentage points) and recycling (37.7 percentage points). Data in the study is extrapolated from 10 retailers and 3 wholesalers, representing 30 percent of those U.S. industries by revenue. The study does not provide information for individual commodities and does not provide information for retailers separately [12].

There are some national studies or sources of data. For example, 2013–2014 supply balance sheet spreadsheets by Statistics Austria provide loss estimates for individual foods available for domestic consumption (e.g., 728 tonnes of cauliflower) and for aggregated food groups (e.g., 168,752 tonnes for all vegetables) though the data do not reveal what share of these losses represents edible food or where the losses occur along the farm-to-fork chain [29]. Greater detail on the extent of food loss and waste by world region can be found in a report by the High Level Panel of Experts on Food Security and Nutrition [30].

Looking more closely at the literature on food loss and waste for fresh fruits and vegetables, Mena *et al.* [24] provide estimates for 'retail loss and waste' as a percent of total weight for 11 select fruits and vegetables categories in the United Kingdom, ranging from 0.5–1 percent for onions to 2.5–5 percent for avocadoes. Meanwhile Eriksson *et al.* [31] estimated that in six Swedish retail stores, fruit and vegetables wasted in relation to quantities delivered was 4.3 percent by mass. Gustavsson and Stage [27] provide retail-level waste estimates for 16 popular fruits and vegetables in Sweden as a percent of sales, and found that small stores are more prone to higher produce waste than larger stores. Lebersorger and Schneider [32] estimate that food loss accounts for 4.2 percent of sales of fruits and vegetables in retail outlets in Austria. (This study also provides a more thorough comparison of retail-level loss estimates worldwide.) Beretta *et al.* [17] (p. 770) estimated that in Switzerland, fruit and vegetable losses in retail stores comprised 8 to 9 percent by calorific content. They also report that less than 1 percent (0.35 to 0.44 percent) of delivered product are lost due to transportation damage, spoilage, and unsatisfactory quality and caution that most substandard produce are sorted out earlier in the food supply chain. Again, the differences in the studies challenge meaningful comparisons among estimates (Eriksson *et al.* (2012), for example, considered whole fruits and vegetables and did not distinguish between edible and non-edible food). Consequently, we are unable to validate our estimates against data from most of the other related peer-reviewed studies. One exception is Buzby *et al.* (forthcoming) [33], which will compare the estimates presented here with the previous estimates for 2005–2006 from Buzby *et al.* [8].

Additionally, there is a lack of literature worldwide about food donations by retailers due to the lack of data [34]. Lebersorger and Schneider [32] estimate that in Austria, 7 percent of food loss in retail stores is donated to social services and that 38 percent of retailers do not donate any food to organizations that serve those in need. Beretta *et al.* [17] cite studies in Switzerland that estimate that food donations comprised 0.15 percent of the amount of food consumed at the retail level, suggesting that there is a great deal of room for improvement. The BSR estimate for the United States was relatively higher at roughly 18 percent of the amount of food waste generated in the retail and wholesale sectors was donated in 2011 [12]. Although food donations are not counted as part of shrink (*i.e.*, they are on a different part of profit and loss balance sheets), it would be helpful to understand the scale of these donations and the ultimate destination (e.g., consumption) because of the implications for the total amounts of food available for consumption.

2. Data and Methods

In late fall 2013, the ERS commissioned Nielsen Perishables Group to repeat the data collection from Buzby *et al.* [8] to obtain the two most recent years of shrink rate data available (*i.e.*, 2011 and 2012) for fresh fruits, vegetables, meat, poultry, and seafood. The primary purpose of the contract was to obtain updated shrink estimates of individual fresh foods for use as food loss assumptions at the retail level in the LAFA data series. In the current article, we focus on fresh fruit and vegetable shrink with a particular emphasis on the use of these estimates to calculate food loss for these products at the retail level in the United States. Greater detail on the foundation study can be found in Buzby *et al.* (forthcoming) [33], which focuses on: (1) comparing the new estimates with the 2005–2006 estimates from Buzby *et al.* [8] as a validation step and to obtain some perspective on changes over time; and (2) using the 2011–2012 estimates to calculate the amount of fresh fruit, vegetables, meat, poultry, and seafood availability (*i.e.*, the amounts of food available for consumption) at the retail level in the United States.

2.1. Commodity Coverage

This study provides shrink estimates for 24 fresh fruits and 31 fresh vegetables in U.S. supermarkets in 2011 and 2012. Processed or preserved fruits and vegetables (*i.e.*, canned, frozen, dried, dehydrated, and juiced) were not included in this study because Nielsen Perishables Group was not able to estimate shrink levels in these fruits and vegetables. In Buzby *et al.* [7], the relative contribution of processed produce to total retail-level lossin the United States in 2010 (*i.e.*, calculated from total poundage) was 3.7 percent for processed fruit (compared to 10.1 percent for fresh fruit) and 4.2 percent for processed vegetables (compared to 12.1 percent for fresh vegetables). Had shrink estimates been available for these processed items, then these estimates could have been used to update the LAFA data series and this would change the resulting estimates of food availability and food loss. The unavailability of these data do not influence the findings in the current study which focuses only on fresh fruits and vegetables. However, if the data were available, they would change the percentages mentioned above in this paragraph.

Data on both random-weight and UPC-coded items were included. Retailer categories were aggregated where necessary to develop data on the fresh fruits and vegetables in this study that match the commodities in the LAFA data series. For example, data on all random-weight apples (e.g., Gala, Granny Smith, McIntosh, Red Delicious, and Golden Delicious *etc.*) were combined with UPC-coded apples to develop the overall category for "fresh apples". UPC-coded apples included packaged apples sold in a bag or "value-added" ones such as pre-sliced apples. Data were only used for apples by themselves, that is, the data did not include caramel or candy-coated apples or sliced apples mixed with other kinds of fruit. Sliced or otherwise minimally processed fresh apples were included if they were not mixed with other fruit and if they were still sold as fresh apples.

Nielsen Perishables Group did not include mixtures of fresh fruit or vegetables (e.g., fruit salad, platters of vegetable sticks, and pre-packaged salads made from both leafy greens and fruit) in their calculations of quantitative data for individual commodities due to the lack of data on the share or weight of the different fruits and vegetables in each mixture. Fresh fruit and vegetable mixtures vary widely from retailer to retailer, and composition shares of the various fruits and vegetables are not consistent. Additionally, separate data categories for mixed fresh fruit and mixed fresh vegetables were not included in the study because the primary purpose of this study was to obtain updated retail-level loss estimates for individual fresh commodities for use in the LAFA data series, which contains only the core individual commodities, not mixtures as that would lead to double counting of product. Therefore, it is not possible to estimate these mixtures' contribution to total produce shrink.

2.2. Method

In order to be used in the analysis sample, data from each store under consideration had to have both weekly shipment data on a particular food commodity (e.g., pounds of fresh cucumber shipments sent from a supplier to the store) and corresponding point-of-sale data on consumer purchases in the Nielsen Perishables Group's proprietary FreshFacts® point-of-sale database (e.g., scanner data showing the pounds of fresh cucumbers sold). Combined, these two types of information enabled Nielsen Perishables Group to aggregate the data across all stores in the sample and match total shipment data to total sales data for each fresh commodity so that amount (*i.e.*, as the residual) and percentage of shrink could be calculated for 2011 and 2012. The average shrink for each commodity for 2011–2012 was then calculated.

Both the shipment data and the point-of-sale data were reported and aligned at the item level for each fresh commodity. For example, a 1-pound bag of fresh carrots would be identified by its UPC-code and its purchase by a particular store. Therefore, the data can be aggregated to the appropriate product level at a particular store and year and a loss rate can be calculated. For example, "Fresh carrots" consist of Universal Product Code (UPC)-coded baby carrots, shredded carrots, and 1-, 3-, and 5-pound bags of whole carrots, plus whole carrots sold by random weight.

The 2011–2012 shrink data for fresh produce are based on true aggregates of each store's data for the individual covered fresh fruits and vegetables, not projections or subjective interviews. Specifically for a given year, the shrink percentage for each commodity was calculated as the total pounds (or other appropriate unit) of a particular product that came into the stores in the sample but were not sold (*i.e.*, the "residual food loss"), divided by the total pounds (or other appropriate unit) of that product that came into the sample stores. This methodology yields actual, as opposed to estimated, shrink rates and provides accurate tracking of shrink trends across fresh produce categories. This methodology and the data sources offer the most accurate possible depiction of retail-level shrink activity for fresh fruit and vegetable items on a national level.

2.3. Supplier Shipment Data

For this study, Nielsen Perishables Group leveraged its existing business relationships to recruit a sample of retail chains in the United States to provide their supplier shipment data on the specific fresh fruits and vegetables covered in the study. Nielsen Perishables Group ensured a good cross section of retailers by developing a sample of retail chains which included: (1) larger national chains and smaller regional chains; (2) retailers from all regions of the country; and (3) traditional supermarkets along with mass merchandisers/super stores. Once Nielsen Perishables Group created the target list, the retailer's main Nielsen contact made the initial contact to request study participation via email to retailers at the corporate level, typically to the Vice President of Produce or Director of Produce for the entire chain. The email explained the purpose of the study and invited them to participate. Retailers were incentivized to participate in the study by a PowerPoint presentation created and delivered to them, which includes national trends found in the study as well as details about how well each retailer's shrink levels ranked compared to the multi-retailer average. Nielsen Perishables Group then conducted phone interviews with those who chose to participate. Participants typically shared their shipment figures for the covered commodities and years by sending a spreadsheet via email. The supermarket retailers ultimately included in the data portion of this study were five large national and regional food retail chains located in all four U.S. regions—East, South, Central, and West—to allow for a geographically-representative sample. We asked a total of 7 retail chains to participate in the shipment data portion of the project and 5 participated, thus we had a 71 percent response rate.

2.4. Point of Sale Data

Nielsen Perishables Group routinely collects point-of-sale data from approximately 18,000 stores in retail supermarkets across the United States and uses this information to develop its proprietary FreshFacts® point-of-sale database. Nielsen Perishables Group sales data tracks weight and package size attributes for produce, meat, poultry, and seafood thus enabling reporting by weight.

In the United States, the food channel (including grocery stores, mass/supercenters and club stores) in total represents 90 percent of the dollar sales for the channels where fresh food is sold. The remaining 10 percent of fresh food sales are from the convenience, dollar, drug and military channels and certain other retail chains. Nielsen Perishables Group used a sample of supermarket retailers in the United States from its proprietary FreshFacts® point-of-sale database to estimate fresh fruit and vegetable shrink at the retail level for 2011–2012. Only stores from the food channel were included, in order to best compare to the sample used in the 2005–2006 food loss analysis documented in Buzby et al. [8]. Specifically, this point-of-sale database includes retail census sales data for key grocery, club and mass/supercenter stores chains in the U.S. with $2 million or more annual all-commodity value (ACV) sales per store. ACV stands for "all-commodity volume", which is a variable on a product's distribution that takes into account differences in the size of a store [35]. The database does not include independent grocers, convenience stores, mom-and-pop grocers, and certain retail chains, such as: Whole Foods, Trader Joe's, Aldi, Costco, HEB, and Hy-Vee (please note this list is not inclusive of all excluded retailers). The sample size was limited by the number of stores providing shipment data, not by the number of retail store coverage in the point-of-sale data.

The study ultimately included 5 national and regional chains with around 2,900 stores in the sample to estimate fresh fruit and vegetable shrink. These stores comprised roughly 16 percent of the approximately 18,000 stores in Nielsen Perishables Group's available universe. Due to data limitations on the entire food sales market, we do not have an estimate of how much of the entire market is covered by the Nielsen Perishables Group proprietary FreshFacts® point-of-sale database or how much is covered by the sample of stores for the shipment and point-of-sales data pairings.

2.5. Data Quality

The shipment data was cleaned and validated for accuracy at the item-level by each retailer. The data input quality varied by retailer, as each had different internal ordering and accounting systems. Data cleaning criteria included aligning weekly sales and procurement receipts for the time period for each individual item sold, and identifying and removing anomalous data points (i.e., shrink levels above 80 percent or below zero). Anomalous data points were primarily driven by the accuracy of the retailer's internal tracking systems for procurement. For example, these outliers can be attributed to procurement receipt reporting discrepancies, longer than expected time in the supply chain, product repurposing in-store (e.g., cutting up fruit to remove blemishes and make fruit salad), or local store procurement outside of the main ordering process. Outliers were found across fruit and vegetable categories, and varied by retailer. After cleaning the data, Nielsen Perishables Group processed the data to meet internal sales data quality standards across all retailer data sets included in the study.

2.6. Food Loss Calculations

Estimates of uneaten food at the retail level in pounds in the United States were first calculated for each covered "whole" (here, meaning edible and nonedible portions combined) fresh fruit and vegetable by multiplying: (1) the per capita retail weight for a given whole commodity in LAFA in 2012; (2) the U.S. population in 2012; and (3) 2011–2012 average shrink estimate for that commodity. We then applied the nonedible share used in the LAFA data series to the resulting amount of uneaten food at the retail-level for each commodity to calculate the amount of food loss in supermarkets after the nonedible portions of fresh fruits or vegetables have been removed. Data on the nonedible share are from the National Nutrient Database for Standard Reference, compiled by USDA's Agricultural Research Service [36] (In the LAFA data series, the nonedible share is removed at the consumer level. It is removed here at the retail level for illustrative purposes).

The resulting food loss estimates may be of interest to the food industry by highlighting possible areas or commodities to target for loss reductions and opportunities for firms to develop and market new loss-reducing technologies, such as innovative packaging for particular types of produce. It is important to note that here, like in Buzby *et al.* [8], we are assuming that the estimated shrink rate equals the actual food loss rate for each fresh fruit and vegetable. This means that the resulting loss estimates are an upper bound since some unknown share of fresh produce go unsold for some reason other than food loss as defined here (e.g., theft).

The food loss estimates may also be of interest for determining the upper bound number of servings that could be donated to feed hungry people.

3. Results

The average annual shrink rates for 2011–2012 for individual fresh vegetables in U.S. supermarkets as estimated by Nielsen Perishables Group varied from 2.2 percent for sweet corn to 62.9 percent for turnip greens. Average fruit shrink was in a narrower range: 4.1 percent for bananas to 43.1 percent for fresh papayas.

3.1. Fresh Fruit

The estimated shrink for fresh fruit at the supermarket level was 13.3 percent in 2011 and 12.3 percent in 2012 (Table 1). This 1-percentage point difference is reasonable given year to year fluctuations in demand and supply of individual commodities. For the bulk of this article, we used the 2011–2012 averages for individual fresh foods because they provide better snapshots of potential trends than using one year of data.

Table 1. Estimated shrink and total loss of select fresh fruit in U.S. supermarkets.

Fresh fruit	U.S. Retail Weight, per capita (Lbs./year)	U.S. Retail Weight, total * (Million lbs./year)	Shrink in 2011 (%)	Shrink in 2012 (%)	Average shrink 2011–2012 (%)	Uneaten Whole Fresh Fruit (Million lbs./year)	Nonedible Share (%)	Fruit Loss (edible) ** (Million Lbs./year)
Papayas	0.9	288	54.8	30.3	43.1	124	33	83
Pineapple	6.1	1,917	30.5	35.8	32.2	617	49	315
Apricots	0.1	29	39.0	28.9	30.0	9	7.0	8
Watermelon	13.3	4,187	23.3	27.7	25.4	1,063	48	553
Honeydew	1.4	444	33.8	18.5	22.5	100	54	46
Mangoes	2.4	743	22.8	20.8	21.1	157	31	108
Apples	15.5	4,856	20.0	19.2	19.2	932	10	839
Avocados	5.0	1,578	25.0	17.2	19.0	300	26	222
Grapefruit	2.3	723	25.2	14.6	18.8	136	50	68
Cantaloupe	7.0	2,196	18.4	17.9	18.2	400	49	204
Peaches	3.7	1,171	13.8	18.8	15.6	183	7	170
Plums	0.6	186	16.4	13.6	15.1	28	6	26
Oranges	10.4	3,261	13.1	15.4	14.8	483	27	352
Kiwi	0.5	156	16.4	13.1	14.7	23	14	20
Pears	2.7	843	12.8	16.7	14.7	124	10	112
Tangerines	3.8	1,192	16.9	14.3	14.7	175	26	130
Strawberries	7.2	2,266	16.4	12.5	14.2	322	6	302
Limes	2.4	763	14.5	13.9	14.0	107	16	90
Cranberries	0.1	20	12.4	12.7	12.7	3	2	3
Cherries	1.4	435	8.0	10.3	10.3	45	9	41
Blueberries	1.2	387	8.3	9.9	8.9	34	5	33
Grapes	7.2	2,251	9.4	7.1	8.7	196	4	188
Lemons	3.7	1,151	8.3	4.9	5.1	59	47	31
Bananas	25.4	7,983	4.1	4.1	4.1	327	36	209
Average	124.2		13.3	12.3	12.6 ***		23.8	
Total		9,028				5,946		4,153

* Assuming U.S. population in 2012 of 314,267,867, which is the U.S. Census Bureau's resident population plus Armed Forces overseas 1 July 2012; ** After removing nonedible share; *** Weighted average; Source: Computed by authors.

The average supermarket shrink rate during 2011–2012 varied considerably across individual fresh commodities. Despite a large decline in shrink between 2011 and 2012 (24.5 percentage points), papayas remained the fresh fruit with the highest 2011–2012 average shrink (43.1 percent). These estimates are in line with the typical range of 10 to 50 percent papaya shrink in the literature with losses up to 80 percent for some shipments [1]. This high loss for papaya may be partly due to consumers' lack of knowledge of when papaya is ripe, how to prepare it, and how to use it as an ingredient. This lack of familiarity with the fruit may mean consumers are more hesitant when deciding whether to purchase papayas at the supermarket and as a result, the loss rate is higher. "Soft fruit" can be caused by crushing and bruising during handling [1] and when the papaya are low in calcium, they are particularly prone to such damage [37]. Additionally, store policies that require the retail chain to carry a range of produce even when sales are minimal may explain some papaya loss [1]. In 2011–2012, pineapples (32.2 percent) and apricots (30 percent) had the second and third highest average shrink.

The largest shrink percentage-point increases between 2011 and 2012 were for pineapple (5.3 percentage points) and peaches (5 percentage points). Peach quality often varies from year to year based on crop conditions and increased peach shrink could be due to a lower quality crop in one of the years.

Table 1 also provides estimated uneaten "whole" fresh fruit (in the sense that it includes both edible and inedible portions) in pounds in 2012 due to shrink in supermarkets using the 2011–2012 average shrink for each fruit and the total retail weight of each fruit in the United States in 2012 from the LAFA data. The top three fruits in terms of amount of uneaten fruit per year are watermelon, apples, and pineapples. After removing the inedible share, the top three fruits in terms of food loss (*i.e.*, edible portions only) are the same but in slightly different order (*i.e.*, apples, watermelon, and pineapple). The reason these differ from the top three fresh fruits in terms of the percent shrink (*i.e.*, papaya, pineapple, and apricots) is the relative weights of the fruit (e.g., watermelon are heavy), their relative importance to U.S. consumption (e.g., apples are one of America's most popular fruits), and perhaps the relative price per pound (e.g., fresh apricot prices are on the higher end of the price spectrum for fresh produce) [38]. In 2012, we estimate that total retail amounts of uneaten fruit for the covered fresh fruits tallied 5.9 billion pounds per year and after removing the nonedible share, this food loss totaled 4.2 billion pounds.

Figure 1 illustrates the point that the relative ranking of fresh fruits differs when viewed in terms of shrink and total amounts of uneaten fruit (both shrink and uneaten fruit here include both edible and inedible portions). The figure contrasts uneaten fresh whole fruit (left axis) ranked in descending order of million pounds per year in supermarkets with the average supermarket shrink rates for 2011–2012 (right axis) for each fruit.

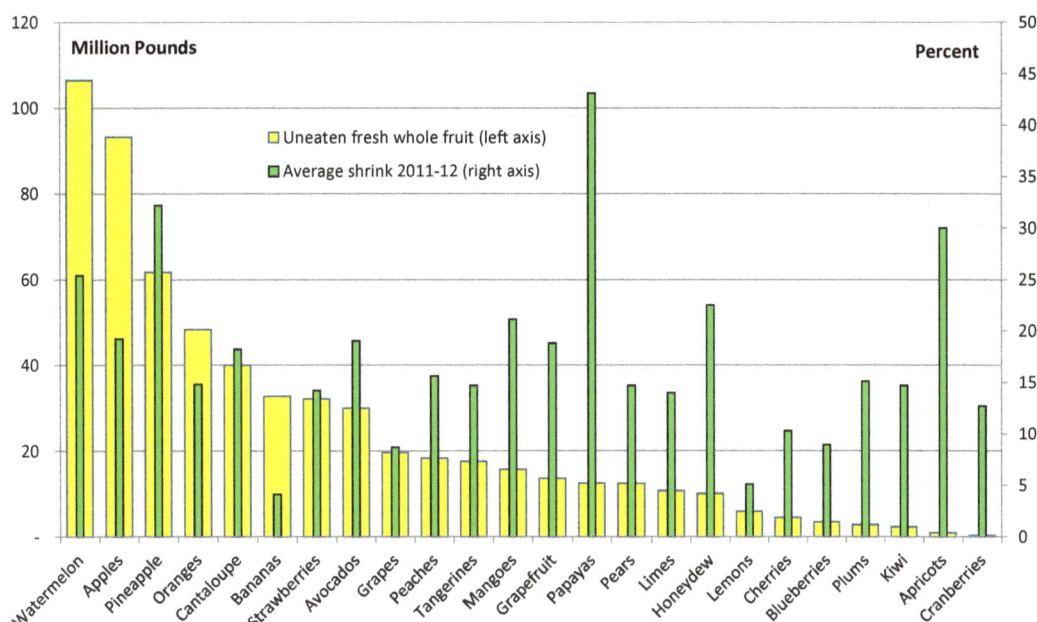

Figure 1. The top fresh fruits in terms of shrink are different than the top fruits in terms of the total amounts of uneaten, whole, fresh fruit in U.S. supermarkets, 2012; Source: computed by authors.

3.2. Fresh Vegetables

Table 2 provides the 2011–2012 estimates of supermarket shrink for the different varieties of fresh vegetables in the LAFA data series. The lowest fresh vegetable shrink was 2.2 percent for sweet corn. The highest average fresh vegetable shrink estimates in 2011–2012 were for fresh turnip greens (62.9 percent), fresh mustard greens (61.1 percent), escarole/endive (47.4 percent), collard greens (43.8 percent), and okra (40.2 percent). Leafy greens are relatively more prone to moisture loss than many other types of produce and this likely contributes to higher shrink. A general lack of consumer knowledge about some of these products and their preparation (e.g., how to cook collard greens) may make some consumers hesitant to purchase collard greens and this could contribute to these high shrink estimates. Additionally, the high shrink for fresh greens can be partly based on the lack of high-quality packaging. These products are typically sold in bunches and are not protected by packaging. In general, greens need to be refrigerated promptly in order to retain their moisture content and stay fresh. Cooking greens have gained in popularity with current juicing trends and focus on healthy eating. According to Nielsen Perishables Group, retailers are increasing the amount of shelf space to display cooking greens and this can lead to higher shrink if the items are not handled optimally and/or if these greens are selling at rates that fall short of expectations.

Table 2 also provides estimated uneaten whole fresh vegetables (both edible and inedible portions) due to shrink in supermarkets using the 2011–2012 average shrink for each fresh vegetable and the total retail weight of each fresh vegetable in the United States in 2012 from the LAFA data. Total estimated uneaten fresh vegetables (*i.e.*, "whole" vegetables including both edible and inedible portions) in U.S. supermarkets for the fresh vegetables covered here tallied 6.1 billion pounds. After removing the nonedible share, this food loss (*i.e.*, edible portions only) totaled 5 billion pounds.

Table 2. Estimated shrink and total loss of select fresh vegetables in U.S. supermarkets in 2012.

Fresh Vegetables	U.S. Retail Weight, per capita (Lbs./year)	U.S. Retail Weight, total * (Million lbs./year)	Shrink in 2011 (%)	Shrink in 2012 (%)	Average Shrink, 2011–2012 (%)	Uneaten Whole Fresh Vegetables (Million lbs./year)	Nonedible Share (%)	Vegetable Loss (edible) ** (Million lbs./year)
Turnip greens	0.4	113	61.7	63.9	62.9	71	30	50
Mustard greens	0.4	124	60.4	61.6	61.1	76	7	70
Escarole/endive	0.3	84	47.9	47.1	47.4	40	14	34
Collard greens	0.8	266	42.5	44.2	43.8	117	43	66
Okra	0.4	121	35.6	53.5	40.2	49	14	42
Kale	0.3	105	30.7	24.2	26.6	28	39	17
Squash	4.3	1,338	24.0	22.9	23.1	309	17	256
Radishes	0.4	128	17.9	27.2	22.7	29	10	26
Snap beans	1.8	581	19.2	23.9	21.9	127	12	112
Artichokes	1.4	433	14.4	30.9	20.8	90	60	36
Eggplant	0.8	248	18.1	22.9	20.6	51	19	41
Romaine and leaf lettuce	10.7	3,373	18.4	21.0	20.2	681	21	540
Spinach	1.4	428	15.1	20.2	18.2	78	28	56
Pumpkins	4.7	1,486	16.5	21.9	18.0	267	30	187
Cauliflower	1.1	340	17.5	17.3	17.3	59	61	23
Mushrooms	2.6	817	19.7	16.5	17.3	141	3	137
Asparagus	1.3	413	12.3	17.0	15.8	65	47	35
Tomatoes	17.3	5,451	11.9	14.7	14.5	790	9	719
Cucumbers	7.1	2,225	12.2	12.1	12.2	271	27	198
Bell peppers	10.7	3,375	8.2	11.3	10.7	361	18	296
Celery	5.5	1,741	6.3	10.0	8.5	148	11	132
Head lettuce	13.2	4,158	9.0	6.4	8.3	345	16	290
Potatoes	34.1	10,715	7.3	8.9	8.3	889	10	800

Table 2. *Cont.*

Fresh Vegetables	U.S. Retail Weight, per capita	U.S. Retail Weight, total *	Shrink in 2011	Shrink in 2012	Average Shrink, 2011–2012	Uneaten Whole Fresh Vegetables	Nonedible Share	Vegetable Loss (edible) **
Cabbage	6.3	1,968	7.0	7.9	7.4	146	20	117
Carrots	7.6	2,384	6.3	8.7	7.2	172	11	153
Broccoli	5.8	1,822	7.0	6.5	6.7	122	39	74
Onions	18.6	5,857	5.3	8.8	6.5	381	10	343
Brussels sprouts	0.3	109	6.1	5.6	5.8	6	10	6
Garlic	1.9	584	1.7	6.1	5.1	30	13	26
Sweet potatoes	6.2	1,948	5.3	4.3	4.4	86	28	62
Sweet corn	9.0	2,828	2.3	2.1	2.2	62	64	22
Average			10.5	12.2	11.6 ***		23.9	
Total	176.8	55,565				6,088		4,968

* Assuming U.S. population in 2012 of 314,267,867, which is the U.S. Census Bureau's resident population plus Armed Forces overseas July 1, 2012; ** After removing nonedible share;

*** Weighted average; Source: Computed by authors.

The top three fresh vegetables in terms of amount of uneaten vegetables per year are potatoes, tomatoes, and "romaine and leaf lettuce". After removing the inedible share, the top three fresh vegetables in terms of food loss remained the same. The main reason these three differ from the top three fresh vegetables in terms of the percent shrink in supermarkets (*i.e.*, fresh turnip greens, fresh mustard greens, and escarole/endive) is likely because potatoes, tomatoes, and "romaine and leaf lettuce" were among the top five fresh vegetables in terms of U.S. retail level availability (*i.e.*, among the most popular for U.S. consumption) (Figure 2).

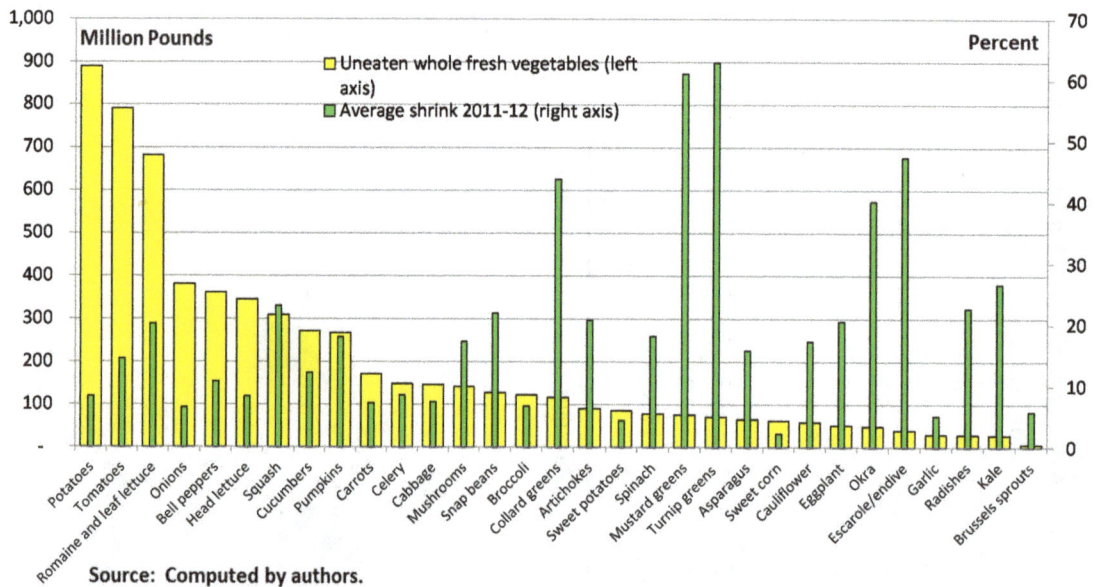

Source: Computed by authors.

Figure 2. The top fresh vegetables in terms of shrink are different than the top vegetables in terms of the total amounts of uneaten, whole, fresh vegetables in U.S. supermarkets, 2012.

4. Discussion and Outlook

The supermarket shrink estimates provided here are high compared to food shrink, loss, and waste estimates reported in other studies both internationally and in the United States. The estimates presented here are higher than the estimates in Mena *et al.* [24]. For example, the estimate for strawberries in the United Kingdom was 2 to 4 percent compared to 14.2 percent in the current study for strawberries in the United States. The estimates presented here are much higher than 4.3 percent of fruits and vegetables by mass reported in Eriksson *et al.* [31] in six Swedish retail stores and slightly higher than the 2005–2006 estimates in Buzby *et al.* [8]. The estimates are not directly comparable to those based on the percent of sales or calories. The overall average supermarket shrink rates here (*i.e.*, weighted by commodity value) were 1.2 percentage point higher for fresh fruit and 2.2 percentage points higher for fresh vegetables than in 2005–2006 [8]. Although, individual estimates often varied considerably from the corresponding average estimates for 2005–2006, the relative position of individual foods in terms of low or high shrink levels remained similar for both fresh fruits and vegetables. For example, estimated shrink for papayas and mustard greens were high in both studies out of all fresh fruits and vegetables respectively.

The LAFA data series continues to be considered *preliminary* or work in progress as there are a series of initiatives underway to improve the data structure and underlying assumptions, including the current study on fresh fruit and vegetable shrink at the retail level. Another recent initiative is that ERS contracted with the National Research Council and the Institute of Medicine of the National Academies of Sciences (NAS) to review the Food Availability Data System (which contains LAFA) and the resulting food availability and loss estimates. This contract led to a workshop held by NAS in April 2014 and a follow-up meeting on food donations in the United States held in March 2015. The proceedings from the NAS workshop has been published and one finding of the workshop was that obtaining a better assessment of the magnitude of food donations, rendering, and transfers to thrift stores would be beneficial [39] (p. 125). It was recognized that the ultimate destination of the product that went unsold in retail stores is important. For example, food donated by retail stores may end up ultimately being consumed by people so the LAFA could take this into account in the future so as to not overestimate loss and underestimate consumption. (The same could be said for theft of food.) (As an aside, some animal feed and other non-food uses during production and manufacturing are removed earlier in the core Food Availability data. Unsold food at the retail level diverted to animal feed could be considered part of food loss by ERS' definition but by some definitions it is not food waste as it has some economic use.) The proceedings from the food donations expert meeting is being developed. The current study improves the understanding of shrink and that component which is food loss but clearly this is an area in need of additional research. For example, more research is needed to determine if fresh produce shrink in supermarkets is dependent on the assortment offered for sale, including the variety of products of a particular fresh fruit or vegetable at different value levels (e.g., lower-, average-, and higher-priced bagged spinach or salad greens).

In this study, we assumed that shrink for fresh fruits and vegetables was food loss. Future research is needed to understand if any of the operational or theft components mentioned previously from the FMI and Retail Control Group study play a notable role. If there are more important factors in produce shrink than food loss, this could partly explain why our estimates are higher than suggested elsewhere in the literature. As previously mentioned, that study does not suggest that theft is a problem for fresh fruits and vegetables but this could be explored further. Future research could also investigate the extent and amount of edible food that supermarkets donate to food banks, soup kitchens, and other places and organizations that serve people in need.

A more thorough investigation of the amounts of and differences in food shrink and loss among retail stores in developed countries could unveil the portion of food shrink that is food loss. Better estimates of shrink and loss could then help identify the potential opportunities to reduce food shrink and loss in retail stores in the future and may spur innovation in food loss-reducing technologies and strategies [40]. Of course, more information on shrink and food loss rates and amounts for particular foods provides information on potential areas to target for shrink and food loss reductions but it does not provide information on what loss-reducing strategies to take or the underlying causes. Supermarkets would need additional information, such as on available and cost-effective loss-reducing technologies and practices (e.g., culling practices for fresh produce displays), to determine which loss-reducing strategies to implement.

Future scientific research questions to be addressed include: How can the scientific community help further food loss prevention in practice? What existing technologies can be refined or transferred to novel applications to reduce food loss? For example, how can new developments in nanotechnology be improved and made more affordable? These nanotechnologies include nanosilver (*i.e.*, has antimicrobial properties) in fresh food packaging containers [41] and nanosensors that can detect storage temperature abuse or trace amounts of gasses or contaminants potentially indicating that food products may not be safe to eat [42]. Also, given that there will always be some level of food loss, what is the next best economic use for uneaten food (e.g., composting, animal feed, and anaerobic digesters) and what cost-effective technologies are needed? Should national or local governments change the economic incentives to reduce food loss or recover uneaten food, such as through taxes or legal liability? How can we better measure food loss on a national level?

There are multiple collaborations under way to improve understanding of food loss and waste and to reduce and prevent food waste. For example, the European Union-funded "Food Use for Social Innovation by Optimising Waste Prevention Strategies" (FUSIONS) alliance with 21 partners across Europe is developing standard approaches for defining food waste [43]. And, the World Resources Institute's Food Loss and Waste Protocol is coordinating the development of a global approach to define and measure food loss and waste [44]. Although food waste is a data-poor area, assessment among some of the larger retailers and food service companies has shown it to be possible [21] (p. 16). Another collaboration is the U.S. Department of Agriculture and the U.S. Environmental Protection Agency's U.S. Food Waste Challenge launched in June 2013, which calls on entities across the food chain–farms, agricultural processors, food manufacturers, grocery stores, restaurants, universities, schools, and local governments–to join efforts to:

Reduce food waste by improving product development, storage, shopping/ordering, marketing, labeling, and cooking methods.

Recover food waste by connecting potential food donors to hunger relief organizations like food banks and pantries.

Recycle food waste to feed animals or to create compost, bioenergy and natural fertilizers.

Retail stores and chains in developed countries vary in their operations (e.g., inventory and pricing reduce food loss and waste would not be appropriate for all stores and chains. Multiple strategies to reduce food loss and waste would likely be needed to make meaningful progress in national or global loss reductions. For example, some issues like waste management priorities, can be addressed at a local level, others require a higher level of consideration, such as the waste of environmental and economic resources represented by food waste [21] (p. 5 and 8). Addressing national and global food loss and waste will likely take the combined efforts of individual stores, retail food chains, consumers, national governments, politico-economic bodies like the European Union and associated agencies, such as the FAO, and international forums like the Nordic Council [45] and Asia-Pacific Economic Cooperation (APEC) [46].

Acknowledgments

The data provided here were developed by Nielsen Perishables Group under a contract with the Economic Research Service of the U.S. Department of Agriculture.

Author Contributions

Jean C. Buzby administered the underlying contract, analyzed the data, and drafted the paper. Jeanine T. Bentley analyzed the use of the retailer data in the LAFA data series. Beth Padera was the lead on the contract for Nielsen Perishables Group and conceived and designed the retailer data collection and analyzed the data. Cara Ammon helped compile and analyze the results. Jennifer Campuzano managed and coordinated retailer engagement to secure data and interviews.

Conflicts of Interest

The views here cannot be attributed to the U.S. Department of Agriculture, the Economic Research Service, MobiSave, Nielsen Perishables Group Inc., or Beacon Research Solutions.

References

1. Paull, R.E.; Nishijima, W.; Reyes, M.; Cavaletto, C. Postharvest handling and losses during marketing of papaya (*Carica papaya* L.). *Postharvest Biol. Technol.* **1997**, *11*, 165–179.
2. Where's My Shrink: Executive Summary. Executive Summary Produced in Conjunction with Food Marketing Institute (FMI) and the Retail Control Group Report. Available online: http://wheresmyshrink.com/executivesummary.html (accessed on 6 April 2015).
3. Kienzlen, M. Where's My Shrink, Surprise, AZ, USA. Personal communication, 2015.
4. ERS. Loss-Adjusted Food Availability (LAFA) Data in the ERS Food Availability Data System (FADS). Available online: http://www.ers.usda.gov/data-products/food-availability-(per-capita)-data-system.aspx (accessed on 6 April 2015).
5. Buzby, J.C.; Hyman, J. Total and per capita value of food loss in the United States. *Food Policy* **2012**, *37*, 561–570.
6. Buzby, J.C.; Hyman, J.; Stewart, H.; Wells, H.F. The value of retail- and consumer-level fruit and vegetable losses in the United States. *J. Consum. Aff.* **2011**, *45*, 492–515.
7. Buzby, J.C.; Wells, H.F.; Hyman, J. *The Estimated Amount, Value, and Calories of Postharvest Food Losses at the Retail and Consumer Levels in the United States*; U.S. Department of Agriculture, Economic Research Service: Washington, DC, USA, 2014; p. 33.
8. Buzby, J.C.; Wells, H.F.; Axtman, B.; Mickey, J. Supermarket Loss Estimates for Fresh Fruit, Vegetables, Meat, Poultry, and Seafood and their Use in the ERS Loss-Adjusted Food Availability Data; Economic Research Service, U.S. Department of Agriculture: Washington, DC, USA, 2009.
9. FAO. Definitional Framework of Food Loss, SAVE FOOD: Global Initiative on Food Loss and Waste Reduction; Food and Agriculture Organization (FAO): Rome, Italy, 2014.
10. Gustavsson, J.; Cederberg, C.; Sonesson, U.; van Otterdijk, R.; Meybeck, A. *Global Food Losses and Food Waste: Extent Causes and Prevention*; Food and Agriculture Organization (FAO) of the United Nations: Rome, Italy, 2011.
11. Parfitt, J.; Barthel, M.; Macnaughton, S. Food waste within food supply chains: Quantification and potential for change to 2050. *Philos. Trans. R. Soc. Biol. Sci. Biol. Sci.* **2010**, *365*, 3065–3081.
12. BSR. Analysis of U.S. *Food Waste among Food Manufactures, Retailers, and Wholesalers*; Prepared for the Food Waste Reduction Alliance; BSR: New York, NY, USA, 2013; p. 24.
13. Kader, A.A. Increasing Food Availability by Reducing Postharvest Losses of Fresh Produce. *Acta Hortic.* **2005**, *682*, 2169–2176.
14. Nahman, A.; de Lange, W. Costs of food waste along the value chain: Evidence from South Africa. *Waste Manag.* **2013**, *33*, 2493–2500.

15. Quested, T.; Parry, A. *New Estimates for Household Food and Drink Waste in the UK; Final Report*, Version 1.1; Waste & Resources Action Programme (WRAP): Oxon, UK, 2011.

16. Whitehead, P.; Parfitt, J.; Bojczuk, K. *Estimates of Waste in the Food and Drink Supply Chain*; Waste Resources Action Programme, Ed.; WRAP: Oxon, UK, 2013.

17. Beretta, C.; Stoessel, F.; Baier, U.; Hellweg, S. Quantifying food losses and the potential for reduction in Switzerland. *Waste Manag.* **2013**, *33*, 764–773.

18. Mena, C.; Adenso-Diazb, B.; Yurt, O. The causes of food waste in the supplier–retailer interface: Evidences from the UK and Spain. *Resour. Conserv. Recycl.* **2011**, *55*, 648–658.

19. Ziegler, G.; Floros, J.D. A Future Perspective to Mitigate Food Losses: The Role of Food Science and Technology. In Proceedings of the IFT 2011 Annual Meeting & Food Expo, New Orleans, LA, USA, 2011.

20. Bio Intelligence Service. *Preparatory Study on Food Waste Across EU 27*; European Commission-Directorate C-Industry, Ed.; European Commission: Paris, France, 2010.

21. House of Lords. *Counting the Cost of Food Waste: EU Food Waste Prevention*; E.U. Committee, Ed.; Authority of the House of Lords, The Stationery Office Limited: London, UK, 2014.

22. Newsome, R.; Balestrini, C.G.; Baum, M.D.; Corby, J.; Fisher, W.; Goodburn, K.; Labuza, T.P.; Prince, G.; Thesmar, H.S.; Yiannas, F. Applications and perceptions of date labeling in food. *Compr. Rev. Food Sci. Food Saf.* **2014**, *13*, 24.

23. Saucede, F.; Fenneteau, H.; Codron, J.-M. Department upkeep and shrinkage control: Two key variables in optimizing the performance of fruit and vegetables departments. *Int. J. Retail Distrib. Manag.* **2014**, *42*, 733–758.

24. Mena, C.; Terry, L.A.; Williams, A.; Ellram, L. Causes of waste across multi-tier supply networks: Cases in the UK food sector. *Int. J. Prod. Econ.* **2014**, *152*, 144–158.

25. Vigneault, C.; Thompson, J.; Wu, S.; Hui, K.P.C.; LeBlanc, D.I. Transportation of fresh horticultural produce. *Postharvest Technol. Hortic. Crops* **2009**, *2*, 1–24.

26. Hodges, R.J.; Buzby, J.C.; Bennett, B. Postharvest losses and waste in developed and developing countries: opportunities to improve resource use. *J. Agric. Sci.* **2010**, *149*, 1–9.

27. Gustavsson, J.; Stage, J. Retail waste of horticultural products in Sweden. *Resour. Conserv. Recycl.* **2011**, *55*, 554–556.

28. Rutten, M.; Nowicki, P.L.; Bogaardt, M.J.; Aramyan, L.H. Reducing Food Waste by Households and in Retail in the EU: A Prioritisation Using Economic, Land Use and Food Security Impacts; LEI Wageningen UR The Hague: Den Haag, The Netherland, 2013.

29. Statistics Austria. *Supply Balance Sheets for the Crop Sector*; Statistics Austria: Vienna, Austria, 2015.

30. HLPE. Food Losses and Waste in the Context of Sustainable Food Systems, High Level Panel of Experts (HLPE) on Food Security and Nutrition of the Committee on World Food Security; Food and Agriculture Organization: Rome, Italy, 2014.

31. Eriksson, M.; Strid, I.; Hansson, P.-A. Food losses in six Swedish retail stores: Wastage of fruit and vegetables in relation to quantities delivered. *Resour. Conserv. Recycl.* **2012**, *68*, 14–20.

32. Lebersorger, S.; Schneider, F. Food loss rates at the food retail, influencing factors and reasons as a basis for waste prevention measures. *Waste Manag.* **2014**, *34*, 1911–1919.

33. Buzby, J.C.; Wells, H.F.; Axtman, B.; Mickey, J. Updated Supermarket Shrink Estimates for Fresh Fruit, Vegetables, Meat, Poultry, and Seafood in the United States and Their Use as Food Loss Assumptions in the ERS Loss-Adjusted Food Availability Data; Economic Research Service, U.S. Department of Agriculture: Washington, DC, USA, 2011–2012.

34. Schneider, F. The evolution of food donation with respect to waste prevention. *Waste Manag.* **2013**, *33*, 755–763.

35. Kilts Center for Marketing. *Nielsen Data Seminar: An Introduction to Scanner Data*; Kilts Center for Marketing: Chicago, IL, USA, 2012.

36. ARS. *National Nutrient Database for Standard Reference*; Agricultural Research Service (ARS), U.S. Department of Agriculture, Ed.; ARS: Washington, DC, USA, 2008.

37. Qiu, Y.X.; Nishina, M.S.; Paull, R.E. Papaya fruit growth, calcium uptake, and fruit ripening. *J. Am. Soc. Hortic. Sci.* **1995**, *120*, 246–253.

38. ERS. *Fruit and Vegetable Prices, Economic Research Service*; U.S. Department of Agriculture: Washington, DC, USA, 2015.

39. NRC and IOM. *Data and Research to Improve the U.S. Food Availability System and Estimates of Food Loss*; The National Academies' National Research Center (NRC), The Institute of Medicine (IOM), Eds.; The National Academies Press: Washington, DC, USA, 2015.

40. Golan, E.; Buzby, J. Innovating to meet the challenge of food waste. *Food Technol.* **2015**, *69*, 20–25.

41. Buzby, J.C. Nanotechnology for food applications: More questions than answers. *J. Consum. Aff.* **2010**, *44*, 528–545.

42. Duncan, T.V. Applications of nanotechnology in food packaging and food safety: Barrier materials, antimicrobials and sensors. *J. Colloid Interface Sci.* **2011**, *363*, 1–24.

43. European Union (EU). About FUSIONS (Food Use for Social Innovation by Optimising Waste Prevention Strategies). Available online: http://www.eu-fusions.org/ (accessed on 15 July 2015).

44. World Resources Institute. Food Loss & Waste Protocol: Addressing the Challenges of Quantifying Food Loss and Waste. Available online: http://www.wri.org/our-work/project/food-loss-waste-protocol (accessed on 12 May 2015).

45. Finnsson, P.T.F. Nordic Countries Play Their Part in Reducing Global Food Waste. Available online: http://nordicway.org/2015/04/nordic-countries-play-their-part-in-reducing-global-food-waste/#.VbhDSPRAV8Y (accessed on 15 July 2015).

46. Asia-Pacific Economic Cooperation. *APEC Action Plan for Reducing Food Loss and Waste*; APEC: Singapore, Singapore, 2014.

3

Effect of Additives and Fuel Blending on Emissions and Ash-Related Problems from Small-Scale Combustion of Reed Canary Grass

Sébastien Fournel [1,2],*, Joahnn H. Palacios [2], Stéphane Godbout [2] and Michèle Heitz [1]

[1] Department of Chemical and Biotechnological Engineering, Université de Sherbrooke, 2500 Université Boulevard, Sherbrooke QC J1K 2R1, Canada; E-Mail: michele.heitz@usherbrooke.ca

[2] Research and Development Institute for the Agri-Environment (IRDA), 2700 Einstein Street, Quebec City QC G1P 3W8, Canada; E-Mails: joahnn.palacios@irda.qc.ca (J.H.P.); stephane.godbout@irda.qc.ca (S.G.)

* Author to whom correspondence should be addressed; E-Mail: sebastien.fournel@usherbrooke.ca

Academic Editor: Stephen R. Smith

Abstract: Agricultural producers are interested in using biomass available on farms to substitute fossil fuels for heat production. However, energy crops like reed canary grass contain high nitrogen (N), sulfur (S), potassium (K) and other ash-forming elements which lead to increased emissions of gases and particulate matter (PM) and ash-related operational problems (e.g., melting) during combustion. To address these problematic behaviors, reed canary grass was blended with wood (50 wt%) and fuel additives (3 wt%) such as aluminum silicates (sewage sludge), calcium (limestone) and sulfur (lignosulfonate) based additives. When burned in a top-feed pellet boiler (29 kW), the four blends resulted in a 17%–29% decrease of PM concentrations compared to pure reed canary grass probably because of a reduction of K release to flue gas. Nitrogen oxides (NO_x) and sulfur dioxide (SO_2) emissions varied according to fuel N and S contents. This explains the lower NO_x and SO_2 levels obtained with wood based products and the higher SO_2 generation with the grass/lignosulfonate blend. The proportion of clinkers found in combustion ash was greatly lessened (27%–98%) with the use of additives, except for lignosulfonate. The positive effects of some additives may allow agricultural fuels to become viable alternatives.

Keywords: agricultural biomass combustion; energy crops; additives; fuel blending; pellets; gas emissions; particulate matter; ash-related problems; melting

1. Introduction

Substituting fossil fuels with renewable forms of energy has become a promising option to face the increase of greenhouse gas concentrations in the atmosphere and the rising cost of oil [1]. This context has motivated the shift to biomass for heat production since it offers many economic, social, and environmental benefits such as financial net saving, local employment opportunities and carbon dioxide (CO_2) emissions reduction compared to petroleum products [2]. In rural areas, there is a growing interest in using agricultural residues and energy crops grown on underutilized lands for heating farm facilities [2–4]. The latter represent moreover several ecological benefits including prevention of soil erosion, limited soil management, and low demand for nutrient inputs [5]. Although combustion is the most mature technology for biomass conversion, emissions from agricultural biomass combustion are generally greater than those from combustion of woody materials, which are the most common solid biofuels. Actually, agricultural biomass burned in small-scale appliances can significantly contribute to higher pollutants release such as particulate matter (PM), nitrogen oxides (NO_x), sulfur dioxide (SO_2), and hydrogen chloride (HCl) [6]. These contaminants can affect air quality and climate by causing respiratory and cardiovascular problems, acid rains, and absorption of solar radiation [7,8].

Comparatively to wood, typical agricultural fuels have higher ash content and higher concentrations of inorganic elements such as nitrogen (N), sulfur (S), chlorine (Cl), potassium (K), and silicon (Si). High amounts of N, S, and Cl in energy crops increase the emissions of NO_x, SO_2, and HCl, respectively. Ash is responsible for dust production and operational problems such as fouling, slagging, and corrosion, which may disturb the burning process, reduce efficiency and lead to unwanted shutdowns and higher levels of compounds from an incomplete combustion including carbon monoxide (CO) and PM [9,10]. Particles consist of aerosol-forming elements like K and Cl, as well as sodium (Na) and S. Boiler corrosion and fouling are also directly related to alkali metals (K and Na) and Cl contents. Chlorine acts as a catalyst, facilitating the movement of iron away from metal surfaces and the deposition of inorganic compounds. Sulfur and Si, in combination with alkali, lead to reactions associated with fouling and slagging in boilers. Potassium and, to a lesser extent, Si, S, and Na, contribute to lower ash melting temperatures in dedicated energy crops [11–15].

Strategies which can be used to reduce pollutants release from agricultural biomass combustion include the use of air staging [16] or flue gas cleaning devices such as filters and electrostatic precipitators [6]. Since the primary cause of emissions is the elemental composition of the feedstock, an alternative, which does not imply possible modifications to the heating system and can act on ash-related problems, is modifying the biomass chemical properties through the use of additives or fuel blending [17]. Additives refer to a group of minerals or products that can alter the ash chemistry, convert problematic species to less troublesome forms and enhance the ash melting temperature in thermal processes. Additives can be introduced before combustion by blending them with the fuel prior to pelletizing the admixture produced [18]. Based on their reactive compounds, additives can be classified as aluminum silicates, calcium, or sulfur based additives [17–19].

Aluminum (Al) silicates based additives, such as kaolin, have been exhaustively studied and have shown an ability to abate particle emissions [11,17,20–24] and ash sintering [25–27] during combustion of agricultural crops and residues. Kaolin mainly acts by binding alkali compounds in ash and by forming K- or Na-Al silicates that have a higher melting temperature than pure K or Na silicates [17–19]. Some works [20,23,28,29] also reported that the addition of kaolin almost eliminated Cl in fly-ash particles whereas HCl levels raised. As clay minerals additives, sewage sludge contains great amounts of Al-Si compounds, can increase ash sintering temperature and can reduce fouling deposition [18,30]. In addition, it has been suggested that S, Ca, and phosphorus (P) comprised in sewage sludge may contribute to the capture and deposition of gaseous alkali chlorides (KCl or NaCl) [30–32]. In fact, these gases can be transformed into sulfates, which are less deleterious deposits [31], or into high melting K- or Na-Ca phosphates [18,19,30]. Additives from waste stream resources such as sewage sludge are of particular interest since they are financially attractive [30]. Calcium based additives, such as lime and limestone, are used for reactions with HCl and SO_2 and have been recognized as well as effective in reducing the slagging tendency in combustion systems by formation of high melting silicates formed of Ca, magnesium (Mg), and alkali [18–20,23,26,27]. Co-firing biomass with calcium based additives actually creates a diluting effect on biomass ash, which restrains physical contact and thus sintering of ash particles [18,19]. Furthermore, lime is already and widely used in agriculture since it is one of the most crucial and beneficial components to successful crop management [33]. Sulfur based additives can decrease the formation of alkali chlorides through different sulfation reactions, as well as increase the melting point of deposits, hence preventing fouling of heat transfer surfaces [18,19]. For instance, the injection of ammonium sulfate greatly reduced gaseous KCl and produced sulfated deposits without any trace of Cl. Concentrations of SO_2 and HCl in flue gas were however higher when ammonium sulfate was added, while nitrogen monoxide (NO) emissions severely dropped because of selective non-catalytic reduction with ammonia (NH_3) [34]. Another option as a sulfur-based additive could be lignosulfonate, which is a by-product of the wood sulfite pulping process. So far, lignosulfonates are used in animal feeds and have been considered as the most effective and popular binding agents for pellets [35]. Their behavior and potential as combustion additives are uncertain since previous experiences showed that pellets with lignosulfonate result in problems with slag formation for wood [36] as well as in an anti-slagging effect for barley straw and husk [37].

Besides the addition of additives, mixing problematic feedstocks with good quality fuels, such as woody materials, may also improve thermal process and reduce emissions. The positive impact may be based on the diluting effect of the fuel having a lower ash content [18,38,39]. The burning of a blend composed of reed canary grass and wood chips only slightly raised fine particles, NO_x and SO_2 releases compared to wood alone, while CO and HCl either decreased or remained unchanged [38]. Nevertheless, different results from Lamberg *et al.* [39] showed elevated levels of incomplete combustion gases using similar wood-grass pellets.

This short review suggests that sewage sludge, limestone, lignosulfonate, and wood could be used as additives for mitigating particulate and gas emissions as well as ash-related operational problems in agricultural biomass heating systems. However, there is currently only a few scientific studies regarding the capacity of these additives to abate pollutant formation and sintering of energy crop ash. The present work was performed with the aim of measuring and comparing PM and gas production and evaluating the ash melting propensity during small-scale combustion of reed canary grass with and without additive (sewage sludge, limestone, lignosulfonate, and wood). This energy crop has a great development potential in the province of Quebec, Canada, but its high concentrations of S, Cl, K, and Si are responsible for increased levels of contaminants and clinkers [40]. The results obtained in this study can provide a better understanding of the effects of biomass-additive and biomass-biomass blending and their potential for controlling emissions and solving ash-related problems.

2. Materials and Methods

2.1. Biomass Fuels, Additives, and Blends

Pellets of reed canary grass and wood were respectively bought from agricultural producers (CLD Du Granit, Lac-Mégantic, QC, Canada) and a pellet mill (Trebio, Portage-du-Fort, QC, Canada). Both biomass fuels were milled (Wiley Mill 1885PL, Thomas Scientific, Swedesboro, NJ, USA) using a 4 mm screen size. Sewage sludge (Osons L'Osier, Rivière-du-Loup, QC, Canada) and lignosulfonate (Granulart, Neuville, QC, Canada) were acquired from research partners. Limestone came from a chemical company (Laboratoires MAT, Quebec City, QC, Canada). Lignosulfonate and limestone were in a powdered form.

The products were weighed on a dry basis and each of the individual blends (Table 1) was mixed manually and then pelletized (GRH200 pelletizer, Granulart, Neuville, QC, Canada). Sewage sludge (SS), limestone (LM) and lignosulfonate (LG) were added to reed canary grass (R) in a percentage of 3 wt% (blends R-SS, R-LM and R-LG, respectively). A review of the literature [10,11,20,22,24–27,29,30] showed that additives are generally blended with biomass in proportions up to 10 wt%. However, satisfactory results were especially obtained by adding 1–5 wt% of additives. By analyzing available data, it seemed that the difference between 1 wt% and 3 wt% was slightly significant, whereas it was negligible between 3 wt% and 5 wt%. Wood (W) was blended with reed canary grass in a 50–50 wt% proportion (blend R-W). This choice of admixture was motivated through theoretical calculations which determined the optimal levels according to the guiding values of Obernberger et al. [9] on major components (N, S, and Cl) in solid biofuels for unproblematic combustion. Furthermore, pure wood and pure reed canary grass pellets were tested to serve as references.

Table 1. Description of the tested blends (expressed in wt% of the different products).

	R	W	R-W	R-SS	R-LM	R-LG
Reed canary grass (R)	100	0	50	97	97	97
Wood (W)	0	100	50	0	0	0
Sewage sludge (SS)	0	0	0	3	0	0
Limestone (LM)	0	0	0	0	3	0
Lignosulfonate (LG)	0	0	0	0	0	3

All blends were experienced only once because the availability of resources (biomass and additives) by the suppliers did not allow realizing more than one replication. Before each experiment, a sample of the tested blend was sent to the Research and Development Institute for the Agri-Environment (IRDA) scientific laboratory (Quebec City, QC, Canada) to determine the physico-chemical properties. A more detailed description of the laboratory methods used can be found in Fournel *et al.* [40].

Additionally, fuel indexes on a molar basis, based on works by Sommersacher *et al.* [27,41] and describing the effect of given elements on alkali release, corrosion risk and ash sintering temperature, were calculated. They correspond respectively to Si/(K + Na), 2S/Cl and (Si + P + K + Na)/(Ca + Mg + Al). Herein, the sum of alkali (K + Na) replaced the K factor in the original indexes to account for possible high Na contents in some admixtures.

2.2. Combustion System

The experimental tests were carried out at a research facility on bioenergy of IRDA (Deschambault, QC, Canada). This facility includes a combustion room in which was installed a commercial 29-kW furnace (BB-100, LEI Products, Madisonville, KY, USA). The BB-100 (Figure 1) is a top-fed, multi-fuel (wood, agricultural crops and residues, waste, *etc.*), hydronic (use of water as the heat-transfer medium), non-catalytic, and non-pressurized boiler.

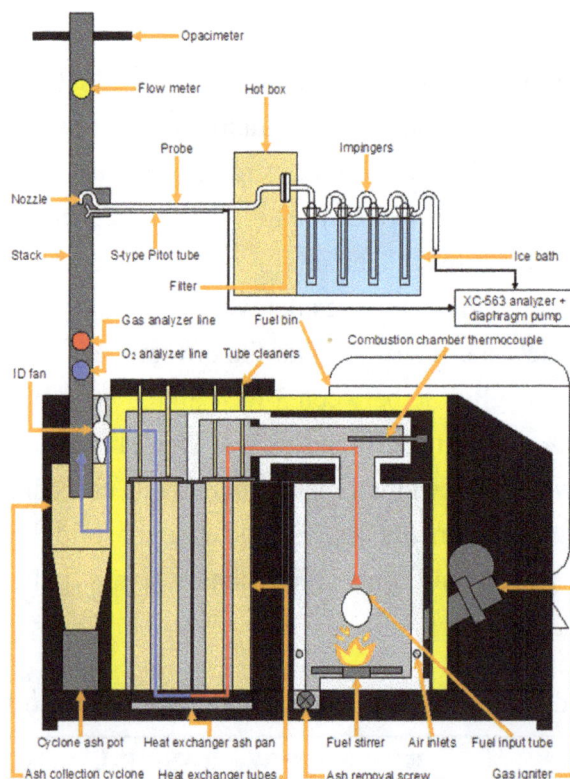

Figure 1. Schematic view of the boiler component parts and main sampling instruments.

The combustion was initiated by using a propane igniter. After reaching the intended temperature, restricted to 675 °C in order to limit the formation of slags, the supply of fuel was instigated. The pellets were continuously supplied to the burning chamber from a storage tank by an auger screw. The overfed material which dropped into the combustion compartment from the fuel input tube was constantly mixed on the ceramic base plate by a fuel stirrer. This apparatus allowed a slow removal of ash to an ash tray in which an auger screw is installed. The air was supplied to the combustion chamber by an induced draft fan, located at the end of the flue gas stream behind the heat exchanger, which pulled up air from the outside inwards. The temperature, the supply rate of fuel and air as well as the frequency of ash removal were controlled by a user interface and regulated for each blend to reach a stable combustion regime, which was then sustained automatically by the boiler's internal computer. The produced heat energy was extracted to the circulating water in a heat exchanger. The feed and hot water temperatures were respectively maintained at 60 °C (± 3 °C) and 70 °C (± 3 °C). Exhaust gases were directed to an exhaust duct via an ash collection cyclone. The boiler finally contains a removable ash pan and pot under the heat exchanger and the cyclone system.

About 25 kg of biomass were burned during a typical 6-h experiment. Each test included a 1-h period for start-up (gas igniter in function), 2 h to reach steady-state combustion (setting of the optimal conditions) and 3 h for measurements and collecting data. All the results presented in the following sections correspond to the data collected during those last three hours.

2.3. Gas and Particulate Measurements

The flue gas was evacuated through a 4.5 m stack composed of double wall stove pipes of 150 mm in diameter. Sampling ports (Figure 1) were fixed along the pipes to install samplers and measuring instruments. The first one is an LC CEM O_2 analyzer (Ametek/Thermox, Pittsburgh, PA, USA) with an internal zirconium oxide cell. It was used to continuously monitor the oxygen (O_2) content of the flue gas. A Fourier transform infrared spectrometer (FTIR; FTLA2000, ABB Bomem, Quebec City, QC, Canada) was then used to constantly analyze concentrations of nine gases (CO_2, CO, CH_4, N_2O, NO, NO_2, NH_3, SO_2, and HCl) from flue gas samples during the experimental combustion tests. The flue gas samples were drawn with a diaphragm pump into a heated stainless steel tube. The IRGAS 100 software (CIC Photonics, Albuquerque, NM, USA) acquired the spectra and quantified the gases each minute. Both instruments were connected to a data logger (CR10X, Campbell Scientific, Edmonton, AB, Canada).

At a distance 1.6 m higher than the FTIR sampling line, the PM sampling train (Figure 1) was inserted. Total PM in the flue gas was sampled according to Method 5H proposed by the United States Environmental Protection Agency. Particles were thereby sampled isokinetically. The PM sampling line included a stainless steel nozzle (12.5 mm in diameter), a stainless steel probe (600 mm long), an S-type Pitot tube, a 75-mm glass fibre filter (Whatman 934-AH, GE Healthcare, Mississauga, ON, Canada) inserted into a Pyrex filter holder installed in a heated compartment maintained at 120 °C, four impingers connected in series in an ice bath, a metering system (XC-563 Digital Meter Console, Apex Instruments, Fuquay-Varina, NC, USA) and a vacuum pump. More details on PM sampling method are given here [42].

An opacimeter (EMS750, Environmental Monitor Service, Yalesville, CT, USA) was installed 0.675 m above the last disturbance to continuously give an indication of opacity in real time. The exhaust gas velocity was monitored by a gas mass flow meter (GF90, Fluid Components Intl., San Marcos, CA, USA; error ±1%).

2.4. Ash Analyses

The ashes collected from the removal screw under the burning chamber, the pan under the heat exchanger, and the pot under the cyclone system were removed the next day of test in the morning after ash had cooled during the night. The three sorts of ash were weighed and sampled. They were analyzed in the same manner than biomass fuels in Section 2.1. Combustion ash was totally sieved (4.75 mm) before sampling to collect clinkers and to calculate the proportion of ash melted, according to the method used by Calvalho et al. [43].

3. Results and Discussion

3.1. Blends Physico-Chemical Properties

The higher heating value (HHV), moisture, ash content and elemental composition of each biomass fuel, additive and blend are presented in Table 2. Reed canary grass contained slightly less carbon (46.1 wt% vs. 50.4 wt%) and high amounts of ash (6.6 wt% vs. 0.8 wt%) compared to wood, which resulted in a lower HHV (17.2 MJ·kg^{-1} vs. 19.5 MJ·kg^{-1}). Main differences in inorganic elements between both fuels were high concentrations of N, S, Cl, K, and Si in reed canary grass. These elevated quantities, combined with relatively low Ca, Mg, and Al contents, can lead to higher levels of NO$_x$ and ash-related operational problems, as suggested by fuel indexes (Table 3). The addition of additives or fuel blending with wood should alter the chemical composition of reed canary grass to limit these inconveniences.

Table 2. Physico-chemical properties of biomass fuels, additives and blends (dry basis for all parameters, except for moisture on wet basis).

	Biomass		Additive			Blend			
	R	**W**	**SS**	**LM**	**LG**	**R-W**	**R-SS**	**R-LM**	**R-LG**
HHV (MJ·kg^{-1})	17.2	19.5	8.0	n.a.	17.0	18.0	17.3	16.7	17.2
Moisture (wt%)	8.9	6.1	59.9	0.2	5.8	8.2	8.6	8.2	7.4
Ash (wt%)	6.6	0.8	36.7	57.8	25.7	4.3	7.2	9.1	7.4
C (wt%)	46.1	50.4	19.0	11.9	42.7	47.8	45.7	45.0	45.8
H (wt%)	6.8	6.8	3.8	0.2	4.9	6.8	6.7	6.5	6.6
O (wt%)	48.8	48.4	14.6	30.1	30.2	49.1	48.2	46.6	46.7
N (wt%)	0.89	0.14	1.65	0.06	1.06	0.61	0.90	0.90	0.90
S (mg·kg^{-1})	1686	256	4269	0	81,219	1013	1553	1582	3805
Cl (mg·kg^{-1})	1226	167	105	32	6035	753	1219	1180	1218
K (mg·kg^{-1})	9099	840	2584	21	754	5291	8511	8535	8813
Na (mg·kg^{-1})	25	82	199	3780	80,885	46	29	121	2384
Si (mg·kg^{-1})	10,696	623	46,765	30	57	6648	11,295	10,818	10,973
P (mg·kg^{-1})	2510	93	26,923	1	30	1370	2528	2315	2404
Ca (mg·kg^{-1})	4053	3252	10,277	388,815	1187	3191	3497	13,729	3995
Mg (mg·kg^{-1})	1575	400	4555	211	226	998	1409	1437	1530
Al (mg·kg^{-1})	281	231	55,194	0	13	235	931	255	268

Notes: n.a., not applicable; HHV, higher heating value.

Table 3. Fuel indexes (mol·mol^{-1}) describing alkali release (Si/(K + Na)), corrosion risk (2S/Cl) and ash sintering temperature ((Si + P + K + Na)/(Ca + Mg + Al)).

	R	W	R-W	R-SS	R-LM	R-LG
Si/(K + Na)	1.63	0.88	1.72	1.84	1.72	1.19
2S/Cl	3.04	3.38	2.98	2.82	2.96	6.91
(Si + P + K + Na)/(Ca + Mg + Al)	3.95	0.47	3.23	3.91	1.66	4.62

Note: alkali release, corrosion risk and ash sintering temperature decrease with increasing index value.

Sewage sludge, limestone, and lignosulfonate respectively comprised high concentrations of Al (55,194 mg·kg^{-1}) and Si (46,765 mg·kg^{-1}), Ca (388,815 mg·kg^{-1}), and S (81,219 mg·kg^{-1}). Sodium amount is also present in a similar quantity than S in lignosulfonate (80,885 mg·kg^{-1}). These characteristics affected fuel indexes (Table 3) as Si/(K + Na) ratio slightly increased for each blend except for R-LG blend, 2S/Cl ratio doubled for R-LG blend and (Si + P + K + Na)/(Ca + Mg + Al) ratio considerably improved for R-LM blend and worsened for R-LG blend. Mixing reed canary grass and wood in equal proportion diluted the problematic elements of the former biomass as R-W blend contained approximately half of S, Cl, K, and Si (Table 2). For this reason, fuel indexes were predominantly improved.

3.2. Gas and Particulate Emissions

The highest PM level was obtained from pure reed canary grass pellets (1182 mg·Nm^{-3}; Table 4). Pure wood pellets, in comparable combustion conditions, produced almost half of this amount (621 mg·Nm^{-3}). Their lower ash content and thus their lower concentrations in ash-forming elements such as K, S, and Cl (Table 2) may be the reason for this reduction in PM. The four blends emitted between 835 mg·Nm^{-3} and 983 mg·Nm^{-3}, signifying that additives allowed a decrease of particles ranging from 17% to 29% compared to pure reed canary grass. These numbers almost correspond with particle drops (31%–57%) obtained by Bäfver et al. [20], Tissari et al. [11] and Carroll and Finnan [17] when kaolin was added (2–5 wt%) to agricultural products (oat grain, miscanthus, or tall fescue). As the latter authors stated, addition of additives with very low concentration of K or high amounts of Al and Si counteracts K volatilization from energy crops during the heating process and thus reduces PM emissions. Since most of the additives used within this study enhanced to some extent the alkali release index (Table 3), K may have been retained in combustion ash (see Section 3.4) rather than been volatilized as fly ash. Besides, PM ensuing from the mixing of reed canary grass with a woody material was only 1.3-fold greater than pure wood. This result is supported by the findings of Kortelainen et al. [38] and Lamberg et al. [39] where aerosol levels from different R-W blends were 1.4 times on average those of wood alone. This indicated that co-combustion of reed canary grass with wood could be an option for small-scale boilers which are capable of operating with fuels comprising moderate quantities of ash [38].

Emissions of CO_2 varied between 137,929 mg·Nm^{-3} and 143,021 mg·Nm^{-3} without particular trend (Table 4). The CO levels were 208 mg·Nm^{-3} for wood pellets, whereas they reached between 356 mg·Nm^{-3} and 431 mg·Nm^{-3} for grass-containing pellets. Methane (CH_4), nitrous oxide (N_2O) and NH_3 were only produced in small amounts (<4 mg·Nm^{-3}) without significant differences between fuels. Sometimes, concentrations were even near the detection limit of the FTIR so that no value was recorded. The measured NO_x emissions were correlated with fuel N, showing that NO_x are mainly formed from the feedstock N as other works revealed [39–41]. Actually, NO_x levels ranged from 63 $mg \cdot Nm^{-3}$ for wood (0.14 wt% N) to 229 $mg \cdot Nm^{-3}$ on average for R, R-SS, R-LM and R-LG blends (0.90 wt% N) with R-W blend (0.61 wt% N) in the middle (185 $mg \cdot Nm^{-3}$). Therefore, the amount of additive (3 wt%) was not high enough to have a real impact on NO_x concentrations. However, mixing two quality-contrasting fuels together (R-W blend) reduced NO_x emissions by almost 20%. Similarly, SO_2 varied according to fuel S as observed in Figure 2. In this figure, the value for R-LM blend slightly deviates from the main linear correlation. In fact, high Ca content can have a strong influence on retention of S in combustion ash since some authors [31,44,45] reported that Ca reacts with SO_2 to form Ca sulfates. This S capture by Ca compounds, which cut SO_2 emissions, was possibly predominant during R-LM burning as the mass balance on S (see Section 3.4) revealed that most of S is indeed comprised in combustion ash. Besides, the addition of lignosulfonate (R-LG blend) radically increased SO_2 levels (Table 2) and the corrosion risk index (Table 3) compared to pure reed canary grass. As mentioned before, the raise of SO_2 was a consequence of the addition of a sulfur based additive which, in return, was supposed to decrease the formation of alkali chlorides through sulfation reactions [18,19]. By analyzing the Cl mass balance (see Section 3.4), less Cl was present in deposits under the heat exchanger and cyclone. Usually, HCl emissions would be also increased [34], but no real HCl values were recorded by the FTIR.

Table 4. Gas and particulate emissions ($mg \cdot Nm^{-3}$ at 13 vol% O_2).

	R	W	R-W	R-SS	R-LM	R-LG
CO	383	208	409	356	431	357
CO_2	140,106	137,929	143,021	138,418	139,257	139,184
CH_4	3.58	3.40	3.30	3.18	3.52	4.33
N_2O	1.64	n.a.	n.a.	2.38	0.77	1.23
NH_3	n.a.	0.15	0.02	0.03	0.06	n.a.
NO_x	222	63	185	221	234	239
SO_2	137	16	66	139	73	423
HCl	n.a.	n.a.	n.a.	n.a.	n.a.	n.a.
PM	1182	621	835	892	955	983

Notes: n.a., not applicable.

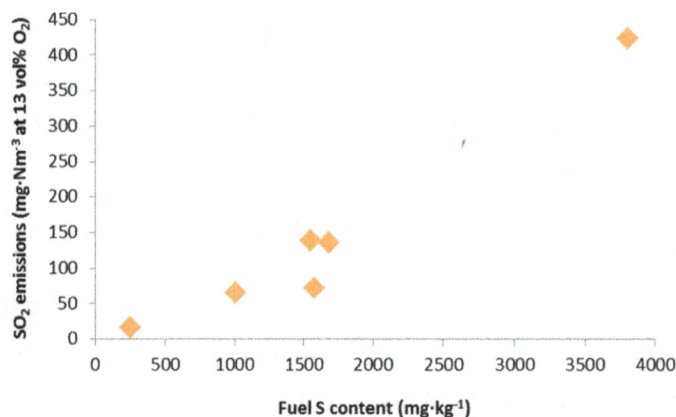

Figure 2. Correlation between SO_2 emissions and fuel S content.

3.3. Ash Melts

Combustion chamber ash sieving allowed the calculation of the proportion of melted ash (Table 5). No sintered ash was collected after wood burning, whereas the reference value obtained with reed canary grass was 3.90 wt%. Mixing this energy crop with wood (2.84 wt%) only reduced ash agglomeration by 27%. The best results were reached with sewage sludge (0.87 wt%) and limestone (0.07 wt%) additions. The 78%–98% sintering reductions when using these additives can be attributed to a surplus of Ca, which contributed to dilute R-LM ash, or a change from relatively low fusion temperature silicates and phosphates to higher fusion temperature silicates and phosphates [24–26,30]. Similar slag formation decreases (51%–67%) were noted by Xiong *et al.* [25] with 3 wt% addition of kaolin and calcite to corn stovers. These results were due to an increase by 100–200 °C of ash melting temperature. Moreover, the $(Si + P + K + Na)/(Ca + Mg + Al)$ ratio (Table 3) serving to estimate the ash sintering temperature was either greatly lessened (R-W and R-LM) or remained unchanged (R-SS) for blends with positive effects. On the contrary, R-LG blend resulted in a severe raise of molten ash proportion (40%). The high concentration of the alkali Na likely led to a melting point decline (Table 3) as Steenari *et al.* [26] experienced with the use of sodium bicarbonate as combustion additive.

Table 5. Proportion of melted ash.

	R	W	R-W	R-SS	R-LM	R-LG
Melted ash (wt%)	3.90	0.00	2.84	0.87	0.07	5.48
Difference with R (%)	n.a.	−100	−27	−78	−98	+40

Notes: n.a., not applicable.

3.4. Ash Analyses

Table 6 presents the content in minor elements of the three sorts of ash. In the case of combustion chamber ash, significant differences were only noticeable regarding Na, Ca, and Al amounts for R-LG (11.3 g·kg^{-1}), R-LM (86.4 g·kg^{-1}) and R-SS (9.3 g·kg^{-1}) blends, respectively. These elevated levels were directly linked with additive addition and can be correlated with the results of Table 5. The greater presence of Ca and Al silicates limited clinkers formation, whereas Na intensified ash agglomeration.

In ash collected under the heat exchanger tubes, S, Cl, and K concentrations drastically increased compared to combustion ash, indicating the importance of alkali volatilization. Excluding pure wood, blends with sewage sludge and limestone were those with the lower K quantities in heat exchanger ash (67–73 g·kg^{-1}) and the higher K levels in combustion ash (44–49 g·kg^{-1}). This showed the impact of Al silicates and Ca on K adsorption. Besides, the effect of lignosulfonate was also very clear since S and Na contents were high within R-LG heat exchanger ash.

In cyclone ash, only few elements were noteworthy. Calcium concentration was unsurprisingly high for R-LM blend while Cl, Na, and Al amounts were high in wood. No particular reason can explain this last result.

The proportion of fuel S, Cl, and K found in the three sorts of ash is illustrated in Figure 3. As explained before, limestone addition allowed a greater retention of S in combustion ash compared to the other blends. Lignosulfonate generated an increase of SO_2 and fused ash, but reduced the presence of Cl in deposits under the heat exchanger and cyclone. Finally, additives adsorbed more K in combustion ash than pure reed canary grass.

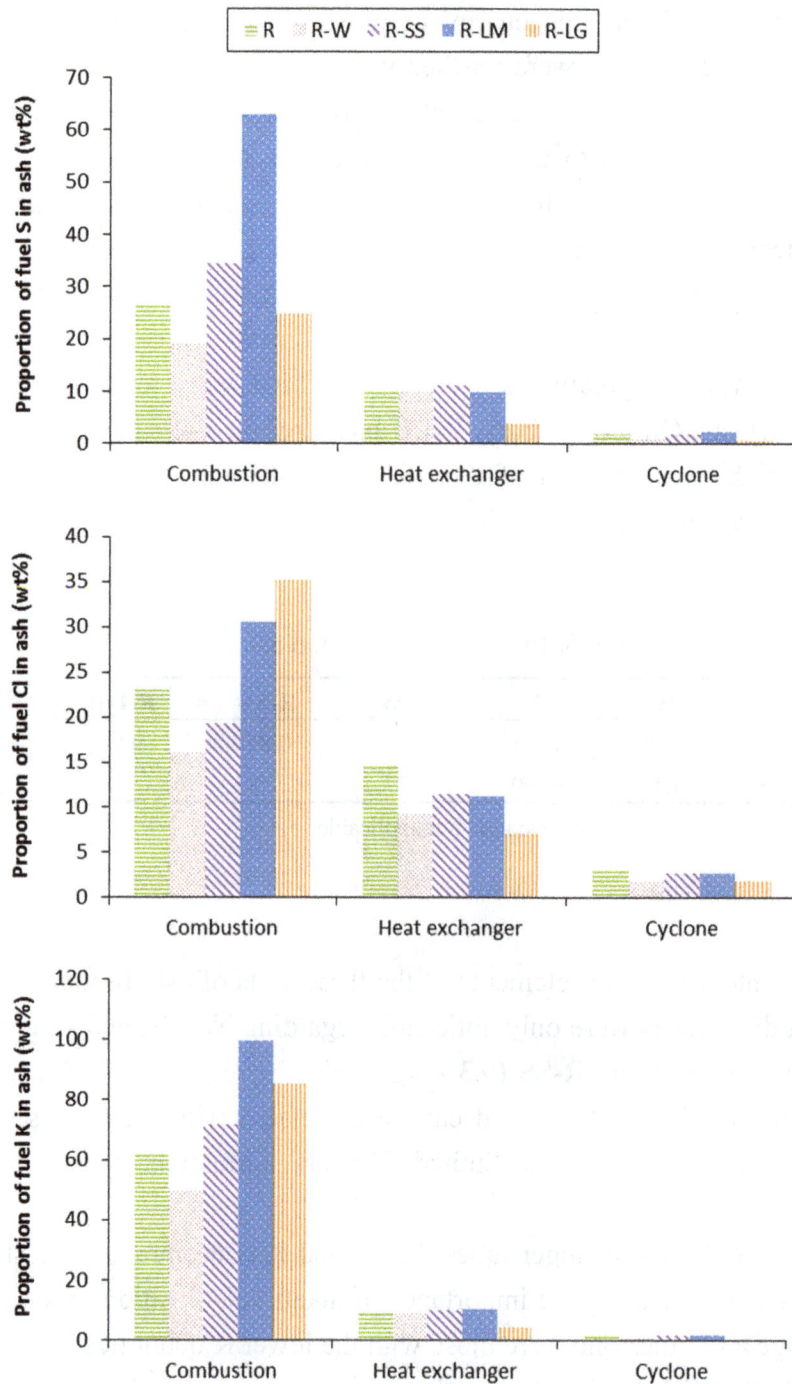

Figure 3. Proportion of fuel S, Cl and K found in combustion, heat exchanger and cyclone ash.

Table 6. Chemical composition of combustion, heat exchanger and cyclone ash (dry basis).

	R	W	R-W	R-SS	R-LM	R-LG
			Combustion Ash			
S (g·kg^{-1})	3.4	1.1	2	3.9	5.8	5.2
Cl (g·kg^{-1})	2.2	0.7	1.2	1.7	2.1	2.4
K (g·kg^{-1})	43.2	12	26.9	44	49.4	41.8
Na (g·kg^{-1})	0.7	1.1	0.6	1.3	1.7	11.3
Si (g·kg^{-1})	13.4	1.6	7.2	15.4	14	12.4
P (g·kg^{-1})	22.1	30.4	18.5	23.1	86.4	21.6
Ca (g·kg^{-1})	8	5.1	5.4	8.3	8.8	7.3
Mg (g·kg^{-1})	2.8	3.8	2.2	9.3	3.1	2.6
Al (g·kg^{-1})	3.4	1.1	2	3.9	5.8	5.2
			Heat Exchanger Ash			
S (g·kg^{-1})	15.2	17.3	16.5	14.8	11.8	29
Cl (g·kg^{-1})	16.4	9.4	11.5	12	9.9	17.6
K (g·kg^{-1})	80.8	33.8	76.7	72.6	66.6	75.5
Na (g·kg^{-1})	2.3	9.4	3.3	3.7	3.1	18.2
Si (g·kg^{-1})	28.8	8.5	27.1	28.3	28.5	25.9
P (g·kg^{-1})	54.9	138.7	79.7	54.3	190.8	71
Ca (g·kg^{-1})	19.6	19.1	21.9	18.6	18.1	16.3
Mg (g·kg^{-1})	7.4	28.5	10.4	12.3	7.1	6.9
Al (g·kg^{-1})	15.2	17.3	16.5	14.8	11.8	29
			Cyclone Ash			
S (g·kg^{-1})	10.6	14.3	10	10.9	10.7	15.3
Cl (g·kg^{-1})	13.3	25.2	12.3	13	9.9	18.2
K (g·kg^{-1})	63.8	38.1	47.9	56.9	44.8	47.8
Na (g·kg^{-1})	2.8	13.1	2.7	5	2.4	8.1
Si (g·kg^{-1})	33	8.4	27.6	31.7	28.3	27.1
P (g·kg^{-1})	77.5	140.7	97.1	89.8	200.8	133.3
Ca (g·kg^{-1})	23.3	20.3	25	22.2	20	19.7
Mg (g·kg^{-1})	9.6	35.8	11	10.9	8.2	8.9
Al (g·kg^{-1})	10.6	14.3	10	10.9	10.7	15.3

4. Conclusions

This study showed the effect of additives such as sewage sludge, limestone, lignosulfonate, and wood on particle and gaseous emissions and on ash sintering during small-scale combustion of reed canary grass. The four created blends resulted in PM decrease due to reduction of K release. Levels of NO$_x$ and SO$_2$ respectively depended on fuel N and S. The proportion of ash melts was greatly lessened with wood blending and the addition of sewage sludge or limestone because of a change of ash chemistry (higher ash sintering temperature compounds). Consequently, blending an energy crop with wood, sewage sludge, or limestone could be a promising strategy to handle problematic properties of agricultural biomass in small-scale heating systems and to help it compete favorably with wood pellets.

Acknowledgments

The authors thank the Mitacs Accelerate program and the "Fonds de recherche du Québec" for their financial contributions. The authors gratefully acknowledge the Research and Development Institute for the Agri-Environment, Granulart, Agriculture and Agri-Food Canada and Université de Sherbrooke which provided in-kind contributions for this study. The authors also recognize the technical and professional support provided by IRDA research staff (Jean-Pierre Larouche, Cédric Morin, Michel Côté, Christian Gauthier and Patrick Dubé).

Author Contributions

All authors conceived and designed the experiments. Sébastien Fournel and Joahnn H. Palacios performed the experimental tests. All authors analyzed the data. Sébastien Fournel wrote the paper. All authors revised the article.

Conflicts of Interest

The authors declare no conflict of interest.

References

1. Dhillon, R.S.; von Wuehlisch, G. Mitigation of global warming through renewable biomass. *Biomass Bioenergy* **2013**, *48*, 75–89.
2. Saidur, R.; Abdelaziz, E.A.; Demirbas, A.; Hossain, M.S.; Mekhilef, S. A review on biomass as a fuel for boilers. *Renew. Sustain. Energy Rev.* **2011**, *15*, 2262–2289.
3. Lewandowski, I.; Scurlock, J.M.O.; Lindvall, E.; Christou, M. The development and current status of perennial rhizomatous grasses as energy crops in the US and Europe. *Biomass Bioenergy* **2003**, *25*, 335–361.
4. Brodeur, C.; Cloutier, J.; Crowley, D.; Desmeules, X.; Pigeon, S.; St-Arnaud, R.M. *La Production de Biocombustibles Solides à partir de Biomasse Résiduelle ou de Cultures Énergétiques*; Ministère de l'Agriculture, des Pêcheries et de l'Alimentation du Québec: Quebec City, QC, Canada, 2008; pp. 1–14.
5. McKendry, P. Energy production from biomass (Part 1): Overview of biomass. *Bioresour. Technol.* **2002**, *83*, 37–46.
6. Nussbaumer, T. Combustion and co-combustion of biomass: Fundamentals, technologies and primary measures for emission reduction. *Energy Fuels* **2003**, *17*, 1510–1521.
7. Williams, A.; Jones, J.M.; Ma, L.; Pourkashanian, M. Pollutants from the combustion of solid biomass fuels. *Prog. Energy Combust.* **2012**, *38*, 113–137.
8. Van Loo, S.; Koppejan, J. *The Handbook of Biomass Combustion and Co-firing*; Earthscan: London, UK, 2008; pp. 291–303.
9. Obernberger, I.; Brunner, T.; Bärnthaler, G. Chemical properties of solid biofuels—Significance and impact. *Biomass Bioenergy* **2006**, *30*, 973–982.
10. Werther, J.; Saenger, M.; Hartgem, E.U.; Ogada, T.; Siagi, Z. Combustion of agricultural residues. *Prog. Energy Combust.* **2000**, *26*, 1–27.

11. Tissari, J.; Sippula, O.; Kouki, J.; Vuorio, K.; Jokiniemi, J. Fine particle and gas emissions from the combustion of agricultural fuels fired in a 20 kW burner. *Energy Fuels* **2008**, *22*, 2033–2042.

12. Jenkins, B.M.; Baxter, L.L.; Miles, T.R., Jr.; Miles, T.R. Combustion properties of biomass. *Fuel Process. Technol.* **1998**, *54*, 17–46.

13. Baxter, L.L.; Miles, T.R.; Miles, T.R., Jr.; Jenkins, B.M.; Milne, T.; Dayton, D.; Bryers, R.W.; Oden, L.L. The behavior of inorganic material in biomass-fired power boilers: Field and laboratory experiences. *Fuel Process. Technol.* **1998**, *54*, 47–78.

14. Cherney, J.H.; Verma, V.K. Grass pellet Quality Index: A tool to evaluate suitability of grass pellets for small scale combustion systems. *Appl. Energy* **2013**, *103*, 679–684.

15. Vassilev, S.V.; Baxter, D.; Vassileva, C.G. An overview of the behaviour of biomass during combustion: Part II. Ash fusion and ash formation mechanisms of biomass types. *Fuel* **2014**, *117*, 152–183.

16. Carroll, J.P.; Finnan, J.M.; Biedermann, F.; Brunner, T.; Obernberger, I. Air staging to reduce emissions from energy crop combustion in small scale applications. *Fuel* **2015**, *155*, 37–43.

17. Carroll, J.P.; Finnan, J.M. The use of additives and fuel blending to reduce emissions from the combustion of agricultural fuels in small scale boilers. *Biosyst. Eng.* **2015**, *129*, 127–133.

18. Wang, L.; Hustad, J.E.; Skreiberg, Ø.; Skjevrak, G.; Grønli, M. A critical review on additives to reduce ash related operation problems in biomass combustion applications. *Energy Procedia* **2012**, *20*, 20–29.

19. Shao, Y.; Wang, J.; Preto, F.; Zhu, J.; Xu, C. Ash deposition in biomass combustion or co-firing for power/heat generation. *Energies* **2012**, *5*, 5171–5189.

20. Bäfver, L.S.; Rönnbäck, M.; Leckner, B.; Claesson, F.; Tullin, C. Particle emission from combustion of oat grain and its potential reduction by addition of limestone or kaolin. *Fuel Process. Technol.* **2009**, *90*, 353–359.

21. Bäfver, L.; Boman, C.; Rönnbäck, M. Reduction of Particle Emissions by Using Additives. Available online: http://www.ieabcc.nl/workshops/task32_2011_graz_aerosols/04_Bafver.pdf (accessed on 30 January 2015).

22. Boman, C.; Boström, D.; Öhman, M. Effect of Fuel Additive Sorbents (Kaolin and Calcite) on Aerosol Particle Emission and Characteristics during Combustion of Pelletized Woody Biomass. Available online: http://pure.ltu.se/portal/files/2208136/22._Effect_of_fuel_additives_on_particle_characteristics_Valencia_2008.pdf (accessed on 30 January 2015).

23. Boström, D.; Grimm, A.; Boman, C.; Björnbom, E.; Öhman, M. Influence of kaolin and calcite additives on ash transformations in small-scale combustion of oat. *Energy Fuels* **2009**, *23*, 5184–5190.

24. Öhman, M.; Hedman, H.; Boström, D.; Nordin, A. Effect of kaolin and limestone addition on slag formation during combustion of wood fuels. *Energy Fuels* **2004**, *18*, 1370–1376.

25. Xiong, S.; Burvall, J.; Örberg, H.; Kalen, G.; Thyrel, M.; Öhman, M.; Boström, D. Slagging characteristics during combustion of corn stovers with and without kaolin and calcite. *Energy Fuels* **2008**, *22*, 3465–3470.

26. Steenari, B.-M.; Lundberg, A.; Pettersson, H.; Wilewska-Bien, M.; Andersson, D. Investigation of ash sintering during combustion of agricultural residues and the effect of additives. *Energy Fuels* **2009**, *23*, 5655–5662.

27. Sommersacher, P.; Brunner, T.; Obernberger, I.; Kienzl, N.; Kanzian, W. Application of novel and advanced fuel characterization tools for the combustion related characterization of different wood/kaolin and straw/kaolin mixtures. *Energy Fuels* **2013**, *27*, 5192–5206.

28. Aho, M. Reduction of chlorine deposition in FB boilers with aluminium-containing additives. *Fuel* **2001**, *80*, 1943–1951.

29. Aho, M.; Silvennoinen, J. Preventing chlorine deposition on heat transfer surfaces with aluminium-silicon rich biomass residue and additive. *Fuel* **2004**, *83*, 1299–1305.

30. Wang, L.; Skjevrak, G.; Hustad, J.E.; Skreiberg, O. Investigation of biomass ash sintering characteriscs and the effet of additives. *Energy Fuels* **2014**, *28*, 208–218.

31. Åmand, L.-E.; Leckner, B.; Eskilsson, D.; Tullin, C. Deposits on heat transfer tubes during co-combustion of biofuels and sewage sludge. *Fuel* **2006**, *85*, 1313−1322.

32. Pettersson, A.; Zevenhoven, M.; Steenari, B.-M.; Åmand, L.-E. Application of chemical fractionation methods for characterisation of biofuels, waste derived fuels and CFB co-combustion fly ashes. *Fuel* **2008**, *87*, 3183−3193.

33. Paradelo, R.; Virto, I.; Chenu, C. Net effect of liming on soil organic carbon stocks: A review. *Agric. Ecosyst. Environ.* **2015**, *202*, 98–107.

34. Kassman, H.; Pettersson, J.; Steenari, B.-M.; Åmand, L.-E. Two strategies to reduce gaseous KCl and chlorine in deposits during biomass combustion—Injection of ammonium sulphate and co-combustion with peat. *Fuel Process. Technol.* **2013**, *105*, 170–180.

35. Tarasov, D.; Shahi, C.; Leitch, M. Effect of additives on wood pellet physical and thermal characteristics: A review. *ISRN Forestry* **2013**, *2013*, 1–6.

36. Nikolaisen, L.; Jensen, T.N.; Hjuler, K.; Busk, J.; Junker, H.; Sander, B.; Baxter, L.; Bloch, L. *Quality Characteristics of Biofuel Pellets*; Danish Technological Institute: Aarhus, Denmark, 2002; p. 24.

37. Skjevrak, G. *Wood Pellets Utilized in the Commercial and Residential Sectors—An In-depth Study of Selected Barriers for Increased Use*; Norwegian University of Science and Technology: Trondheim, Norway, 2013; p. 130.

38. Kortelainen, M.; Jokiniemi, J.; Nuutinen, I.; Torvela, T.; Lamberg, H.; Karhunen, T.; Tissari, J.; Sippula, O. Ash behaviour and emission formation in a small-scale reciprocating-grate combustion reactor operated with wood chips, reed canary grass and barley straw. *Fuel* **2015**, *143*, 80–88.

39. Lamberg, H.; Tissari, J.; Jokiniemi, J.; Sippula, O. Fine particle and gaseous emissions from a small-scale boiler fueled by pellets of various raw materials. *Energy Fuels* **2013**, *27*, 7044–7053.

40. Fournel, S.; Palacios, J.H.; Morissette, R.; Villeneuve, J.; Godbout, S.; Heitz, M.; Savoie, P. Influence of biomass properties on technical and environmental performance of a multi-fuel boiler during on-farm combustion of energy crops. *Appl. Energy* **2015**, *141*, 247–259.

41. Sommersacher, P.; Brunner, T.; Obernberger, I. Fuel Indexes: A novel method for the evaluation of relevant combustion properties of new biomass fuels. *Energy Fuels* **2012**, *26*, 380–390.

42. Fournel, S.; Palacios, J.H.; Morissette, R.; Villeneuve, J.; Godbout, S.; Heitz, M.; Savoie, P. Particulate concentrations during on-farm combustion of energy crops of different shapes and harvest seasons. *Atmos. Environ.* **2015**, *104*, 50–58.

43. Carvalho, L.; Wopienka, E.; Pointner, C.; Lundgren, J.; Verma, V.K.; Haslinger, W.; Schmidl, C. Performance of a pellet boiler fired with agricultural fuels. *Appl. Energy* **2013**, *104*, 286–296.

44. Theis, N.; Skrifvars, B.J.; Zevenhoven, M.; Hupa, M.; Tran, H. Fouling tendency of ash resulting from burning mixtures of biofuels. Part 2: Deposit chemistry. *Fuel* **2006**, *85*, 1992–2001.

45. Pisupati, S.V.; Bhalla, S. Influence of calcium content of biomass-based materials on simultaneous NO_x and SO_2 reduction. *Environ. Sci. Technol.* **2008**, *42*, 2509–2514.

4

Temperature Impact on the Forage Quality of Two Wheat Cultivars with Contrasting Capacity to Accumulate Sugars

Máximo Lorenzo [1], Silvia G. Assuero [2,*] and Jorge A. Tognetti [2,3]

[1] INTA, Estación Experimental Balcarce, C.C. 276, Balcarce 7620, Argentina;
E-Mail: lorenzo.maximo@inta.gob.ar

[2] Laboratorio de Fisiología Vegetal, Facultad de Ciencias Agrarias, Universidad Nacional de Mar del Plata, C.C. 276, Balcarce 7620, Argentina

[3] Comisión de Investigaciones Científicas de la Provincia de Buenos Aires, La Plata 1900, Argentina;
E-Mail: jtognetti2001@yahoo.com.ar

* Author to whom correspondence should be addressed; E-Mail: assuero.silvia@inta.gob.ar

Academic Editor: Cory Matthew

Abstract: Wheat is increasingly used as a dual-purpose crop (for forage and grain production) worldwide. Plants encounter low temperatures in winter, which commonly results in sugar accumulation. High sugar levels might have a positive impact on forage digestibility, but may also lead to an increased risk of bloat. We hypothesized that cultivars with a lower capacity to accumulate sugars when grown under cold conditions may have a lower bloat risk than higher sugar-accumulating genotypes, without showing significantly lower forage digestibility. This possibility was studied using two wheat cultivars with contrasting sugar accumulation at low temperature. A series of experiments with contrasting temperatures were performed in controlled-temperature field enclosures (three experiments) and growth chambers (two experiments). Plants were grown at either cool (8.1 °C–9.3 °C) or warm (15.7 °C–16.5 °C) conditions in field enclosures, and at either 5 °C or 25 °C in growth chambers. An additional treatment consisted of transferring plants from cool to warm conditions in the field enclosures and from 5 °C to 25 °C in the growth chambers. The plants in the field enclosure experiments were exposed to higher irradiances (*i.e.*, 30%–100%) than those in the growth chambers. Our results show that (i) low temperatures led to an increased hemicellulose content, in parallel with sugar accumulation; (ii) low temperatures produced negligible changes in *in vitro* dry matter digestibility while leading to a higher *in vitro* rumen gas production, especially in the higher sugar-accumulating cultivar; (iii) transferring plants from cool to warm conditions led to a sharp decrease in *in vitro* rumen gas production in both cultivars; and (iv) light intensity (in contrast to temperature) appeared to have a lower impact on forage quality.

Keywords: *Triticum aestivum* L.; dual purpose; cellulose; hemicellulose; lignin; crude protein; *in vitro* rumen gas production; *in vitro* dry matter digestibility

---

1. Introduction

Wheat is increasingly cultivated as a dual-purpose crop in several main wheat areas of the world, including the USA southern Great Plains [1,2], Australia [3,4], China [5] and the Argentinean Pampas region [6–8]. The reasons for this expansion are mainly the capacity of wheat to provide forage early in winter without excessively decreasing grain production. This practice increases the profitability at the whole-farm system level and additionally reduces the risk associated with both price and climate variability [9–15].

Wheat is often considered a high quality, cool season forage when consumed at earlier developmental stages due to the high digestibility of young leaf blades, which is in turn associated with a low lignin content [16]. In general, forage is considered high quality when the *in vitro* digestibility of dry matter (IVDMD) is higher than 600 g·kg^{-1} DM [17]. Accordingly, values higher than 800 g·kg^{-1} DM IVDMD have been reported for wheat at the pre-stem elongation stage [16,18]. Nevertheless, several reports have related the intake of wheat and other annual winter grasses to bloat risk due to high levels of rapidly fermentable components (*i.e.*, soluble protein and sugars [19–21]). Pasture bloat takes place when the grazing animal's capacity to expel gases produced by fermentation is exceeded [22], and gases become trapped in bio-film complexes [21,23].

In vitro rumen gas production has been positively correlated with plant protein fractions and IVDMD when incubated with mixed rumen microorganisms [24]. However, this correlation is not necessarily straightforward. The concentration of soluble protein and sugars in wheat leaves may vary, depending on genotypic and environmental conditions [25–28]. Exposure of grasses to low temperature induces a steady accumulation of both components, while reversion to non-chilling conditions determines a very rapid decline in their concentration [25,29]. Considerable variation in the capacity to accumulate sugars and proteins exists among wheat cultivars: cultivars which undergo deeper cold-acclimation (winter hardy cultivars) are able to accumulate substantially higher amounts of compatible solutes in their cells compared with less hardy cultivars [26–28]. Because of the transient nature of solute accumulation under cold conditions, the ratio between rapidly fermentable non-structural carbohydrates and proteins, and structural components of grass cells may vary with temperature, and thus wheat pastures might present a variable bloat risk while maintaining a constantly high IVDMD.

In addition to temperature, light intensity may also play a role in determining IVDMD and bloat risk. A reduction in light intensity has been associated with reduced forage quality in some evergreen species [30]. However, there are conflicting reports regarding the influence of light intensity on lignin content, even though most studies suggest that higher intensities favor an increase in lignin levels [31].

In the present work, we studied the forage composition of two wheat cultivars with contrasting capacity to accumulate solutes when grown under cold conditions, in parallel with IVDMD and *in vitro* gas rumen production, as affected by temperature and light intensity. A set of experiments with contrasting temperatures was conducted in both field enclosures (high irradiance, three experiments) and growth chambers (low irradiance, two experiments) to test the following hypotheses: (i) low temperature increases the concentration of soluble and structural components of wheat leaf blades; (ii) low temperature increases *in vitro* rumen gas production, without a significant effect on forage digestibility; (iii) the effect of low temperature on *in vitro* rumen gas production is stronger in a cultivar with a higher capacity to accumulate solutes; and (iv) low temperature effects on *in vitro* rumen gas production and IVDMD are enhanced under higher light intensity conditions.

2. Experimental Section

2.1. Plant Material

Two wheat (*Triticum aestivum* L.) cultivars were selected for their contrasting morpho-physiological responses to low temperature, which have been described elsewhere [29,32,33]. Briefly, ProINTA Pincén is a winter hardy wheat that reduces its growth more and accumulates higher sugar concentration than Buck Patacón under low temperatures. In all experiments, seeds were soaked in tap water for 24 h at ambient temperature prior to sowing.

2.2. Experimental Layout

2.2.1. Field Enclosure Experiments

Three experiments were conducted in field enclosures during the winter seasons of 2005, 2006 and 2008 at the Facultad de Ciencias Agrarias campus (Universidad Nacional de Mar del Plata, Balcarce, Argentina, 37°45′47.94″ S, 58°17′38.82″ W, 130 m a.s.l.) under a natural photoperiod. The enclosures were constructed of pipe structures covered with polyethylene film (100 μm thick) (Figure S1). Plants were grown up to the fourth fully expanded leaf stage in polyethylene containers with a 0.1-m diameter and a 0.6-m height filled with a uniform mixture of soil (topsoil of a Typic Argiudol) and vermiculite (1:1 v/v) located in an excavation within the enclosures in order to maintain the top of the containers at soil level. Twenty-four containers were placed in each enclosure (12 for each cultivar, from which three were monitored during development and harvested, three were used for water status determination and the rest were used as borders). The substrate was saturated at sowing with ½-strength, and irrigated daily thereafter with ¼-strength Hoagland's solution [34]. Seeds (12 per container) were germinated and seedlings were thinned to 6 plants per container after emergence. Two electrical fan heaters with a thermostatic control (set to turn on under 16 °C) were located at opposite corners of one of the enclosures (warm treatment) at sowing. Accordingly, two electrical fans were located at opposite corners of the other enclosure (cool treatment) where the roof permanently covered the plants while the sidewalls were opened during the diurnal period and closed during the night. Air temperature was measured using thermistors and recorded using a data logger (Meteo, Cavadevices, Buenos Aires, Argentina) every 30 min. Thermistors were protected by shields to prevent absorption of solar radiation. The fourth channel of the data logger was used to record the photosynthetically active radiation. In the three experiments, the mean temperatures measured in the cool environment were very similar (ranging between 8.1 °C–9.3 °C), as were those of the warm enclosures (15.7 °C–16.5 °C) (Table 1). Daily mean air temperature dynamics during the 2006 field enclosure experiment, as well as the air temperatures recorded on one typical day of the same experiment, are shown in Figure S2 to illustrate the temperature conditions in the field enclosures. The average photosynthetic daily light integral (DLI) values diverged between cool and warm because of the different duration of the growing periods (Table 1). In 2008, a third treatment consisting of transferring plants from cool to warm conditions at the third leaf stage was applied. Plants were harvested early in the morning following a developmental criterion (*i.e.*, when 100% of plants attained the third fully expanded leaf stage for the cool and warm treatments, or the fourth fully expanded leaf stage for the cool-warm treatment); therefore, the harvest dates differed between the treatments.

Table 1. Average ($\pm$ SD) daily mean air temperature and photosynthetic daily light integral (DLI) in the cool and warm field enclosures for the 2005 (sown on 20 June), 2006 (sown on 19 June) and 2008 (sown on 12 June) experiments. The cool-warm (C-W) data correspond to the post-transferred period only.

Field Enclosure	Mean air temperature (°C)			DLI (mol photons $m^{-2} \cdot day^{-1}$)		
	Cool	C-W	Warm	Cool	C-W	Warm
2005	8.5 ± 2.8	–	16.5 ± 1.7	12.7 ± 5.7	–	17.8 ± 7.8
2006	9.3 ± 3.2	–	15.7 ± 2.2	12.7 ± 3.4	–	11.1 ± 4.1
2008	8.7 ± 3.3	16.6 ± 2.3	16.3 ± 1.9	14.0 ± 5.3	12.2 ± 3.9	11.9 ± 3.6

2.2.2. Growth Chamber Experiments

Two complete independent experiments were carried out in growth chambers at either 5 °C $\pm$ 0.5 °C or 25 °C $\pm$ 1 °C, under otherwise similar environmental conditions: 200 µmol photon $m^{-2} \cdot s^{-1}$ (photosynthetically active radiation, PAR) at the canopy level provided by fluorescent lamps (Osram Lumilux 21–840), 50% $\pm$ 10% relative humidity and a 12-h photoperiod. The average photosynthetic DLI (for both experiments) was 8.5 ± 0.1 and 8.6 ± 0.1 (mol photons $m^{-2} \cdot day^{-1}$) at 5 °C and 25 °C, respectively. Twenty-four plastic containers (0.1-m diameter, 0.3-m depth) filled with vermiculite and saturated with ½-strength Hoagland's solution [34] were placed in the chamber (12 for each cultivar, from which three were monitored during development and harvested, three were used for water status determination and the rest were used as borders). Seeds (twelve per container) were germinated and seedlings were thinned to 6 plants per container after emergence. Plants were harvested as described above for the field experiment.

The time in days and the thermal time from sowing to harvest for both field enclosures and growth chamber experiments are shown in Table S1.

2.3. Determinations

2.3.1. Plant Development

The number of fully expanded leaves was recorded at least twice a week.

2.3.2. Relative Water Content

The relative water content (RWC) at harvest was determined on the youngest fully expanded leaf of the mainstem as described by Equiza *et al.* [32]. Sampling was performed early in the morning in parallel with harvesting for other determinations. In all experiments (the field enclosures and the growth chambers), irrespective of cultivars and temperature treatments, the RWC values at harvest were higher than 96%. Therefore, differences in the concentrations of the cell components, when expressed per unit of fresh mass, are not attributable to variation in water status among cultivars or treatments.

2.3.3. Dry matter content and sugar concentration

The dry matter content was expressed as $g \cdot DM \cdot kg^{-1}$ FM. The total sugar concentration (TSC) in the leaf blades (mainly fructan, sucrose and monosaccharides) was quantified spectrophotometrically according to the phenol-sulfuric acid procedure [35]. Briefly, oven-dried leaf blades were ground, weighed and extracted in boiling distilled water ($10 \ mg \cdot DM \cdot mL^{-1}$) for 10 min. The mixtures were centrifuged at 1000 g, and supernatants were used for analysis. The reaction mixture contained 0.57 mL of a 5% phenol solution and 2.85 mL of H_2SO_4 in a total volume of 4 mL. The mixture was stirred and incubated for 20 min in a bath at 25 °C, agitated and after 15 min at ambient temperature, the absorbance at 490 nm was read using a UV-1700 PharmaSpec spectrophotometer (Shimadzu Corp., Kyoto, Japan). A glucose solution was used as a standard. All samples were run in duplicate, and the values are expressed on a fresh mass (FM) basis.

2.3.4. Cell Wall Components

The neutral detergent fiber (NDF) and acid detergent fiber (ADF) contents were determined using F57 filter bags (ANKOM A200, ANKOM Technology Corp., Fairport, NY, USA) according to Komareck *et al.* [36] and Komareck *et al.* [37], respectively. The lignin (ADL) content was determined using the acid detergent fiber permanganate lignin method [38]. The cellulose (ADF-ADL) and hemicellulose (NDF-ADF) contents were estimated by the difference. All values are expressed on a FM basis.

2.3.5. Crude Protein

The crude protein (CP) concentration was determined from the nitrogen levels (CP = 6.25 × N) using a LECO FP-528 (LECO Corporation, St. Joseph, MI, USA) nitrogen auto-analyzer [39]. The values are expressed on a FM basis.

2.3.6. True *in Vitro* Dry Matter Digestibility (IVDMD)

This procedure followed the ANKOM-DAISY procedure [40]. Samples (0.5 g DW) were weighed directly into F57 filter bags that were sealed with a heater and placed in a Daisy[II] Incubator (ANKOM Technology Corp., Fairport, NY, USA) digestion jar. Buffered rumen fluid was prepared according to Goering and Van Soest [38] and transferred into the jars containing the bags. The jars were then placed in the Daisy[II] Incubator at 39 °C, with continuous rotation. After 48 h of incubation in buffered rumen fluid, the bags were gently rinsed under cold tap water and placed in an ANKOM[200] Fiber Analyzer to remove microbial debris and any remaining soluble fractions using neutral detergent solution so that true digestibility could be determined. Incubations were performed in duplicate.

2.3.7. *In Vitro* Rumen Gas Production

Fresh wheat leaf blade samples were cut into 5-mm long pieces prior to all *in vitro* experiments. *In vitro* rumen gas production was determined following the general procedure described previously by Fay *et al.* [41] and Min [21,24], with modifications. The method consisted of measuring a syringe plunger displacement (ml) in 0–6-h incubation periods over a period of 28 h. Total *in vitro* rumen gas production was corrected to blank incubations (*i.e.*, no ruminal fluid). The rumen fluid was collected from a cannulated steer continuously receiving an alfalfa diet, mixed and strained through four layers of cheesecloth and flushed with CO_2 gas for *in vitro* rumen incubation. The *in vitro* rumen incubation procedure consisted of placing 2.5 g of minced fresh forage in 100-mL volumetric flasks containing 50 mL of rumen fluid diluted with artificial saliva [42], buffered to pH 6.8, saturated with CO_2 gas and maintained at 39 °C. Luer-type syringes (30 mL) with a 50/18 hypodermic needle, previously lubricated with distilled water to ensure consistent plunger resistance and movement to avoid gas losses, were inserted into the flask rubber stoppers. All gases were collected from the *in vitro* rumen incubation for gas production analyses. *In vitro* incubation was undertaken in duplicate.

2.4. Experimental Design and Statistics

A completely randomized design with three replicates (containers) per combination of two cultivars and two or three growth temperatures (depending on the experiment) was used. The temperature effect on plant carbon status (dry matter content and total sugar concentration), cell wall components (cellulose, hemicellulose and lignin), crude protein and *in vitro* rumen gas production at 28 h was analyzed using two-way ANOVA (Statistica 7, StatSoft Inc., Tulsa, OK, USA). Means were separated using Tukey's test at a significance level of 5%. No attempt was made to compare the effect of light intensity because of differences in temperature conditions between the field enclosure and the growth chamber experiments.

3. Results

3.1. Forage Composition

3.1.1. Forage Dry Matter Content

Significantly higher leaf blade dry matter content (DMC) values were found in the cool than in the warm environments for both cultivars (Table 2); the increase induced by lower temperatures was more pronounced in winter hardy Pincén than in Patacón (between 23%–29% and 15%–19%, respectively, for field enclosures, and averaging 47% and 16%, respectively, for growth chambers). Accordingly, Pincén had a higher DMC in cool environments. Conversely, no significant differences were found between the two cultivars under warm growing conditions. Transferring the plants from cool to warm conditions resulted in a significant decrease in DMC in Pincén but not in Patacón for both the field enclosures and growth chamber experiments.

Similar DMC values between the field enclosures and the growth chamber experiments were found for Pincén, while in Patacón, the growth chamber values were approximately 12% lower than their counterparts in the field enclosures.

It is well known that during cold acclimation of grasses, cellular dry matter content increases due to a transient deposition of many solutes, including non-structural carbohydrates, proteins, amino acids, *etc.*, while the cell water content may not be affected [43]. Since the RWC in our experiments was close to saturation (*i.e.*, higher than 96%) and was unaffected by temperature treatments or cultivars, the changes in DMC reflected the variation in the C concentration, not the plant water status. Because not all components are accumulated in the same proportion, the concentration of a component that accumulates less than the average could be seen as diminishing when expressed on a dry matter basis. For this reason, the concentrations of the different forage components listed below are expressed on a fresh mass basis, as in similar experiments reported elsewhere [44].

3.1.2. Total Sugar Concentration (TSC)

In general, the TSC results were similar to those of the DMC, with higher leaf blade TSC in cool than in warm treatments for both cultivars in all experiments (Figure 1), but the cold-induced increases were larger than for the DMC (between 155%–167% and 83%–100%, for Pincén and Patacón, respectively). Within each experiment and under cool conditions, Pincén showed the highest values.

For both cultivars, similar TSC values between the field enclosures and the growth chamber experiments were attained under the cooler environments. On the other hand, plants grown in growth chambers under warm conditions had TSC values that were approximately 40% lower than their counterparts in the field enclosures.

Table 2. Dry matter content (DMC, g kg^{-1} FM) of leaf blades of wheat cv. Pincén and cv. Patacón grown in cool (8.1 °C–9.3 °C) or warm (15.7 °C–16.5 °C) field enclosures in the 2005, 2006 and 2008 experiments, and in growth chambers at either 5 °C or 25 °C. Plants were harvested at the 3rd fully expanded leaf stage, or at the 4th fully expanded leaf stage for plants that were transferred from cool to warm and from 5 °C to 25 °C in the 2008 field enclosure experiment and in the second growth chamber experiment, respectively. Values are the means (± SE) of three replicates. Within each experiment, different letters indicate significant differences ($p < 0.05$).

	Pincén			Patacón		
Field enclosure	**Cool**	**Cool-Warm**	**Warm**	**Cool**	**Cool-Warm**	**Warm**
2005	179 ± 6.9 a	N.D.	146 ± 7.3 c	161 ± 8.3 b	N.D.	140 ± 6.7 c
2006	176 ± 7.1 a	N.D.	141 ± 3.6 c	159 ± 8.5 b	N.D.	140 ± 6.0 c
2008	175 ± 2.3 a	162 ± 6.7 b	136 ± 2.9 c	159 ± 6.1 b	155 ± 4.6 b	134 ± 8.9 c
Growth Chamber	5 °C	5 °C–25 °C	25 °C	5 °C	5 °C–25 °C	25 °C
Experiment 1	189 ± 10.3 a	N.D.	128 ± 7.3 c	141 ± 8.7 b	N.D.	123 ± 8.6 c
Experiment 2	183 ± 9.0 a	156 ± 5.6 b	125 ± 6.1 d	142 ± 2.3 c	133 ± 8.9 cd	121 ± 8.9 d

N.D.: Not determined.

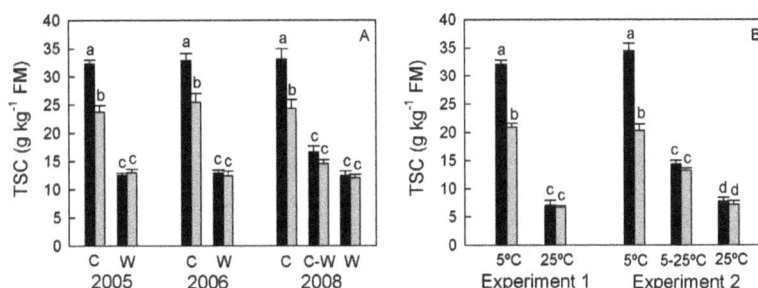

Figure 1. Total sugar concentration (TSC, g·kg^{-1} FM) of leaf blades of wheat cv. Pincén (black bars) and cv. Patacón (grey bars). (**A**) plants grown in cool (C, 8.1 °C–9.3 °C) or warm (W, 15.7 °C–16.5 °C) field enclosures in the 2005, 2006 and 2008 experiments. (**B**) plants grown in growth chambers at 5 °C or 25 °C. Plants were harvested at the 3rd fully expanded leaf stage, or at the 4th fully expanded leaf stage for plants that were transferred from cool to warm (C-W) and from 5 °C to 25 °C in the 2008 field enclosure experiment and in the second growth chamber experiment, respectively. Vertical bars indicate SE (n = 3). Within each experiment, different letters indicate significant differences ($p < 0.05$).

3.1.3. Structural Carbohydrates and Lignin

Cellulose and hemicellulose were the main cell wall components, ranging between 18 and 37 g·kg^{-1} FM, and 12 and 49 g·kg^{-1} FM, respectively (Figure 2). The lignin content was generally low, ranging between 1.2 and 3.5 g·kg^{-1} FM.

For both cultivars, the cellulose content of the leaf blades increased slightly under cooler conditions (between 13% and 25%, and 8% and 45% for cool *vs.* warm conditions for Pincén and Patacón, respectively, Figure 2A,B) except in the 2005 field enclosure experiment when no significant differences were found between temperatures. In the field enclosures, transferring plants from cool to warm conditions resulted in a slight (4%–7%) but significant decrease in the cellulose content of the leaf blades. The values obtained in the growth chamber experiments tended to be similar or higher than their counterparts in the field enclosures.

Hemicellulose varied most among the temperature treatments and cultivars (Figure 2C,D). Similar to cellulose, the hemicellulose values were generally higher under cool conditions (between 51% and 177%, and between 24% and 95% higher than the warm condition values for Pincén and Patacón, respectively) with the sole exception of Patacón in the 2006 experiment (−11%). The hemicellulose values of the transferred plants decreased and approached those of the warm-grown plants. The hemicellulose values in the growth chamber experiments tended to be lower than those in the field enclosures, particularly under warm conditions.

The lignin content of the leaf blades was higher under cool conditions (between 19% and 53%, and 12% and 44% higher than the warm condition values for Pincén and Patacón, respectively, Figure 2E,F). Transferring plants from cool to warm conditions resulted in a 14%–58% reduction in lignin concentrations, which approached the values of warm-grown plants or were even lower in one case (Figure 2F). The lignin values in the growth chamber were similar to those for the field enclosures, with the exception of the transferred plants of Patacón in Experiment 2, which, for unknown reasons, presented a rather low value.

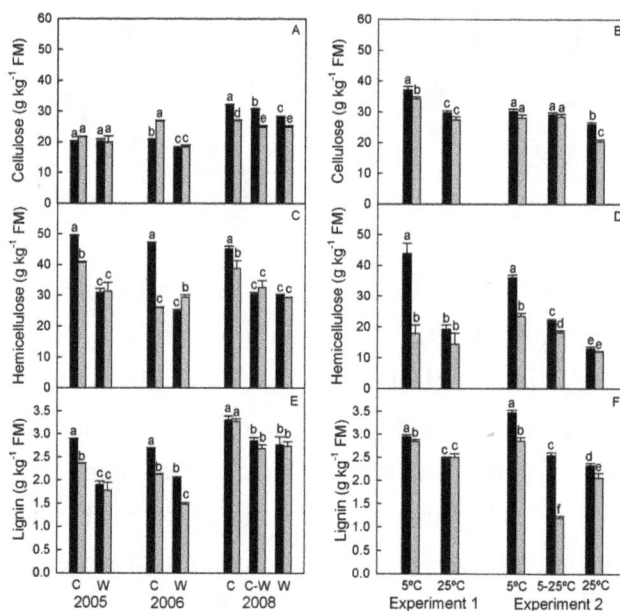

Figure 2. Cellulose (g·kg^{-1} FM, **A,B**), hemicellulose (g·kg^{-1} FM, **C,D**) and lignin (g·kg^{-1} FM, **E,F**) contents of leaf blades of wheat cv. Pincén (black bars) and cv. Patacón (grey bars). (**A,C,E**): plants grown in cool (C, 8.1 °C–9.3 °C) or warm (W, 15.7 °C–16.5 °C) field enclosures in the 2005, 2006 and 2008 experiments. (**B,D,F**): plants grown in growth chambers at 5 °C or 25 °C. Plants were harvested at the 3rd fully expanded leaf stage, or at the 4th fully expanded leaf stage for plants that were transferred from cool to warm (C-W) and from 5 °C to 25 °C in the 2008 field enclosure experiment and the second growth chamber experiment, respectively. Vertical bars indicate SE ($n = 3$). Within each experiment, different letters indicate significant differences ($p < 0.05$).

3.1.4. Crude Protein Concentration

The crude protein (CP) concentration in the leaf blades ranged between 37 and 54 $g \cdot kg^{-1}$ FM, and 32 and 48 $g \cdot kg^{-1}$ FM for Pincén and Patacón, respectively. The values for winter hardy Pincén were significantly higher under cooler conditions in both the field enclosures and the growth chamber experiments, except for the 2006 experiment when the difference was not significant (Figure 3). In contrast, no cold-induced increase in CP concentration was observed in Patacón except for the 2006 experiment. Transferring Pincén plants from cool to warm conditions did not significantly modify the CP concentration in either the field enclosures or the growth chamber experiments. No straightforward trend was observed for the CP concentration in Patacón. In general, similar values were found in the field enclosure and the growth chamber experiments, except that for Experiment 2, somewhat higher values were observed, especially for Pincén.

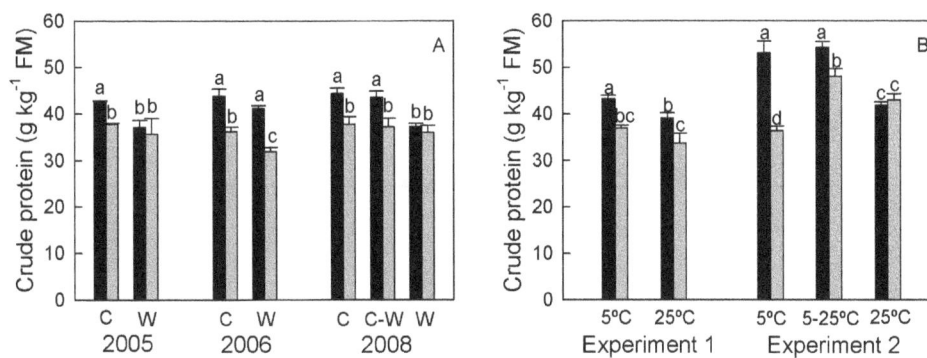

Figure 3. Crude protein content ($g \cdot kg^{-1}$ FM) of the leaf blades of wheat cv. Pincén (black bars) and cv. Patacón (grey bars). (**A**) plants grown in cool (C, 8.1 °C–9.3 °C) or warm (W, 15.7 °C–16.5 °C) field enclosures in the 2005, 2006 and 2008 experiments. (**B**) plants grown in growth chambers at 5 °C or 25°C. Plants were harvested at the 3rd fully expanded leaf stage, or at the 4th fully expanded leaf stage for plants that were transferred from cool to warm (C-W) and from 5 °C to 25 °C in the 2008 field enclosure experiment and in the second growth chamber experiment, respectively. Vertical bars indicate SE ($n = 3$). Within each experiment, different letters indicate significant differences ($p < 0.05$).

3.2. Forage Quality

3.2.1. True *in Vitro* Dry Matter Digestibility (IVDMD)

The IVDMD values were consistently high (above 75%, Table 3) irrespective of temperature, light environment, and cultivar. Although in some experiments significant differences were found among treatments, a straightforward pattern was not observed. In addition, the actual differences were small, even between the most contrasting treatments (approximately 70 and 40 $g \cdot kg^{-1}$ DM for the field enclosures and growth chambers, respectively).

Table 3. True *in vitro* dry matter digestibility (IVDMD, g·kg^{-1} DM) of leaf blades of wheat cv. Pincén and cv. Patacón grown in cool (8.1 °C–9.3 °C) or warm (15.7 °C–16.5 °C) field enclosures in the 2005, 2006 and 2008 experiments, and in two growth chamber experiments at 5 °C or 25 °C. Plants were harvested at the 3rd fully expanded leaf stage, or at the 4th fully expanded leaf stage for plants that were transferred from cool to warm and from 5 °C to 25 °C in the 2008 field enclosure experiment and in the second growth chamber experiment, respectively. Values are the means (± SE) of three replicates. Within each experiment, different letters indicate significant differences ($p < 0.05$).

Field Enclosure	Pincén			Patacón		
	Cool	Cool-Warm	Warm	Cool	Cool-Warm	Warm
2005	913 ± 1.6 a	N.D.	917 ± 13.4 a	894 ± 4.9 a	N.D.	912 ± 14.0 a
2006	951 ± 33.2 a	N.D.	868 ± 13.4 a	942 ± 28.4 a	N.D.	905 ± 23.4 a
2008	754 ± 38.4 b	783 ± 4.3 ab	820 ± 3.9 ab	842 ± 7.5 a	830±2.6 ab	838 ± 2.9 ab
Growth chamber	5 °C	5 °C–25 °C	25 °C	5 °C	5 °C–25 °C	25 °C
Experiment 1	941 ± 1.6 c	N.D.	972 ± 2.1 a	953 ± 1.8 bc	N.D.	962 ± 3.4 ab
Experiment 2	975 ± 4.6 a	963±1.4 ab	936 ± 2.3 d	954 ± 2.1 bc	910±3.5 e	946 ± 3.1 cd

N.D.: Not determined.

3.2.2. *In Vitro* Rumen Gas Production

In vitro rumen gas production analysis of leaf blades was performed for the 2008 field enclosure experiment and in the growth chamber Experiment 2. Curvilinear relationships were obtained between cumulative gas production and time up to 28 h of incubation irrespective of temperature, light environment and cultivar for plants grown at constant temperature (Figure 4).

Different gas production profiles were observed between temperatures and cultivars under constant temperature. After the first 4–5 h, the cumulative gas production from plants grown under cool conditions was approximately 65%–70% (field enclosures) or 55%–78% (growth chambers) higher than the values observed in warm environments for both cultivars. Moreover, Pincén cumulative gas production after this period was always higher than that observed for Patacón (approximately 35% in the field enclosures and 17% in the growth chambers).

Linear relationships best fit the data of plants transferred from cool to warm conditions. Cumulative gas production was considerably reduced in transferred plants, *i.e.*, to values markedly lower than those of plants grown under warm conditions (−30% and −56% for Pincén, and −15% and −44% for Patacón for field enclosures and growth chamber experiments, respectively, at 28 h of incubation).

On the other hand, slight differences were found between the light environments for plants of either cultivar grown at constant temperatures (*i.e.*, at 28 h, the field enclosure values were between −3% and +11% of the equivalent treatments in the growth chambers).

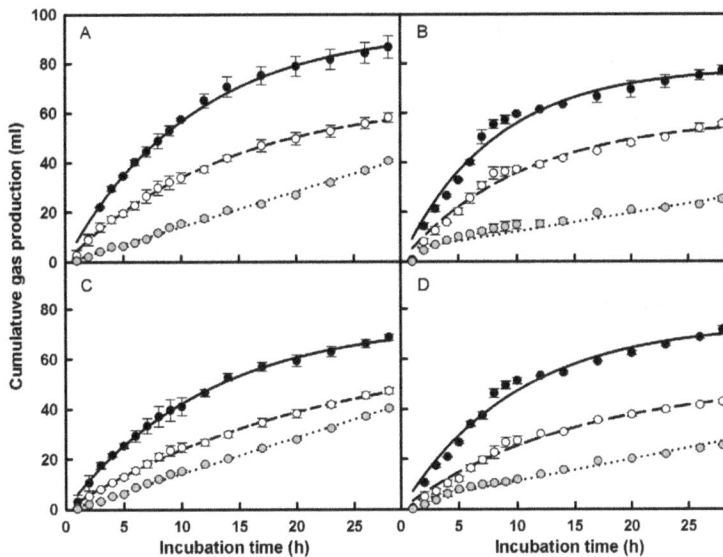

Figure 4. Cumulative gas production (ml) of leaf blades of wheat cv. Pincén (**A,B**) and cv. Patacón (**C,D**). (**A,C**): plants were grown in cool (8.1 °C–9.3 °C, black symbols, solid line) or warm (15.7 °C–16.5 °C, white symbols, dashed line) field enclosures in the 2008 experiment. (**B,D**): plants were grown in growth chambers at 5 °C (black symbols, solid line) or 25 °C (white symbols, dashed line) in Experiment 2. Plants were harvested at the 3rd fully expanded leaf stage except those that were transferred (grey symbols, dotted line) from cool to warm conditions in the 2008 field enclosure experiment (**A,C**) or from 5 °C to 25 °C (**B,D**) in the second growth chamber experiment, in which the plants were harvested at the 4th fully expanded leaf stage. Values correspond to the incubation of 2.5 g of minced fresh forage (mean ± SE of three replicates). The fitted models are $y = a \times (1 - \exp(-b \times t))$ [45], where y is the volume of gas produced at time t, a is the final asymptotic gas volume, and b is the fractional rate of gas production and, for transferred plants, $y = a + b \times t$, where a is the y-axis value at $t = 0$ and b is the rate of gas production.

4. Discussion

Plant components that affect forage quality, including readily digestible ones (soluble carbohydrates and crude protein, which are associated with bloat risk [19–21]) and non-digestible ones (lignin) generally increased under cooler conditions. This led to an increase in the dry matter content of the leaf blades, which was more pronounced in Pincén than in Patacón (Table 2). In the present work, soluble carbohydrates showed the most important changes while cellulose showed the least pronounced ones (Figure 1). In addition, changes were generally more marked in winter hardy cv. Pincén than in cv. Patacón. An increase in soluble carbohydrate concentration in cold temperatures is a very well known phenomenon in temperate grasses, which results from photosynthesis being less affected than growth by cold [46]. This response, which is associated with the capacity of plants to develop freezing tolerance [47–49], has been reported previously for the wheat cultivars studied here [25,32]. Transferring plants from cool to warm conditions led to a sharp decrease in TSC (Figure 1). This response may be the consequence of either a decreased TSC concentration in pre-existing leaves, or of a low concentration in the leaves developed under warm conditions, or more likely a combination of both. In any case, preliminary observations indicate that a peak in respiration within the first 24 h after transfer occurs in Pincén but not in Patacón [50].

Much less is known about the changes in the concentration of structural components due to temperature. An increase in cell wall thickness in grasses acclimated to cold has been reported [32,51]. Our finding of increased hemicellulose and lignin and, to a lesser extent, cellulose concentrations under cooler conditions (Figure 2) might be associated with this anatomical response. In plants transferred from cool to warm environments, the concentration of the different compounds tended to decrease toward values close to those of warm-grown plants. Since the proportion of the different components of leaf blades remained unmodified by ontogeny (at least up to the fifth leaf stage; data not shown) under continuous environmental conditions, the observed changes in forage composition in transferred plants could be solely attributable to the effect of temperature.

On the other hand, the light environment appeared to have only a marginal effect on the concentration of forage constituents. It has been argued that because of energy balance factors, low temperature and high light environments modify grass morpho-physiological characters in a similar manner (i.e., promoting compactness of growth habit [52], and a comparable redox state of photosystem II [53]). Nevertheless, it has been noted that freezing tolerance (measured as LT50) depends on low temperature exposure that is independent of irradiance levels [54]; therefore, the observed marked increase in TSC due to cold rather than to light is in agreement with the corresponding winter hardiness of Pincén and Patacón. While cold promotes a strong C accumulation due to an altered balance between growth and C utilization [46], it has been shown that light intensity not only favors assimilate availability but also plant growth (i.e., higher shoot fresh mass at harvest in the field enclosures compared with the growth chambers [28]), thus preventing a substantial carbohydrate accumulation. Consequently, within each temperature treatment (i.e., cool and 5 °C or warm and 25 °C), plants grown under higher irradiances (field enclosures) were larger than those grown under low irradiances (growth chambers), as shown previously [28].

Both parameters of forage quality, i.e., IVDMD and gas production, had markedly different responses to growing conditions and cultivar (Table 3 and Figure 4). A low correlation between both parameters was reported previously [55]. A first possible reason for the divergence between the two parameters could be the different incubation times. IVDMD was assessed at the end of a 48 h-incubation period while cumulative gas production assessed the dynamics of forage degradation up to 28 h, which is sensitive to changes in the proportion of forage constituents that differ in digestion rates. A 48 h-incubation period exceeds the ruminal retention time of high quality feeds in animals of high potential production (e.g., < 24 h) [56]. However, a second major factor is that fresh tissue was used for analyzing gas production, in contrast with IVDMD, which uses dry matter. Thus, variation in the dry matter content of the tissue is likely to modify only in vitro rumen gas production. There is strong evidence indicating that voluntary consumption of fresh forage by ruminants is closely related to forage volume (which in turn is related to fresh mass), and not to dry mass [57,58]. Therefore, in vitro rumen gas production from fresh forage incubation may more closely reflect what is actually happening in the animal rumen.

Despite large variations in the concentration of the forage components, IVDMD was almost unaffected by changes in temperature, light environment or cultivar (Table 3). There are reports indicating that cool temperatures may increase IVDMD. For example, this response has been found in tall fescue [59], timothy [60,61], and six other temperate grasses [61]. However, in other cases, no consistent effect of temperature on IVDMD was found [62]. A possible explanation for the lack of IVDMD increase at low temperature here is that even under warm environments, the values were high (ranging from 820 to 917 $g \cdot kg^{-1}$ DM in the field enclosures and 936 to 972 $g \cdot kg^{-1}$ DM in the growth chambers). The high values of IVDMD were expected since we evaluated the forage quality of the leaf blades of very young wheat plants in this study. It is well known that IVDMD decreases with leaf age [63,64].

Contrasting results have been reported with respect to the effect of light on IVDMD [65]. It has been suggested that only shade-intolerant species have their quality reduced by shade, mainly because of a decrease in total soluble carbohydrate concentration [66]. In our experiments, plants from the growth chambers (*i.e.*, lower light intensity) tended to show higher IVDMD values than those from the field enclosures (Table 3).

On the other hand, *in vitro* rumen gas production was largely modified by temperature and, to a lesser extent, by cultivars (Figure 4). In general, variation in gas production was in agreement with the changes in TSC (Figure 1). This is expected since gas is produced as the result of fermentation by ruminal fluid. The only exception was observed in transferred plants, which exhibited the lowest gas production despite showing TSC values that were higher than those of warm grown plants. In this sense, it has been shown that a high proportion of TSC in cold-acclimated plants consists of fructans that are inserted into the lipid headgroup region of the plasma membrane and help to stabilize it under freezing stress [67–69]. Given the fact that fructans may represent as much as 80% of TSC in cold-acclimated wheat [70], a hypothesis for further study is that part of the fructans inserted in the plasma membranes during growth at low temperature could remain there for a certain time after plants have been subjected to warmer conditions. This category of fructans might be less easily fermented by ruminal enzymes, but still captured in the chemical quantification of TSC. Because gas production is associated with bloat risk, information from further research into these points could be useful for grazing management of dual-purpose wheat crops grown in environments with changing temperatures in the autumn-winter period, conditions commonly found in the Argentinean Pampas.

It is well known bloat risk is tightly related to high sugar levels [19–21], but in turn, the latter are required for high freezing tolerance [47–49]. Consequently, it appears that simultaneously improving wheat for both low bloat risk and freeze hardening may be difficult. However, if the hypothesis that fructan inserted into the plasma membranes is less readily fermented in the rumen is supported, then studying genotype-associated variation in fructan partitioning between soluble and membrane-bound fractions could provide useful information for breeding purposes.

5. Conclusions

In parallel with the expected increase in sugar accumulation, low temperatures led to an increase in hemicellulose and crude protein concentration. This response was more marked in the hardy cultivar Pincén. While negligible changes in response to temperature were observed in *in vitro* dry matter digestibility, *in vitro* rumen gas production was much higher at cooler temperatures, especially in the higher sugar accumulating cultivar. This effect was rapidly reversed in plants transferred from cool to warm conditions. In contrast to temperature, light intensity appeared to have a lower impact on forage quality. Future experiments should focus on the remobilization of cool-induced membrane-bound fructans and their association with rumen gas production in animals.

Acknowledgments

This work, which is part of Máximo Lorenzo's PhD thesis at Universidad Nacional de Mar del Plata (UNMdP), Argentina, is supported by a grant from UNMdP (AGR-360). The authors thank Patricio Fay for his assistance with the *in vitro* rumen gas production analyses.

Author Contributions

All co-authors contributed equally to this work.

Conflicts of Interest

The authors declare no conflict of interest.

References

1. Hossaina, I.; Epplin, F.M.; Krenzer, E.G. Planting date influence on dual-purpose winter wheat forage yield, grain yield, and test weight. *Agron. J.* **2003**, *95*, 1179–1188.
2. Butchee, J.D.; Edwards, J.T. Dual-purpose wheat grain yield as affected by growth habit and simulated grazing intensity. *Crop Sci.* **2013**, *53*, 1686–1692.
3. Dove, H.; McMullen, G. Diet selection, herbage intake and liveweight gain in young sheep grazing dual-purpose wheats and sheep responses to mineral supplements. *Anim. Prod. Sci.* **2009**, *49*, 749–758.
4. Kelman, W.M.; Dove, H. Growth and phenology of winter wheat and oats in a dual-purpose management system. *Crop Pasture Sci.* **2009**, *60*, 921–932.
5. Tian, L.H.; Bell, L.W.; Shen, Y.Y.; Whish, J.P.M.; Nan, Z.B. Dual-purpose use of winter wheat in western China: Cutting time effects on forage production and grain yield. *Crop Pasture Sci.* **2012**, *63*, 520–528.
6. Arzadun, M.J.; Arroquy, J.I.; Laborde, H.E.; Brevedan, R.E. Grazing pressure on beef and grain production of dual-purpose wheat in Argentina. *Agron. J.* **2003**, *95*, 1157–1162.
7. Arzadun, M.J.; Arroquy, J.I.; Laborde, H.E.; Brevedan, R.E. Effect of planting date, clipping height, and cultivar on forage and grain yield of winter wheat in Argentinean Pampas. *Agron. J.* **2006**, *98*, 1274–1279.
8. Peralta, N.; Abbate, P.E.; Marino, A. Effect of the defoliation regime on grain production in dual purpose wheat. *Agriscientia* **2011**, *28*, 1–11.
9. Morant, A.E.; Merchán, H.D.; Lutz, E.E. Comparación de la producción forrajera de cultivares de trigo para doble propósito. *Rev. Argent. Prod. Anim.* **1998**, *18*, 213–214.
10. Lutz, E.E.; Merchán, H.D.; Morant, A.E. Carne y grano de un trigo doble propósito en condiciones semiá-ridas. *Phyton (Buenos Aires)* **2000**, *67*, 195–200.

11. Lutz, E.E.; Merchán, H.D.; Morant, A.E. Estado de desarrollo de la planta de trigo (var. ProINTA Pincén) al momento de la última defoliación y su rendimiento en grano. *Phyton (Buenos Aires)* **2000**, *68*, 83–87.

12. Bainotti, C.T.; Gomes, D.; Masiero, B.; Salines, J.; Fraschina, J.; Bertram, N.; Navarro, C. Evaluación de Cultivares de trigo como Doble Propósito. Available online: http://agrolluvia.com/wp-content/uploads/2010/05/INTA-Marcos-Ju%C3%A1rez-Evaluaci%C3%B3n-de-cultivares-de-trigo-como-doble-prop%C3%B3sito1.pdf (accessed on 18 May 2015).

13. Morant, A.E.; Merchán, H.D.; Lutz, E.E. Evaluación de genotipos de trigos para doble propósito. Fecha de siembra y producción de grano. *Rev. Argic. Prod. Anim.* **2003**, *23*, 222–223.

14. Lutz, E.; Merchán, H.; Morant, A. Mezcla de variedades de trigo para doble propósito. *Phyton (Buenos Aires)* **2008**, *77*, 217–223.

15. Bell, L.W.; Moore, A.D. Mixed Crop-livestock Businesses Reduce Price- and Climate-induced Variability in Farm Returns: A Model-derived Case Study. Available online: http://aciar.gov.au/files/node/13992/mixed_crop_livestock_businesses_reduce_price_and__20972.pdf (accessed on 3 August 2015).

16. Walker, D.W.; West, C.P.; Bacon, R.K.; Longer, D.E.; Turner, K.E. Changes in forage yield and composition of wheat and wheat-ryegrass mixtures with maturity. *J. Dairy Sci.* **1990**, *73*, 1296–1303.

17. Paterson, J.A.; Bowman, J.P.; Belyea, R.L.; Kerley, M.S.; Williams, J.E. The impact of forage quality and supplementation regimen on ruminant animal intake and performance. In *Forage Quality, Evaluation, and Utilization*; Fahey, G.C., Ed.; American Society of Agronomy, Crop Science Society of America, Soil Science Society of America: Madison, WI, USA, 1994; pp. 59–114.

18. Kelman, W.M.; Dove, H.; Flint, P. The Potential of Winter Wheat Cultivars and Breeding Lines for Use in Dual-purpose (Grain and Graze) Systems. Available online: http://www.regional.org.au/au/asa/2006/poster/systems/4613_kelmanw.htm (accessed on 18 May 2015).

19. Howarth, R.E.; Horn, G.W. Wheat pasture bloat of stocker cattle: A comparison with legume pasture bloat. In Proceedings of the National Wheat Pasture Symposium; Division of Agriculture, Oklahoma State University: Stillwater, OK, USA, 1984; pp. 24–25.

20. Horn, G.W. Growing cattle on winter wheat pasture: Management and herd health considerations. *Vet. Clin. North Am. Food A* **2006**, *22*, 335–356.

21. Min, B.R.; Pinchak, W.E.; Mathews, D.; Fulford, J.D. *In vitro* rumen fermentation and *in vivo* bloat dynamics of steers grazing winter wheat to corn oil supplementation. *Anim. Feed Sci. Technol.* **2007**, *133*, 192–205.

22. Mayland, H.F.; Cheeke, P.R.; Majak, W.; Goff, J.P. Forage-induced animal disorders. In *Forages*, 6th ed.; Nelson. C.J., Moore. K.M., Collins. M., Eds.; Blackwell Publication: Ames, IA, USA, 2007; Volume 2, pp. 687–707.

23. Malinowski, D.P.; Pitta, D.W.; Pinchak, W.E.; Min, B.R.; Emendack, Y.Y. Effect of nitrogen fertilisation on diurnal phenolic concentration and foam strength in forage of hard red wheat (*Triticum aestivum* L.) cv. Cutter. *Crop Pasture Sci.* **2011**, *62*, 656–665.

24. Min, B.R.; Pinchak, W.E.; Fulford; J.D.; Puchala, R. Wheat pasture bloat dynamics, *in vitro* ruminal gas production, and potential bloat mitigation with condensed tannins. *J. Anim. Sci.* **2005**, *83*, 1322–1331.

25. Tognetti, J.A.; Calderón, P.L.; Pontis, H.G. Fructan metabolism: Reversal of cold acclimation. *J. Plant Physiol.* **1989**, *134*, 232–236.

26. Tognetti, J.A.; Salerno, C.L.; Crespi, M.D.; Pontis, H.G. Sucrose and fructan metabolism of different wheat cultivars at chilling temperatures. *Physiol. Plant.* **1990**, *78*, 554–559.

27. Equiza, M.A.; Miravé, J.P.; Tognetti, J.A. Differential root *versus* shoot growth inhibition and its relationship with carbohydrate accumulation at low temperature in different wheat cultivars. *Ann. Bot.* **1997**, *80*, 657–663.

28. Lorenzo, M.; Assuero, S.G.; Tognetti, J.A. Low temperature differentially affects tillering in spring and winter wheat in association with changes in plant carbon status. *Ann. App. Biol.* **2015**, *166*, 236–248.

29. Equiza, M.A.; Tognetti, J.A. Morphological plasticity of spring and winter wheats under changing temperatures. *Funct. Plant Biol.* **2002**, *29*, 1427–1436.

30. Blair, R.M.; Alcaniz, R.; Harrell, A. Shade intensity influences the nutrient quality and digestibility of southern deer browse leaves. *J. Range Manag.* **1983**, *36*, 257–264.

31. Moura, J.C.; Bonine, C.A.; de Oliveira Fernandes Viana, J.; Dornelas, M.C.; Mazzafera, P. Abiotic and biotic stresses and changes in the lignin content and composition in plants. *J. Integr. Plant Biol.* **2010**, *52*, 360–376.

32. Equiza, M.A.; Miravé, J.P.; Tognetti, J.A. Morphological, anatomical and physiological responses related to differential shoot *vs.* root growth inhibition at low temperature in spring and winter wheat. *Ann. Bot.* **2001**, *87*, 67–76.

33. Assuero, S.G.; Lorenzo, M.; Pérez, N.M.; Velázquez, L.; Tognetti, J.A. Tillering promotion by paclobutrazol in wheat and its relationship with plant carbohydrate status. *N. Z. J. Agric. Res.* **2012**, *55*, 347–358.

34. Hoagland, D.R.; Arnon, D.I. The water-culture method for growing plants without soil. *Calif. Agric. Exp. Stn. Circ.* **1950**, *347*, 1–32.

35. Dubois, M.; Gilles, K.A.; Hamilton, J.K.; Rebers, P.A.; Smith, F. Colorimetric method for determination of sugars and related substances. *Anal. Chem.* **1956**, *28*, 350–356.

36. Komareck, A.R.; Robertson, J.B.; van Soest, P.J. Comparison of the filter bag technique to conventional filtration in the Van Soest NDF analysis of 21 feeds. Proceedings of the National Conference on Forage Quality, Evaluation and Utilization, Lincoln, NE, USA, 13–15 April 1994; Fahey, G.C., Ed.; Nebraska University: Lincoln, NE, USA, 1994.

37. Komareck, A.R.; Robertson, J.B.; Van Soest, P.J. A comparison of methods for determining ADF using the filter bag technique *versus* conventional filtration. *J Dairy Sci.* **1993**, *77*, 24–26.

38. Goering, H.K.; van Soest, P.J. Forage fiber analyses (Apparatus, Reagents, Procedures and Some Applications). In *USDA-ARS Agricultural Handbook 379*; US Government Printing Office: Washington, DC, USA, 1970; p. 20.

39. Horneck, D.A.; Miller, R.O. Determination of total nitrogen in plant tissue. In *Handbook of Reference Methods for Plant Analysis*; Kalra, Y.P., Ed.; CRC Press: London, UK, 1998; pp. 75–83.

40. ANKOM Tecnology. Analytical Methods *in vitro* True Digestibility Method (IVTD-Daisy). Available online: https://ankom.com/sites/default/files/document-files/Method_3_Invitro_0805 _D200%2CD200I.pdf (accessed on 3 August 2015).

41. Fay, J.P.; Cheng, K.-J.; Hanna, M.R.; Howarth, R.E.; Costerton, J.W. *In vitro* digestion of boat-safe and boat-causing legumes by rumen microorganisms: Gas and foam production. *J Dairy Sci.* **1980**, *63*, 1273–1281.

42. McDougall, E.I. Studies on ruminant saliva. 1. The composition and output of sheep's saliva. *Biochem. J.* **1948**, *43*, 99–109.

43. Tanino, K.; Weiser, C.J.; Fuchigami, L.H.; Chen, T.H. Water content during abscisic acid induced freezing tolerance in bromegrass cells. *Plant Physiol.* **1990**, *93*, 460–464.

44. Wanner, L.A.; Junttila, O. Cold-induced freezing tolerance in Arabidopsis. *Plant Physiol.* **1999**, *120*, 391–400.

45. Ørskov, E.R.; McDonald, I. The estimation of protein degradability in the rumen from incubation measurements weighted according to rate of passage. *J. Agric. Sci.* **1979**, *92*, 499–503.

46. Pollock, C.J. The response of plants to temperature change. *J. Agric. Sci.* **1990**, *115*, 1–5.

47. Levitt, J. Responses of plants to environmental stress. *Chilling, Freezing, and High Temperature Stresses*, 2nd ed.; Academic Press: New York, NY, USA, 1980; p. 447.

48. Tarkowski, Ł.P.; van den Ende, W. Cold tolerance triggered by soluble sugars: A multifaceted countermeasure. *Front. Plant Sci.* **2015**, *6*, 203.

49. Van den Ende, W. Multifunctional fructans and raffinose family oligosaccharides. *Front. Plant Sci.* **2013**, *4*, 247.

50. Panelo, J.S.; Redi, W.I.; Lorenzo M.; Tognetti, J. Efecto del Incremento de la Temperatura Sobre la Fotosíntesis y la Respiración en Plantas de Trigo Aclimatadas a Bajas Temperaturas. Available online: http://fisiologiavegetal.org/fv2014/abstract-index/abstracts/#905 (accessed on 21 September 2014).

51. Huner, N.P.A.; Palta, J.P.; Li, P.H.; Carter, J.V. Anatomical changes in leaves of Puma rye in response to growth at cold-hardening temperatures. *Bot. Gaz.* **1981**, *142*, 55–62.

52. Huner, N.P.A.; Oquist, G.; Sarhan, F. Energy balance and acclimation to light and cold. *Trends Plant Sci.* **1998**, *3*, 224–230.

53. Ndong, C.; Danyluk, J.; Huner, N.P.; Sarhan, F. Survey of gene expression in winter rye during changes in growth temperature, irradiance or excitation pressure. *Plant Mol. Biol.* **2001**, *45*, 691–703.

54. Gray, G.R.; Chauvin, L.P.; Sarhan, F.; Huner, N.P. Cold acclimation and freezing tolerance (A complex interaction of light and temperature). *Plant Physiol.* **1997**, *114*, 467–474.

55. Getachew, G.; Robinson, P.H.; DePeters, E.J.; Taylor, S.J. Relationships between chemical composition, dry matter degradation and *in vitro* gas production of several ruminant feeds. *Anim. Feed Sci. Technol.* **2004**, *111*, 57–71.

56. Lopez-Guisa, J.M.; Satter, L.D. Effect of forage source on retention of digesta markers applied to corn gluten meal and brewers grains for heifers. *J. Dairy Sci.* **1991**, *74*, 4297–4304.

57. John A.; Ulyatt M.J. Importance of dry matter content to voluntary intake of fresh grass forages. *Proc N. Z. Soc. Anim. Prod.* **1987**, *47*, 13–16.

58. Cabrera Estrada, J.I.; Delagarde, R.; Faverdin, P.; Peyraud, J.L. Dry matter intake and eating rate of grass by dairy cows is restricted by internal, but not external water. *Anim. Feed Sci. Technol.* **2004**, *114*, 59–74.

59. Allinson, D.W. Influence of photoperiod and thermoperiod on the IVDMD and cell wall components of tall fescue. *Crop Sci.* **1971**, *11*, 456–458.

60. Bertrand, A.; Tremblay, G.F.; Pelletier, S.; Castonguay, Y.; Bélanger, G. Yield and nutritive value of timothy as affected by temperature, photoperiod and time of harvest. *Grass Forage Sci.* **2008**, *63*, 421–432.

61. Thorvaldsson, G.; Tremblay, G.F.; Tapani Kunelius, H. The effects of growth temperature on digestibility and fibre concentration of seven temperate grass species. *Acta Agric. Scand. Sect. B* **2007**, *57*, 322–328.

62. Crasta, O.R.; Cox, W.J.; Cherney, J.H. Factors affecting maize forage quality development in the northeastern USA. *Agron. J.* **1997**, *89*, 251–256.

63. Agnusdei, M.G.; di Marco, O.N.; Nenning, F.R.; Aello, M.S. Leaf blade nutritional quality of rhodes grass (*Chloris gayana*) as affected by leaf age and length. *Crop Pasture Sci.* **2012**, *62*, 1098–1105.

64. Di Marco, O.N.; Harkes, H.; Agnusdei, M.G. Calidad de agropiro alargado (*Thinopyrum ponticum*) en estado vegetativo en relación con la edad y longitud de las hojas. *RIA* **2013**, *39*, 105–110.

65. Reynolds, S.G. *Pasture-Cattle-Coconut Systems*; FAO RAPA Publication: Bangkok, Thailand, 1995; p. 668.

66. Samarakoon, S.P.; Wilson, J.R.; Shelton, H.M. Growth, morphology and nutritive quality of shaded *Stenotaphrum secundatum*, *Axonopus compressus* and *Pennisetum clandestinum*. *J Agric. Sci.* **1990**, *114*, 161–169.

67. Livingston, D.P., III; Hincha, D.K.; Heyer, A.G.; Norio, S.; Noureddine, B.; Shuichi, O. The relationship of fructan to abiotic stress tolerance in plants. In *Recent Advances in Fructooligosaccharides Research*; Norio, S., Noureddine, B., Shuichi, O., Eds.; Research Signpost: Kerala, India, 2007; pp. 181–199.

68. Valluru, R.; van den Ende, W. Plant fructans in stress environments: Emerging concepts and future prospects. *J. Exp. Bot.* **2008**, *59*, 2905–2916.

69. Livingston, D.P., III; Hincha, D.K.; Heyer, A.G. Fructan and its relationship to abiotic stress tolerance in plants. *Cell Mol. Life Sci.* **2009**, *66*, 2007–2023.

70. Vágújfalvi, A.; Kerepesi, I.; Galiba, G.; Tischner, T.; Sutka, J. Frost hardiness depending on carbohydrate changes during cold acclimation in wheat. *Plant Sci.* **1999**, *144*, 85–92.

5

Variation in Response to Moisture Stress of Young Plants of Interspecific Hybrids between White Clover (*T. repens* L.) and Caucasian Clover (*T. ambiguum* M. Bieb.)

Athole H. Marshall *, Matthew Lowe and Rosemary P. Collins

Institute of Biological, Environmental and Rural Sciences, Aberystwyth University, Gogerddan, Aberystwyth, Ceredigion SY233EE, UK; E-Mails: mjl@aber.ac.uk (M.L.); rpc@aber.ac.uk (R.P.C.)

* Author to whom correspondence should be addressed; E-Mail: thm@aber.ac.uk

Academic Editor: Cory Matthew

Abstract: Backcross hybrids between the important forage legume white clover (*Trifolium repens* L.), which is stoloniferous, and the related rhizomatous species Caucasian clover (*T. ambiguum* M. Bieb), have been produced using white clover as the recurrent parent. The effect of drought on the parental species and two generations of backcrosses were studied in a short-term glasshouse experiment under three intensities of drought. Plants of Caucasian clover maintained a higher leaf relative water content and leaf water potential than white clover at comparable levels of drought, with the response of the backcrosses generally intermediate between the parents. Severe drought significantly reduced stolon growth rate and leaf development rate of white clover compared to the control, well-watered treatment, whilst differences between these two treatments in the backcross hybrids were relatively small. The differences between parental species and the backcrosses in root morphology were studied in 1m long vertical pipes. The parental species differed in root weight distribution, with root weight of Caucasian clover significantly greater than white clover in the 0.1 m to 0.5 m root zone. The backcrosses exhibited root characteristics intermediate between the parents. The extent to which these differences influence the capacity to tolerate drought is discussed.

Keywords: white clover; interspecific hybrids; drought; leaf development rate; root weight distribution

1. Introduction

Changing climatic conditions mean that the growing demand for meat and milk based products must be met against a backdrop of rising global temperatures and changing patterns of precipitation [1]. Extreme weather events, including periods of drought, will increasingly become a major factor limiting crop productivity in many parts of the world, including the UK [2]. Adaptation of agriculture to predicted climate change scenarios is essential, with the development of improved plant varieties better able to tolerate periods of drought [1] increasingly a key objective of many plant breeding programmes [3]. Selection criteria that will lead to new improved varieties of wheat [4,5] and grain legumes [6,7] better able to cope with drought are being developed. Grassland systems face similar challenges from climate change, therefore the development of new varieties of forage grasses and legumes better able to tolerate periods of drought is crucial.

The most important forage legume component of temperate pastures is white clover (*Trifolium repens* L.) [8], a nitrogen fixing species that produces forage of high quality. It is an outbreeding, highly heterozygous allotetraploid ($2n = 4x = 32$) species and the wide genetic variation within its gene pool has been used successfully in the production of new varieties with improvements in many traits. Less variation has been identified for traits such as drought tolerance, which have proved difficult to improve significantly by conventional selection methods [9]. Although some authors [10] showed differences between ten white clover cultivars with respect to their response to drought, others [11] found little variation in response to a drought stress gradient between six lines (three cultivars and three germplasm accessions). Selection for deeper, more extensive root systems has been recommended for better tolerance to intermittent drought [12]. Selection for thicker roots as an indirect selection criterion has, however, been unsuccessful [13], although selection for increased root weight ratio (proportion of total plant DM allocated to roots) was found to improve the growth and survival of white clover in drought prone environments [14].

Introgression of genes from closely related species has been used successfully to introduce desirable traits into white clover [15–18] including improved drought tolerance [19]. Caucasian or Kura Clover (*Trifolium ambiguum* M. Bieb) is a strongly rhizomatous perennial legume species with good drought tolerance and persistence [20]. It is considered to have a wider range of adaptation than white clover [21], although slow seedling establishment tends to reduce its competitiveness with grasses in mixtures [22]. The extensive root and rhizome system is thought to act as a nutrient store that can be remobilised and used for growth, thus allowing this species to persist under stressful conditions [23]. Hybrids have been developed between white clover and Caucasian clover with the objective of introgressing the rhizomatous trait from Caucasian clover into white clover [16] as a strategy for improving drought tolerance whilst retaining the desirable agronomic traits associated with the latter species. Fertile backcrosss (BC) hybrids (derived from backcrossing to white clover) have been produced and these are essentially like white clover, but with rhizomes as well as stolons. A drought experiment comparing the BC1 and BC2 hybrids with the white clover and Caucasian clover parents in deep soil bins [16] showed that the backcross hybrids maintained lower values of leaf relative water content (RWC) and leaf water potential than Caucasian clover, but higher levels than white clover at comparable levels of drought. The mechanism by which Caucasian clover maintains a higher leaf RWC is not known, nor is the extent to which this mechanism operates within the hybrids. However, previous studies have shown that the hybrids allocate a higher proportion of their total DM yield to roots than white clover *i.e.*, they maintain a higher root to shoot ratio [16]. Previous studies on white clover have shown that stolon growth and leaf development rate (LDR) are reduced by drought [24,25], but little is known about the effect of drought on these growth parameters in the backcross hybrids.

This study had the following objectives: firstly, to quantify the response of the backcross hybrids to drought; and secondly, to identify the extent to which ability to withstand drought may be related to differences in root depth distribution.

2. Materials and Methods

2.1. Experiment 1

2.1.1. Plant Material and Experimental Treatments

The *T. ambiguum* (Caucasian clover) accession Ah1254, collected in Turkey in 1971, and the *T. repens* (white clover) medium-leaved variety Menna were used in the hybridization programme. Fertile F1 plants were used as the basis for two generations of backcrossing to white clover as the recurrent parent. Details of the development of these backcrosses including methods of embryo rescue used in the development of the original hybrids and their morphological characterisation have been described previously [8,16]. Four genotypes within each of the white clover, Caucasian clover, BC1 and BC2 populations, selected based on their use in the development of the backcross populations, were cloned to provide six-plants of each genotype so that there were two clonal plants of each genotype available for each of three drought regimes. The genotypes of the BC1 and BC2 were selected on the basis of the presence of rhizomes and had been used in previous studies on forage yield and quality [16]. Clonal plants were obtained by removing a growing point with three nodes and planting in multi-compartment trays containing John Innes No. 3 compost. When they had produced at least three trifoliate leaves they were transplanted into 25 cm diameter × 27 cm deep pots filled with John Innes No. 3 compost. No rhizobia were added to the soil however nodules were observed on plant roots.

2.1.2. Drought Tolerance

There were three treatments: control (C) plants maintained at field capacity; moderate drought (M) plants maintained at 80% field capacity; severe drought (S) plants maintained at 65% field capacity. Field capacity was defined as the volume of water required for the soil within the pot to be saturated and was determined daily on the control plants. The M and S plants received 80% and 65% respectively of the quantity of water required by the C plants to maintain them at field capacity. This was repeated daily throughout the course of the experiment.

The experiment began when the plants were 3 months old, when they were cut to a height of 3cm above ground level. At 21 and 35 days after the start of the experiment, pre-dawn leaf water potential was measured. Two leaflets were sampled per plant and leaf water potential measured using a pressure bomb (Portable plant moisture system SKPM 1400/40; Skye Instruments Ltd. Llandrindod Wells, UK) using the method described previously [25]. After 21 and 35 days, leaf relative water content (RWC) was determined on three leaves per plant as described [16] using the formula

$$RWC = ((FW - DW) / (RW - DW)) \times 100$$

where FW = fresh weight, RW = rehydrated weight and DW = dry weight.

2.1.3. Plant Growth

Non-destructive measurements of stolon length and leaf development rate (LDR) were carried out on one rando mLy selected stolon per plant. At the beginning of the experiment the selected stolon was marked with an acrylic paint dot behind the youngest fully expanded leaf. After 7, 14, 21 and 28 days, stolon length from the tip of the growing point to the paint mark was measured and leaf development recorded using the criteria established by Carlson [26]: all leaves produced after the paint mark were given a score using the Carlson visual scale for leaf development, where 1.0 indicates a fully expanded leaf and 0.1 indicates a leaf just visible as it emerges. The sum of these scores was calculated for the measured stolon. The absence of stolons in Caucasian clover and the difficulty of measuring LDR in this species meant that this part of the experiment only compared white clover with the BC1 and BC2 hybrids. Thirty five days after the start of the experiment all plants were cut to a height of 3 cm above soil level. The leaf area of three leaves per plant was measured using a Delta-T-Devices leaf area meter and the dry weight of above ground material determined by drying for 12 h at 80 °C in a forced draught oven.

2.2. Experiment 2

Root Depth Distribution

Four clonal plants of each of the four genotypes of the populations used in Experiment 1 were obtained as described previously and planted into multi-compartment trays containing John Innes No. 3 compost. When they had produced three trifioliate leaves, they were transplanted into 1 m deep × 15 cm diameter plastic pipes with several drainage holes drilled in the base, into which was inserted a polythene tube filled with vermiculite. The pipes were placed vertically on a gravel bed in a glasshouse maintained at ambient temperature. The plants received 100 mL water daily and once a week received an additional 50 mL of a standard full-nutrient solution [27]. After ten weeks the polythene tube was removed from the pipe and the above ground foliage cut to ground level with hand held shears. The root column was removed and separated into 10 cm deep horizontal sections. The roots within each section were removed by washing under running water. The dry weight of the above ground biomass and root biomass within each section were determined after drying at 80 °C for 24 h in a forced draught oven.

2.3. Data Analysis

Experiment 1 was established as a split-plot design with two replicate blocks, comprising drought treatments as whole plots and genotypes as sub-plots. Growth parameters (leaf water potential, leaf relative water content, leaf development rate, stolon growth rate, dry matter yield and leaf size) were analysed by analysis of variance (ANOVA) using GenStat® (VSN International, Hemel Hempstead, UK) Release 13 [28] to determine significant effects of population, genotype within population and drought, and their interactions. Experiment 2 was established as a split-plot design with four replicate blocks, comprising populations as whole plots and genotypes as sub-plots. Root dry weight at each depth was analysed separately by ANOVA as above to determine significant effects of population and genotype within population.

3. Results

3.1. Experiment 1

3.1.1. Overall analysis

For most of the growth parameters measured there were significant effects of population and drought, and significant population × drought interactions, but no significant differences between genotypes within populations (Table 1). Consequently for all growth parameters only the population × drought means are presented.

Table 1. Significance levels for effect of drought, population and their interaction on plant growth parameters.

Treatment	LWP		Leaf RWC		SGR	LDR	DM Yield	Leaf Area
	21 Day	35 Day	21 Day	35 Day				
Drought (D)	NS	***	**	*	*	*	**	*
Population (P)	NS	***	***	***	***	***	***	***
D × P	***	***	***	***	NS	NS	***	***

NS, not significant; * $p < 0.05$; ** $p < 0.01$; *** $p < 0.001$; Key to abbreviations: LWP—leaf water potential, Leaf RWC—leaf relative water content, SGR—stolon growth rate, LDR—leaf development rate, DM yield—dry matter yield.

3.1.2. Plant Water Status

Results for the effects of the drought treatments and population on leaf water potential (LWP) are presented in Table 2. Twenty one days after the start of the experiment overall values of LWP were not affected by drought treatment, nor was there a difference between populations. However, there was a significant drought treatment × population interaction, such that LWP in white clover decreased more under the S drought treatment compared with LWP in Caucasian clover and the backcross hybrids. Thirty five days after the start of the experiment, drought treatment had a significant effect on overall values of LWP, which were greatly reduced under treatment S, followed by treatment M, and both were less than under the well-watered control treatment C. There was also a significant difference between populations and a significant drought × population interaction. As a result, LWP in white clover was significantly lower than in the other populations, and the magnitude of this reduction was greatest under the most severe drought treatment. Leaf RWC was significantly influenced by drought, population and there was a significant drought × population interaction when measured 21 and 35 days after the start of the experiment (Table 3). After 21 days, leaf RWC was lower under S than under M and C and in Caucasian clover was greater than that of white clover and the BC1 and BC2 hybrids. Leaf RWC of Caucasian clover was unaffected by moisture stress however in white clover and the BC1 and BC2 hybrids the leaf RWC was significantly lower under S than M and C. A similar result was observed after 35 days with the leaf RWC of Caucasian clover unaffected by moisture stress but the leaf RWC of white clover and the BC1 and BC2 hybrids significantly reduced under S in comparison with M and C.

Table 2. Leaf water potential (MPa) of Caucasian clover, white clover, BC1 and BC2 hybrids after 21 and 35 days at three levels of drought. C—control treatment, M—moderate moisture stress, S—severe moisture stress.

| Population | Days after Start of Drought | | | | | |
| | 21 | | | 35 | | |
	C	M	S	C	M	S
Caucasian Clover	−0.47	−0.58	−0.76	−0.50	−0.73	−0.76
White Clover	−0.27	−0.51	−1.35	−0.39	−0.85	−2.00
BC1	−0.32	−0.57	−0.80	−0.30	−0.69	−1.69
BC2	−0.26	−0.41	−0.94	−0.32	−0.62	−1.49
S.e.d.						
Drought (D)	0.270 NS			0.037 ***		
Population (P)	0.072 NS			0.059 ***		
D × P	0.291 *** (0.124 ***)			0.092 *** (0.097 ***)		

NS not significant; *** $p < 0.001$; S.e.d in brackets to be used when comparing means with same level of drought.

Table 3. Leaf relative water content (%) of Caucasian clover, white clover, BC1 and BC2 hybrids after 21 and 35 days at three levels of drought. C—control treatment, M—moderate moisture stress, S—severe moisture stress.

| Population | Days after Start of Drought | | | | | |
| | 21 | | | 35 | | |
	C	M	S	C	M	S
Caucasian Clover	93.1	94.3	93.1	92.2	92.9	92.0
White Clover	91.2	90.9	69.3	91.6	92.9	71.3
BC1	94.1	93.2	76.5	92.7	92.5	76.4
BC2	93.1	93.1	69.9	92.3	93.1	68.4
S.e.d.						
Drought (D)	1.23 **			1.96 *		
Population (P)	1.61 ***			1.63 ***		
D × P	2.72 *** (2.79 ***)			3.13 *** (2.82 ***)		

S.e.d. in brackets to be used when comparing means with same level of drought. * $p < 0.05$; ** $p < 0.01$; *** $p < 0.001$.

3.1.3. Plant Growth

Stolon growth rate (SGR) and leaf development rate (LDR) were influenced by drought and population but there was no significant interaction (Table 4). Drought reduced SGR, and generally the

SGR of white clover was significantly higher than the BC2 and both were higher than in the BC1. The LDR of white clover was significantly greater than the backcross hybrids which were not significantly different from each other. Drought treatment reduced LDR but only under S; under M and C it did not differ significantly. Leaf area was significantly influenced by drought, differed between populations and there was a significant drought × population interaction. Generally leaf area was reduced by drought and the leaf area of white clover was greater than the BC1 and BC2 hybrids with the leaf area of Caucasian clover smallest. Leaf area of white clover and Caucasian clover was reduced by the M treatment and the leaf area of white clover further significantly reduced under the S treatment, unlike Caucasian clover which showed no further reduction in leaf area. The BC1 and BC2 hybrids exhibited a similar response to the S treatment as white clover.

Table 4. Stolon growth rate (mm/7 days), leaf development rate (quantified using Carlson Scale) and leaf area (mm^2) of Caucasian clover, white clover, BC1 and BC2 hybrids after 35 days at three levels of drought. C—control treatment, M—moderate moisture stress, S—severe moisture stress.

Population	Stolon Growth Rate			Leaf Development Rate			Leaf Area (mm^2)		
	C	M	S	C	M	S	C	M	S
Caucasian Clover	-	-	-	-	-	-	358.2	303.0	184.9
White Clover	3.8	4.0	1.1	9.9	10.0	5.6	280.4	241.9	178.6
BC1	2.1	1.6	0.4	6.1	7.0	4.8	309.8	265.4	257.0
BC2	3.1	2.3	0.6	6.4	7.0	4.3	322.2	248.3	220.8
Drought (D)	0.39 *			0.38 *			10.90 *		
Population (P)	0.32 ***			0.43 ***			10.19 ***		
D × P	0.65 NS (0.55 NS)			0.73 NS (0.75 NS)			18.77 *** (17.64 ***)		

NS, not significant; * $p < 0.05$; *** $p < 0.001$; S.e.d. in brackets to be used when comparing means with same level of drought.

Overall DM yield per plant was greater under C than in M and both greater than under the S treatment (Table 5). DM yield of white clover was significantly greater than the BC1 and BC2 hybrids and all had DM yields significantly greater than Caucasian clover reflecting the slow establishment of this species. There was also a significant drought × population interaction as drought had no significant effect on the DM yield of Caucasian clover but the DM yield of white clover and the BC1 and BC2 hybrids was significantly reduced by drought stress but white clover was reduced by a greater amount than the hybrids.

3.2. Experiment 2

There was a significant difference between populations in root dry weight to depths of 0.5 m and significant differences between genotypes within populations (Table 6). However, at depths below 0.5 m differences between populations were small and insignificant and are not shown. Root dry weight of white clover and Caucasian clover in the 0 to 0.1 m root zone was comparable (Figure 1). However, in subsequent zones, up to a depth of 0.5 m, the root dry weight of Caucasian clover was significantly greater than that of white clover (Figure 1). Apart from the 0 to 0.1 m root zone where the BC2 had the greatest root dry weight, the root dry weight of the BC1 and BC2 hybrids were not significantly different and were generally intermediate between the two parental species. Differences in root dry weight between genotypes of white clover, BC1 and BC2 hybrids were observed at depths of 0.1–0.4 m but no significant differences between genotypes of Caucasian clover were observed.

Table 5. Dry matter yield (g/plant) of Caucasian clover, white clover, BC1 and BC2 hybrids after 35 days at three levels of drought. C—control treatment, M—moderate moisture stress, S-severe moisture stress.

Population	Moisture Level		
	C	M	S
Caucasian Clover	3.3	2.1	1.5
White Clover	30.8	20.2	4.4
BC1	22.7	17.7	4.9
BC2	26.9	16.1	4.4
Drought (D)	0.98 **		
Population (P)	0.75 ***		
P × S	1.49 *** (1.30 ***)		

** $p < 0.01$; *** $p < 0.001$; S.e.d. in brackets to be used when comparing means with same level of drought.

Table 6. Significance levels for effect of population and genotype within population on root dry weight at different depths.

Significance	Root Depth (m)				
	0.1	0.2	0.3	0.4	0.5
Population	*	**	***	***	***
Genotypes within Population	***	**	**	**	NS

NS, not significant; * $p < 0.05$; ** $p < 0.01$; *** $p < 0.001$.

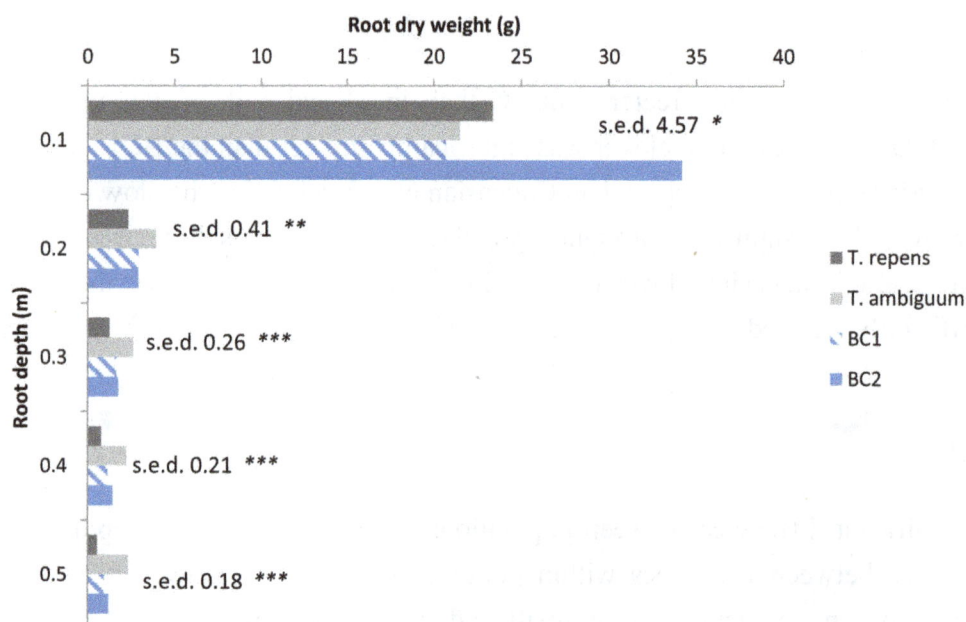

Figure 1. Root dry weight in 0.1 m sections of soil columns containing Caucasian clover, white clover, BC1 and BC2 hybrids. NS, not significant; * $p < 0.05$; ** $p < 0.01$; *** $p < 0.001$.

4. Discussion

4.1. Plant Water Status in Response to Drought

It has been predicted that climate change will affect the distribution patterns of rainfall [29] and that this may have a negative impact on grassland productivity [30]. Development of forage varieties with a greater ability to tolerate drought is therefore an increasingly important target for forage plant breeding programmes [31]. Temperate pasture species such as white clover, whose agronomic yield consists of foliage, are considered to be more susceptible to drought stress than cereals or grain legumes [32]. One strategy to improve the drought tolerance of white clover is the development of interspecific hybrids between white clover and the drought tolerant, rhizomatous species Caucasian clover. Although the value of Caucasian clover as a drought tolerant species has been recognised in many Mediterranean areas, it has not been fully exploited due to its slow establishment [33].

Interspecific hybrids have been developed to introgress the rhizomatous trait into white clover without compromising the DM yield and forage quality of white clover [34,35]. Confirmation that Caucasian clover has a greater tolerance of drought than white clover has been clearly provided by the present experiment. Here, plants of Caucasian clover were found to maintain a higher leaf RWC and LWP than white clover under moderate and severe drought, confirming results from previous experiments carried out in deep soil bins [16]. Improved drought tolerance was also evident in the BC1 and BC2 hybrids, which maintained higher values of LWP and RWC than white clover, although lower than Caucasian clover. Consequently, after 35 days of the experiment, the plants of white clover were visibly wilting whilst those of the BC1 and BC2 hybrids were still turgid. In this experiment, plants were grown individually in large pots with limited opportunity for development of adventitious roots and rhizomes and were relatively juvenile plants when measured. Therefore care is necessary in extrapolating results directly to a field situation. Nevertheless, the results are comparable with previous experiments where genotypes of these populations were grown in deep soil bins and growth was typical of that observed in a previous a field plot experiment, in which adventitious roots were able to develop [16].

4.2. Root Depth Distribution

The objective of this experiment was to determine whether differences in response to drought in this germplasm might be associated with variation in root depth distribution. Plants tolerate drought through a range of mechanisms that include enhanced capture of soil moisture, limiting water loss and retention of cellular rehydration [36]. Rooting depth is one factor that influences soil moisture uptake. Drought resistant cultivars of species such as bean (*Phaseolus vulgaris* L.) [37] are those with thicker and deeper penetrating roots, and a higher proportion of total dry matter allocated to roots. In white clover, significant genetic variation has been observed for many root characters [38], some of which might be expected to confer improved drought tolerance. Some authors [39] for example, found significant genetic variation in seedling root system depth, and a high heritability for this character, within a collected population of white clover. These differences were related to dry matter yield under drought conditions, with genotypes from the "long" selection group producing higher yields than those with shorter roots. However, selection for thicker roots in white clover had varied success [13,14]. In the present experiment, Caucasian clover had a greater root weight than white clover at depths below 0.2 m. The BC1 and BC2 hybrids also had a greater root weight than white clover at depths below 0.2 m. This suggests that the introgression approach increased the allocation of resources to roots at lower depths thereby contributing to the improved drought tolerance of the hybrids compared to white clover.

When considering the effect of root system type on plant persistence in conditions of drought it is important to take into account both the timing of soil water deficits and the use of stored soil water in relation to crop phenology [40]. Thus, where soil water reserves exist at depth, the ability of plants to produce deep, extensive root systems is likely to be advantageous [41]. Conversely, where moisture reserves are confined to the upper layers of soil, then rooting depth becomes less important than the ability of plants to produce an efficient extraction system in these surface layers [41]. Although depth of rooting depends on soil type, cultivar and management, white clover is generally considered to be a shallow-rooted species, with most roots distributed in the top 0.1–0.2 m of soil [38]. This distribution pattern is likely to be detrimental to the ability of the species to persist under drought and suggests why the hybrids, with greater root weight than white clover below 0.2m, are more drought tolerant. In addition, there is evidence that the growth of white clover root systems *per se* is adversely affected by drought conditions. For example, the production of new roots of white clover has been found to be greatly reduced under drought stress [42,43]. Studies [44] comparing root system development in seedlings of three legume species, *T. repens*, *Lotus corniculatus* and *Medicago lupulina*, under different levels of drought found that the depth of penetration of the root system of white clover was considerably more reduced by drought than that of the other species. Consequently, in deep, moist soils the shallow root system of white clover was found to be adequate to sustain vegetative growth and seed production, but in conditions of drought an inability to penetrate the soil profile effectively was found not only to reduce dry matter yield but also to lead to lower seed production per plant and, ultimately, to plant mortality [44]. The effect of drought on the root growth of the BC hybrids was not included in the present study but could be the focus of future work, particularly including an analysis of root growth in different soil types.

4.3. Plant Growth and Development

Persistence and dry matter yield of white clover within mixed swards is determined by stolon growth and development and the rate at which leaves are produced at stolon nodes [45]. In white clover, stolon growth and leaf development rate have been shown to be reduced by drought stress [24,25] In one study [25], a moderate drought reduced leaf development rate by 40% and also reduced individual leaf area, effects that would have a considerable detrimental effect on DM yield if replicated in a mixed sward. A similar response to drought by white clover was observed in the present experiment, as severe drought significantly reduced stolon growth rate and production of new leaf area. Although the stolon growth, LDR and leaf area of the BC1 and BC2 hybrids were also reduced by severe drought the difference between growth in the well-watered control and the severe drought was, unlike white clover, relatively small. Both the moderate and severe drought treatment reduced the DM yield of white clover, BC1 and BC2 hybrids in comparison with the well-watered control. However, the greater drought tolerance in the hybrids was only reflected in a slight reduction in DM yield of the BC1 under moderate drought compared with white clover and the BC2. The reduction in DM yield of white clover, BC1 and BC2 hybrids was comparable under the severe drought treatment. Some similarity between white clover and the BC2 is not surprising since they are very closely related but it is possible that greater differences would have been observed had the experiment continued for longer than 35 days.

Although the BC1 and BC2 hybrids are essentially white clover-like in appearance, one of the surprising results was the lower stolon growth rate and of LDR both BC hybrids in the well-watered treatment compared with the white clover parent. No evidence of differences in stolon growth rate has been found during characterisation of this germplasm in spaced plant nurseries [8]. Further agronomic studies, carried out in field plots, have shown that the DM yield of the BC1 and BC2 hybrids when grown with *L. perenne* was slightly lower than in white clover in the first harvest year [34]. However, DM yield did improve in subsequent years and was comparable with white clover in the 2nd and 3rd harvest years, suggesting that any differences in stolon growth may not be significant for long term pasture performance. Caucasian clover is a persistent species, but slow to establish with consequently low initial DM yields and grows poorly in pastures compared with white clover [21], largely due to its rhizomatous growth. The low stolon growth rate and DR may be a consequence of the introgression of the rhizomatous trait from Caucasian clover and further selection for improved stolon growth rate may be required.

5. Conclusions

Analysis of plant water status (leaf relative water content, leaf water potential) and plant growth and development in response to different levels of drought revealed significant variation between Caucasian clover and white clover, with an intermediate response of backcross hybrids between these two species. The parental species differed in root depth distribution and in root weight distribution, with root weight of Caucasian clover significantly greater than white clover in the 0.1 m to 0.5 m root zone with the backcross hybrids exhibiting root characteristics intermediate between the parental species. It is suggested that differences in root distribution is a factor influencing the extent to which plants are able to tolerate drought, however, further studies are required to quantify the impact of the differences in root distribution and rhizomes in mature plants.

Acknowledgments

This research was funded by the Department for the Environment, Food and Rural Affairs through the Sustainable Livestock LINK programme.

Author Contributions

Athole Marshall contributed to the experimental analysis and writing of the manuscript, aided by Rosemary Collins. Matthew Lowe was responsible for management of the experiments and data collection.

Conflicts of Interest

The authors declare no conflict of interest.

References

1. Foresight. *The Future of Food and Farming. Final Project Report*; The Government Office for Science: London, UK, 2011.
2. Hopkins, A.; Del Prado, A. Implications of climate change for grassland in Europe: Impacts, adaptations and mitigation options: A review. *Grass Forage Sci.* **2007**, *62*, 118–126.
3. Abberton, M.T.; MacDuff, J.H.; Marshall, A.H.; Humphreys, M.W. The genetic improvement offorage grasses and legumes to enhance adaptation of grasslands to climate change. In Proceedings of the United National Climate Change Conference, Nusa Dua, Indonesia, 3–14 December 2007.
4. Foulkes, M.J.; Sylvester-Bradley, R.; Weightman, R.; Snape, J.W. Identifying physiological traitsassociated with improved drought resistance in winter wheat. *Field Crops Res.* **2007**, *103*, 11–14.
5. Dodd, I.C.; Whalley, W.R.; Ober, E.S.; Parry, M.A.J. Genetic and management approaches to boost UK winter wheat yields by ameliorating water deficits. *J. Exp. Bot.* **2011**, *62*, 5241–5248.
6. Lizana, C.; Wentworth, M.; Martinez, J.P.; Villegas, D.; Meneses, R.; Murchie, E.H.; Pastenes, C.; Lercari, B.; Vernieri, P.; Horton, P.; *et al.* Differential adaptation of two varieties of common bean to abiotic stress. I. Effects of drought on yield and photosynthesis. *J. Exp. Bot.* **2006**, *57*, 685–697.
7. Martinez, J.P.; Silva, H.; Ledent, J.F.; Pinto, M. Effect of drought stress on the osmotic adjustment, cell wall elasticity and cell volume of six cultivars of common beans (*Phaseolus vulgaris* L.). *Eur. J. Agron.* **2007**, *6*, 30–38.
8. Abberton, M.T.; Michaelson-Yeates, T.P.T.; Marshall, A.H.; Holdbrook-Smith, K.; Rhodes, I. Morphological characteristics of hybrids between white clover, *Trifolium. repens* L. and Caucasian clover, *Trifolium. ambiguum* M. Bieb. *Plant Breed.* **1998**, *117*, 494–496.
9. Abberton, M.T.; Marshall, A.H. Progress in breeding perennial clovers for temperate agriculture. *J. Agric. Sci.* **2005**, *143*, 117–135.
10. Barbour, M.; Caradus, J.R.; Woodfield, D.R.; Silvester, W.B. Water stress and water use efficiency of ten white clover cultivars. In *White Clover: New Zealand's Competitive Edge*, Woodfield, D.R., Ed.; Grassland Research and Practice Series No. 6; New Zealand Grassland Association: Palmerston North, New Zealand, 1996; pp. 159–162.
11. Brink, G.E.; Pederson, G.A. White clover response to a water application gradient. *Crop Sci.* **1998**, *38*, 771–775.
12. Collins, R.P. The effect of drought stress and winter stress on the persistence of white clover. In *Lowland Grasslands of Europe: Utilization and Development*; Fisher, G., Frankow-Lindberg, B.E., Eds.; REUR Technical Series No. 64; FAO: Rome, Italy, 2002; pp. 17–32.

13. Annicchiarico, P.; Piano, E. Indirect selection for root development of white clover and implications for drought tolerance. *J. Agron. Crop Sci.* **2004**, *190*, 28–34.

14. Caradus, J.R.; Woodfield, D.R. Genetic control of adaptive root characteristics in white clover. *Plant Soil* **1998**, *200*, 63–69.

15. Hussain, S.W.; Williams, W.M.; Mercer, C.F.; White, D.W.R. Transfer of clover cyst nematode resistance from *Trifolium. nigrescens* Viv. to *T. repens* by interspecific hybridisation. *Theor. Appl. Genet.* **1997**, *95*, 1274–1281.

16. Marshall, A.H.; Rascle, C.; Abberton, M.T.; Michaelson-Yeates, T.P.T.; Rhodes, I. Introgression as a route to improved drought tolerance in white clover (*Trifolium. repens* L.). *J. Agron. Crop Sci.* **2001**, *187*, 11–18.

17. Williams, W.M.; Hussain, S.W. Development of a breeding strategy for interspecific hybrids between Caucasian clover and white clover. *NZ J. Agric. Res.* **2008**, *51*, 115–126.

18. Williams, W.M. Trifolium interspecific hybridisation: Widening the white clover gene pool. *Crop Pasture Sci.* **2014**, *65*, 1091–1106.

19. Nichols, S.N., Hofman, R.W., Williams, W.M. Drought resistance of Trifolium repens × *Trifolium uniflorum* interspecific hybrids. *Crop Pasture Sci.* **2014**, *65*, 911–921.

20. Coolbear, P.; Hill, M.J.; Efendi, F. Relationships between vegetative and reproductive growth in a four year old stand of Caucasian clover (*Trifolium. ambiguum* M Bieb.) cv. Monaro. *Proc. Agron. Soc. N. Z.* **1994**, *24*, 77–82.

21. Taylor, N.L.; Smith, R.R. Kura clover (*Trifolium. ambiguum* M.B.) breeding, culture and utilization. *Adv. Agron.* **1998**, *63*, 153–178.

22. Black, A.D.; Moot, D.J.; Lucas, R.J. Development and growth characteristics of Caucasian and white clover seedlings, compared with perennial ryegrass. *Grass Forage Sci.* **2006**, *61*, 442–453.

23. Fu, S.M.; Hill, M.J.; Hampton, J.G. Root system development in Caucasian clover cv. Monaro and its contribution to seed yield. *N. Z. J. Agric. Res.* **2001**, *44*, 23–29.

24. Turner, L.B. The effect of water stress on the vegetative growth of white clover (*T. repens* L.): comparison of long-term water deficit and a short-term developing drought. *J. Exp. Bot.* **1991**, *42*, 311–316.

25. Belaygue, C.; Wery, J.; Cowan, A.A.; Tardieu, F. Contribution of leaf expansion, rate of leaf appearance and stolon branching to growth of plant leaf area under water deficit in white clover. *Crop Sci.* **1996**, *36*, 1240–1246.

26. Carlson, G.E. Growth of clover leaves developmental morphology at ten stages. *Crop Sci.* **1996**, *6*, 293–294.

27. Hoagland, D.R.; Snyder, W.C. Nutrition of strawberry under controlled conditions: (a) Effects of deficiencies of boron and certain other elements: (b) Susceptibility to injury from sodium salts. *Proc. Am. Soc. Hortic. Sci.* **1933**, *30*, 288–296.

28. Payne, R.W.; Murray, D.A.; Harding, S.A.; Baird, D.B.; Soutar, D.M. *Introduction to GenStat® for Windows™*; VSN International: Hemel Hempstead, UK, 2010.

29. Humphreys, M.W.; Yadav, R.S.; Cairns, A.J.; Turner, L.B.; Humphreys, J.; Skøt, L. A changing climate for grassland research. *New Phytol.* **2006**, *169*, 9–26.

30. Humphreys, M.O. Grass roots for improved soil structure and hydrology. *IBERS Knowl.-Based Innov.* **2011**, *2011*, 21–25.

31. Humphreys, M.O. Genetic improvement of forage crops—Past, present and future. *J. Agric. Sci.* **2005**, *143*, 441–448.

32. Turner, N.C.; Begg, J.E. Responses of pasture plants to water deficits. In *Plant Relations in Pastures*; Wilson, J.R., Ed.; CSIRO: Melbourne, Australia, 1978; pp. 50–66.

33. Widdup, K.H.; Knight, T.L.; Waters, C.J. Genetic variation for rate of establishment in Caucasian clover. *Proc. Agron. Soc. N. Z.* **1998**, *60*, 213–217.

34. Marshall, A.H.; Williams, T.A.; Abberton, M.T.; Michaelson-Yeates, T.P.T.; Powell, H.G. Dry matter production of white clover (*Trifolium. repens* L.), Caucasian clover (*T. ambiguum* M. Bieb) and their associated hybrids when grown with a grass companion over three harvest years. *Grass Forage Sci.* **2003**, *59*, 91–99.

35. Marshall, A.H.; Williams, T.A.; Abberton, M.T.; Michaelson-Yeates, T.P.T.; Olyott, P.; Powell, H.G. Forage quality of white clover (*Trifolium. repens* L.) × Caucasian clover (*T. ambiguum* M. Bieb) hybrids when grown with a grass companion over three harvest years. *Grass Forage Sci.* **2004**, *59*, 91–99.

36. Blum, A. Drought resistance, water-use efficiency, and yield potential—Are they compatible, dissonant or mutually exclusive? *Aust. J. Agric. Res.* **2005**, *56*, 1159–1168.

37. Sponchiado, B.N.; White, J.W.; Castillo, J.A.; Jones, P.G. Root growth of four common bean cultivars in relation to drought tolerance in environments with contrasting soil types. *Exp. Agric.* **1989**, *25*, 249–257.

38. Caradus J.R. The structure and function of white clover root systems. *Adv. Agron.* **1990**, *43*, 1–46.

39. Ennos, R.A. The significance of genetic variation for root growth within a natural population of white clover (*Trifolium. repens* L.). *J. Ecol.* **1985**, *73*, 615–624.

40. Clarke, J.M.; McCaig, T.N. Breeding for efficient root systems. In *Plant breeding—Principles and Prospectsi*; Hayward, M.D., Bosemark, N.O., Romagosa, I., Eds.; Chapman & Hal: London, UK, 1993; pp. 485–499.

41. Wilson, D. Breeding for morphological and physiological traits. In *Plant Breeding II*; Frey, K.J, Ed.; Iowa State University Press: Ames, IA, USA, 1981; pp. 233–290.

42. Thomas, H. Effects of drought on growth and competitive ability of perennial ryegrass and white clover. *J. Appl. Ecol.* **1984**, *21*, 591–602.

43. Stevenson, C.A.; Laidlaw, A.S. The effect of moisture stress on stolon and adventitious root development in white clover (*Trifolium. repens* L.). *Plant Soil* **1985**, *85*, 249–257.

44. Fouldes, W. Response to soil moisture supply in three leguminous species. I. Growth, reproduction and mortality. *New Phytol.* **1978**, *80*, 535–545.

45. Elgersma, A.; Fengrui, L. Effects of cultivar and cutting frequency on dynamics of stolon growth and leaf appearance rate in white clover grown in mixed swards. *Grass Forage Sci.* **1997**, *3*, 370–380.

6

A Modified Thermal Time Model Quantifying Germination Response to Temperature for C_3 and C_4 Species in Temperate Grassland

Hongxiang Zhang [1], Yu Tian [2] and Daowei Zhou [1,*]

[1] Northeast Institute of Geography and Agroecology, Chinese Academy of Sciences, Changchun 130012, China; E-Mail: zhanghongxiang@iga.ac.cn

[2] Animal Science and Technology College, Jilin Agricultural University, Changchun 130118, China; E-Mail: tiany0115@163.com

* Author to whom correspondence should be addressed; E-Mail: zhoudaowei@iga.ac.cn

Academic Editor: Cory Matthew

Abstract: Thermal-based germination models are widely used to predict germination rate and germination timing of plants. However, comparison of model parameters between large numbers of species is rare. In this study, seeds of 27 species including 12 C_4 and 15 C_3 species were germinated at a range of constant temperatures from 5 °C to 40 °C. We used a modified thermal time model to calculate germination parameters at suboptimal temperatures. Generally, the optimal germination temperature was higher for C_4 species than for C_3 species. The thermal time constant for the 50% germination percentile was significantly higher for C_3 than C_4 species. The thermal time constant of perennials was significantly higher than that of annuals. However, differences in base temperatures were not significant between C_3 and C_4, or annuals and perennial species. The relationship between germination rate and seed mass depended on plant functional type and temperature, while the base temperature and thermal time constant of C_3 and C_4 species exhibited no significant relationship with seed mass. The results illustrate differences in germination characteristics between C_3 and C_4 species. Seed mass does not affect germination parameters, plant life cycle matters, however.

Keywords: germination rate; base temperature; thermal time constant; seed size

1. Introduction

Temperature not only affects seed formation and development, but also influences seed germination and seedling establishment [1,2]. Fastest germination usually occurs at optimal temperatures [3] or over an optimal temperature range [4]. Seeds germinate at lower percentages and rates at temperatures lower or higher than the optimum [5]. Extreme high temperature will kill seeds [6], while extreme low temperature impedes the start of germination-physiological processes [7].

The rate of germination (defined as the reciprocal of the time taken for 50% seeds to germinate) usually increases linearly with temperature in the suboptimal range and then decreases linearly [8–10]. Garcia-Huidobro *et al.* [11] developed a linear thermal time model (TT model) to calculate the cardinal temperatures and the thermal time constant at suboptimal ($\theta_1(g)$) and supraoptimal temperatures ($\theta_2(g)$) of different subpopulations (germination fractions/percentiles) g in a seed lot. The two equations are:

$$GR_g = 1/t_g = (T - T_b(g)) / \theta_1(g) \quad T < T_o \tag{1}$$

$$GR_g = 1/t_g = (T_c(g) - T) / \theta_2(g) \quad T > T_o \tag{2}$$

For any given subpopulation, germination rate can be described by two straight lines. The slopes of the two lines are $\theta_1(g)$ and $\theta_2(g)$ with the intersection of the two lines defined as T_o. The two points where germination percentages equal zero were defined as the base, $T_b(g)$, and maximal temperature, $T_c(g)$, respectively [11].

Recently, we showed that for ryegrass and tall fescue species, germination rate was not significantly different over an optimal temperature range, thus we proposed a modified thermal time model (MTT model), with equations as follows [4]:

$$GR_g = 1/t_g = (T - T_b(g)) / \theta_1(g) \quad T < T_{ol}(g) \tag{3}$$

$$GR_g = 1/t_g = K \quad T_{ol}(g) \leq T \leq T_{ou}(g) \tag{4}$$

$$GR_g = 1/t_g = (T_c(g) - T) / \theta_2(g) \quad T > T_{ou}(g) \tag{5}$$

Where T_{ol} is the lower limit of the optimum temperature range and T_{ou} is the upper limit of the optimum temperature range. Different subpopulations in a seed population may have different T_{ol} and T_{ou} values. K is the average value of T_{ol} to T_{ou} for a given subpopulation.

The base temperature and thermal time constant in the model have great significance, and can be used to compare germination timing between different species or for the same species in different habitats or climatic conditions [12,13]. However, most studies use the thermal time model to investigate intraspecific variation of germination or differences between several species [3,14–16]. However, comparison of thermal time model parameters between large numbers of species and between different functional groups is lacking [17,18]. Knowing and comparing the base temperature and thermal time constant at the species level can increase our ability to predict species distribution shift under climate change. It may also provide useful information for plant breeding purposes.

It is widely accepted that high temperature favours C4 species while low temperature favours C3 species. Physiological models predict that the C3 *vs.* C4 crossover temperature of net assimilation rates (*i.e.*, the temperature above which C4 plants have higher net assimilation rates than C3 plants) is approximately 22 °C [19]. However, it remains unclear whether there are significant differences in germination base temperature and thermal time constant between the two groups.

Seed mass is one of the most important functional traits, which affects many aspects of species' regeneration processes [20], including germination. Compared with small seeded species, large seeded species generally germinate better under drought [21], shade [22] and salt conditions [23]. The relationship between seed mass and thermal time parameters has not been tested.

In this study, we used the modified thermal time model to calculate the base temperatures and thermal time constants of different C3 and C4 species in the Songnen grassland. We had two main objectives: (1) to compare the difference of germination response and model parameters between C3 and C4 species; (2) to test the relationship between model parameters and seed size within the two group species.

2. Materials and Methods

2.1. Plant Materials and Habitats

Twenty seven species were used in this study, among which *Plantago asiatica*, *Saussurea glomerata*, *Lactuca indica*, *Cynanchum sibiricum*, *Dracocephalum moldavica*, *Cynanchum chinense*, *Allium odorum*, *Convolvulus arvensis*, *Pharbitis purpurea*, *Bidens parviflora*, *Achillea mongolica*, *Potentilla chinensis*, *Stipa baicalensis*, *Lappula echinata*, *Incarvillea sinensis* were C3 species, *Kochia prostrate*, *Artemisia anethifolia*, *Salsola collina*, *Portulaca oleracea*, *Setaria viridis*, *Chenopodium album*, *Amaranthus retroflexus*, *Amaranthus blitoides*, *Chloris virgata*, *Eriochloa villosa*, *Euphorbia humifusa*, *Echinochloa crusgalli* were C4 species [24,25]. Species information was given in Table 1. Mature seeds were collected in autumn from wild populations in Changling, Jilin Province of China. The seeds were stored in cloth bags in a fridge at 4 °C until used. Mean seed mass was calculated by weighing 30 seeds of each species on a microbalance, with five replicates.

2.2. Germination Tests

The experiments were conducted in programmed chambers (HPG-400HX; Harbin Donglian Electronic and Technology Co. Ltd., Harbin, China) under a 12-h light/12-h dark photoperiod, with light at approximately 200 $\mu mol \cdot m^{-2} s^{-1}$ supplied by cool white fluorescent lamps (Sylvania). Eight constant temperature treatments from 5 °C to 40 °C at 5 °C intervals were set in different chambers. There were four replicates at each temperature. For each replicate, 100 seeds were germinated on two layers of filter paper in Petri dishes (10 cm in diameter). The filter paper was kept moistened with distilled water. Seeds were considered to have germinated when the radicle emerged, and germinated seeds were removed. Germination was recorded every 8 h in the first week, every 12 h in the second week and then once a day as germination rates decreased. Germination tests were terminated when no seeds had germinated for 3 consecutive days.

Table 1. Photosynthetic-type (P), family, life cycle, single seed weight (calculated from 30 seeds, $n = 5$) of 27 wild species in this study.

P	Species	Family	Life Cycle	Seed Weight (mg)
	Kochia prostrata	Amaranthaceae	Annual	0.762 ± 0.013
	Chenopodium album	Amaranthaceae	Annual	0.579 ± 0.006
	Salsola collina	Amaranthaceae	Annual	1.632 ± 0.064
	Amaranthus blitoides	Amaranthaceae	Annual	0.965 ± 0.019
	Amaranthus retroflexus	Amaranthaceae	Annual	0.502 ± 0.006
C$_4$	*Setaria viridis*	Poaceae	Annual	0.815 ± 0.007
	Chloris virgata	Poaceae	Annual	0.629 ± 0.025
	Echinochloa crusgalli	Poaceae	Annual	1.836 ± 0.028
	Eriochloa villosa	Poaceae	Annual	3.549 ± 0.353
	Portulaca oleracea	Portulacaceae	Annual	0.134 ± 0.003
	Euphorbia humifusa	Euphorbiaceae	Annual	0.434 ± 0.007
	Artemisia anethifolia	Compositae	Biennial	1.019 ± 0.012
	Lappula echinata	Boraginaceae	Annual	2.170 ± 0.052
	Incarvillea sinensis	Bignoniaceae	Annual	0.660 ± 0.010
	Dracocephalum moldavica	Labiatae	Annual	1.892 ± 0.031
	Bidens parviflora	Compositae	Annual	5.530 ± 0.139
	Saussurea glomerata	Compositae	Perennial	2.843 ± 0.077
	Lactuca indica	Compositae	Perennial	1.031 ± 0.028
	Achillea mongolica	Compositae	Perennial	0.030 ± 0.001
C$_3$	*Allium odorum*	Liliaceae	Perennial	2.187 ± 0.017
	Convolvulus arvensis	Convolvulaceae	Perennial	31.82 ± 0.131
	Pharbitis purpurea	Convolvulaceae	Perennial	28.55 ± 0.442
	Cynanchum sibiricum	Asclepiadaceae	Perennial	5.973 ± 0.124
	Cynanchum chinense	Asclepiadaceae	Perennial	4.217 ± 0.070
	Potentilla chinensis	Rosaceae	Perennial	0.411 ± 0.010
	Stipa baicalensis	Poaceae	Perennial	7.980 ± 0.194
	Plantago asiatica	Plantaginaceae	Perennial	0.229 ± 0.002

2.3. Data Analysis

Germination data were arcsine transformed before being subjected to statistical analysis. For modeling purposes, a seed population was considered to be composed of subpopulations defined by differences in their relative germination rates (Garcia-Huidobro *et al.*, 1982 [11]). In this study, the 1st and 50th germination percentiles were used to calculate thermal time model parameters, as they represent first germination and half of the seeds germination. Germination rates were defined as the reciprocal of 1% and 50% germination times. The differences between germination rates at different constant temperatures were tested by One-Way ANOVA ($p < 0.05$). The base temperature (T_b) and thermal time constant (θ_1) at suboptimal temperatures of each species were predicted by the modified thermal time model (Equation (3) in the introduction). Differences in T_b and θ_1 of C$_3$ and C$_4$ species were examined using Independent-Samples T test ($p < 0.05$). Linear regression was used to test the relationship between T_b, θ_1 and seed mass of C$_3$ and C$_4$ species. Data transformation and analysis of variance were carried out in SPSS (version 13.0, SPSS Inc., Chicago, IL, USA). Regression and calculation of model parameters were carried out in SigmaPlot (version 10.0, Systat Software Inc., Richmond, CA, USA).

3. Results and Discussion

3.1. Germination Responses of C_3 and C_4 Species to Temperature

The twelve C_4 species exhibited a variety of responses to constant temperatures (Figure 1). Seeds of *Kochia prostrata, Salsola collina, Chloris virgata, Echinochloa crusgalli* and *Artemisia anethifolia* germinated well (>80%) at a wide range of temperatures from 5–35 °C or 10–35 °C. The germination percentages of *Amaranthus retroflexus, Eriochloa villosa* and *Portulaca oleracea* increased with temperature until 30 °C, and then kept constant or decreased slightly. The germination percentages of *Chenopodium album* and *Euphorbia humifusa* increased with temperature, then decreased greatly above 30 °C. For *Amaranthus blitoides* and *Setaria viridis*, more than half of the seeds did not germinate at most temperatures.

For most C_3 species, the relationship between germination percentage and temperature resembled an upside-down "U" or "V" (Figure 2). Only three species *Dracocephalum moldavica, Bidens parviflora* and *Lactuca indica* had more than 90% seed germination at all temperatures from 5 °C until 30 °C or 35 °C.

The germination rate either increased with temperature until 40 °C, or increased until an optimal temperature, then decreased, irrespective of whether they were C_3 or C_4 species (Figures 3 and 4). The trends of germination rate change with temperature were similar for 1% and 50% germination percentiles of each species. The germination rates of C_4 species were generally higher than those of C_3 species, with *S. collina* most rapid, and *C. virgata* next.

C_3 and C_4 species are classified by their photosynthetic pathway. C_3 species are mainly distributed to high latitude regions with cooler climate, while C_4 species are generally found at low latitudes with warmer climate and strong light [26]. From our study, the two types of species also had different germination responses to temperature [27]. The twelve C_4 species used were all annual or biennial and distributed widely in the research region; More than half these species had a wide optimal temperature range. At 5 °C, seven species exhibited no seed germination or lower than 10 percent of seeds germinated; however, all the C_4 species could germinate at 40 °C (Figure 1). By contrast, twelve of fifteen C_3 species could germinate at 5 °C and ten of the fifteen species could not germinate at 40 °C (Figure 2). Seeds of C_4 species germinated faster than those of C_3 species at the optimal 30 °C ($p < 0.05$). Plant responses to temperature reflect the environments in which those species live, thus the differences in germination optima between species may have ecological significance [12].

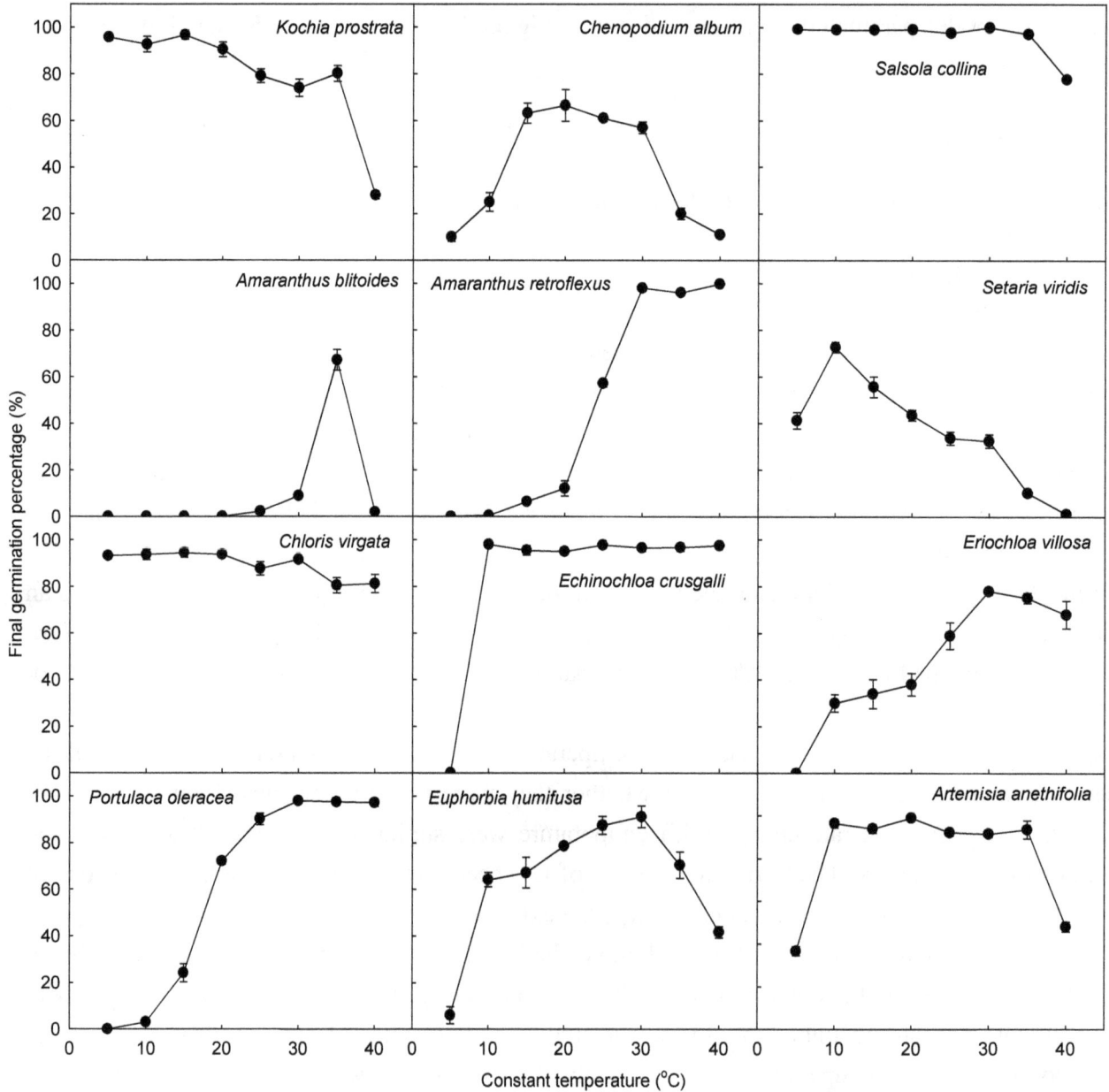

Figure 1. Final germination percentages of C4 species at a range of constant temperatures from 5 °C to 40 °C. Bars represent ±SE (*n* = 4).

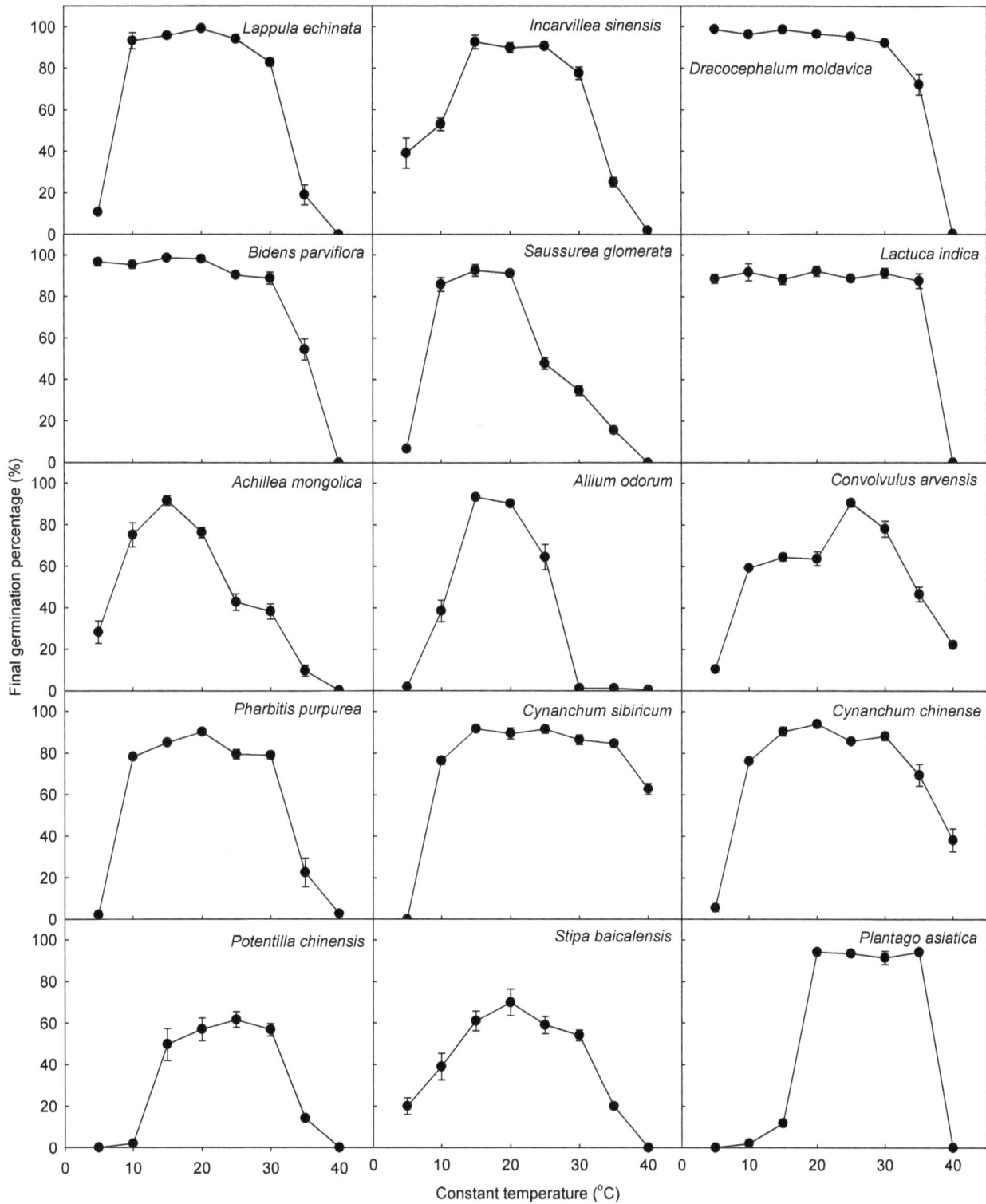

Figure 2. Final germination percentages of C3 species at a range of constant temperatures from 5 °C to 40 °C Bars represent ±SE (*n* = 4).

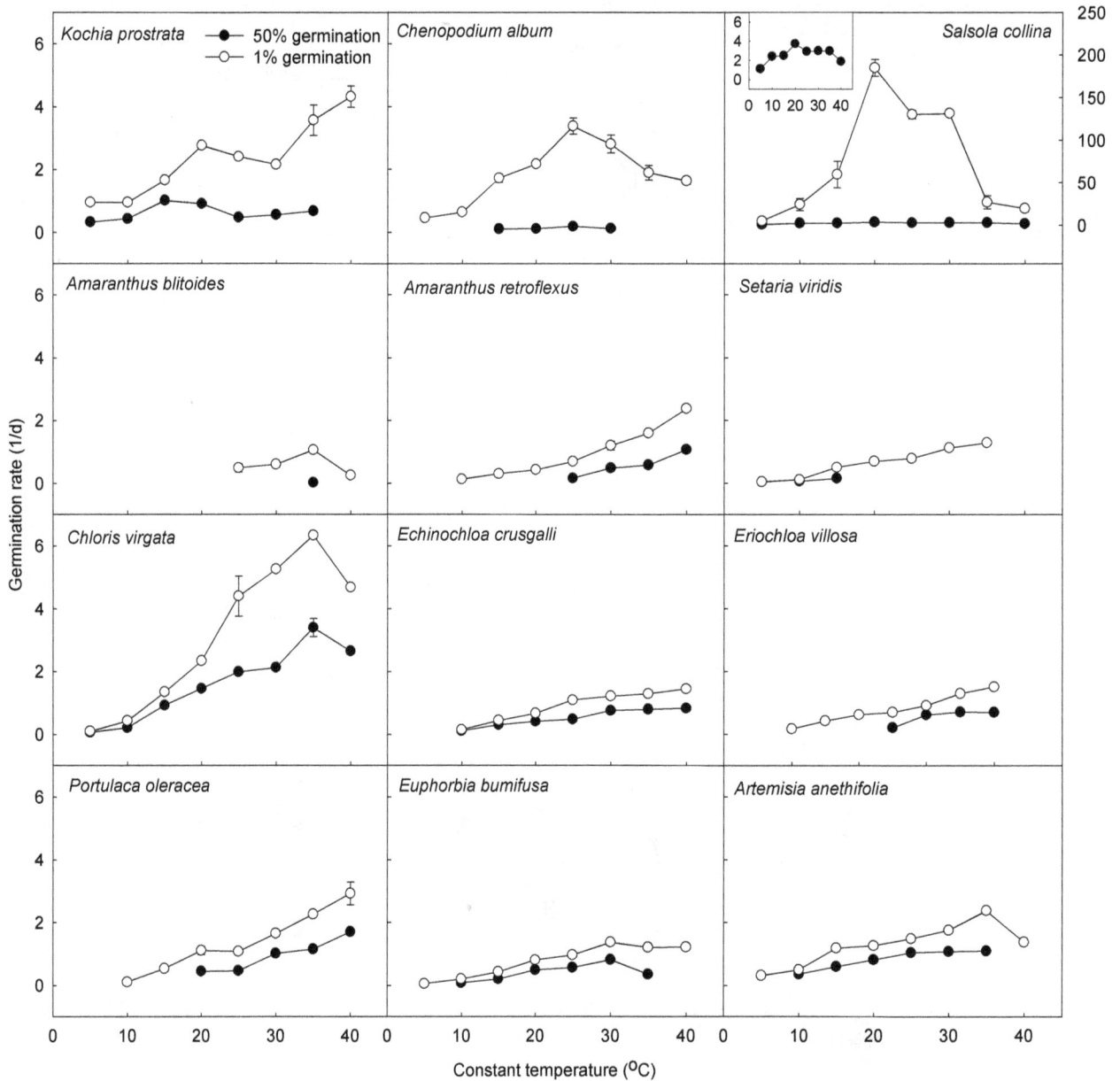

Figure 3. Germination rates of C4 species for 1% (○) and 50% (●) germination percentiles at a range of constant temperatures from 5 °C to 40 °C. Bars represent ±SE (*n* = 4). For *Salsola collina*, scaling of y axis was given on the right-hand side; the enlarged figure was for 50% germination percentile.

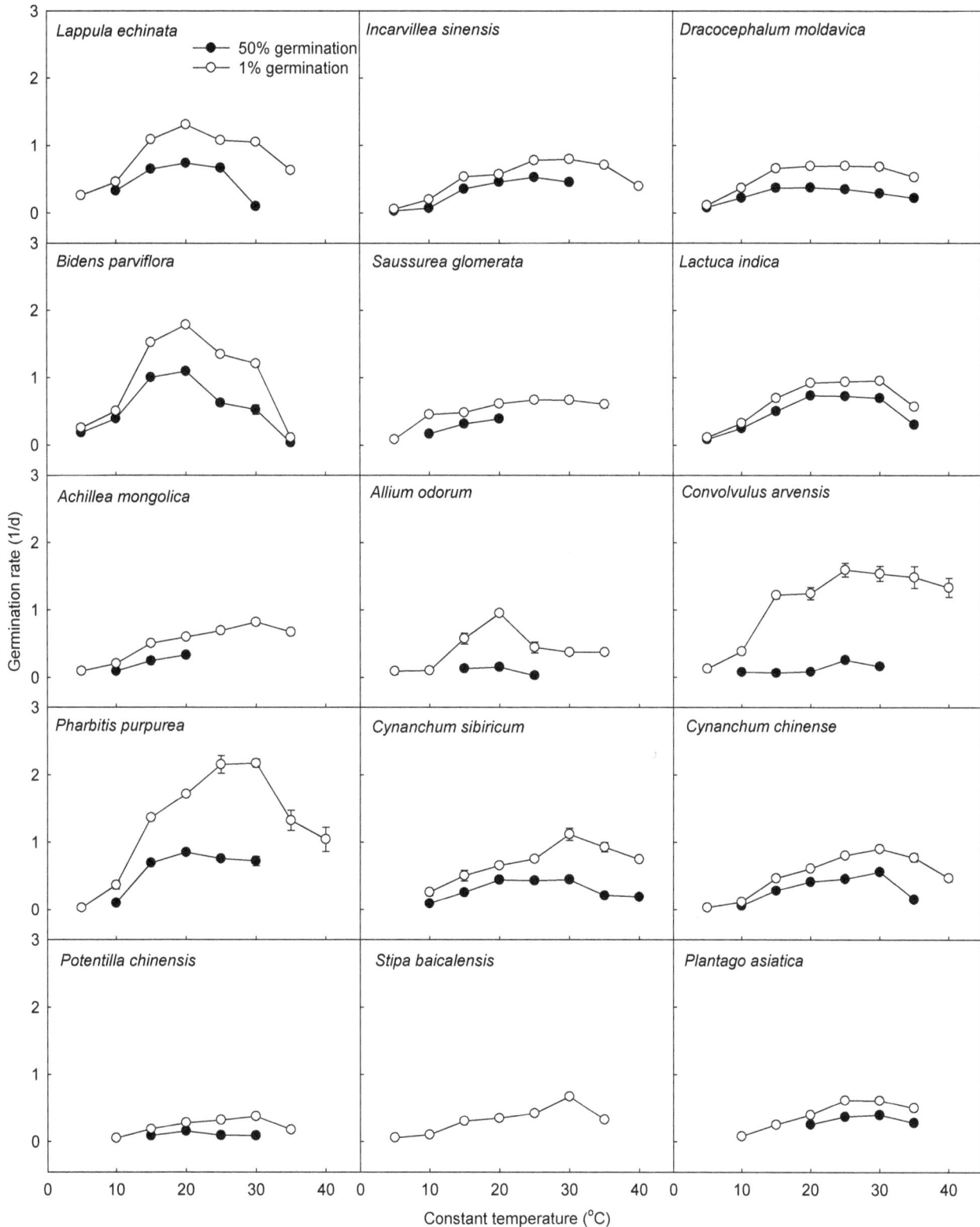

Figure 4. Germination rates of C_3 species for 1% (○) and 50% (●) germination percentiles at a range of constant temperatures from 5 °C to 40 °C. Bars represent ±SE ($n = 4$).

3.2. Comparison of Model Parameters between C_3 and C_4 Species

We used the modified thermal time model to predict base temperature and thermal time constant of a range of C_4 and C_3 species (Tables 2 and 3). As a whole, the estimation was accurate ($p < 0.05$), except the 50% germination percentile for several species (e.g., *K. prostrata*, *Lappula echinata*) and 1% germination percentile for *S. collina*, *A. blitoides*, and *Saussurea glomerata*. The poor fit for these species was due to the lower number of regression points (three or four). The average base temperature of C_3 species for the 1% germination percentile ($T_b = 3.8 \pm 0.4$ °C, $n = 14$) was lower than that for C_4 species ($T_b = 4.9 \pm 1.4$ °C, $n = 10$), with the difference approaching significance ($p = 0.074$). The average thermal time constant of C_3 species for 1% germination percentile ($\theta_1 = 24.7 \pm 4.0$ °C·d, $n = 14$) was higher than that of C_4 species ($\theta_1 = 15.2 \pm 2.2$ °C·d, $n = 10$) ($p = 0.45$). The differences between model parameters of C_3 and C_4 species for 50% germination percentile ($n = 5$ for C_3, $n = 7$ for C_4) were similar, with a significant difference in θ_1 ($p < 0.05$). Among the 27 species in this study, the differences in base temperature were not significant between annuals and perennials, but the differences in thermal time constant were significant between annuals and perennials (16.0 °C·d and 26.2 °C·d, respectively; $p < 0.01$). This result was consistent with previous research [17], which indicated that germination responses to temperature was related to plant life cycle.

Table 2. Estimated parameters of thermal time model for 1% and 50% germination percentiles (G) of C_4 species at suboptimal temperatures.

Species	G	N	T_b (°C)	θ_1 (°C·d)	R^2	p
Kochia prostrata	1%	8	−3.6	11.1	0.85	0.001
	50%	3	1.4	14.6	0.86	0.24
Chenopodium album	1%	5	3.7	6.8	0.95	0.0044
	50%	3	1.7	126.6	0.81	0.28
Salsola collina	1%	4	6.5	0.09	0.84	0.08
	50%	4	−2.8	6.3	0.92	0.04
Amaranthus blitoides	1%	3	17.3	17.6	0.89	0.21
	50%					
Amaranthus retroflexus	1%	7	11.6	13.9	0.92	0.0006
	50%	4	22.3	17.7	0.94	0.031
Setaria viridis	1%	7	4.8	23.3	0.98	< 0.0001
	50%	2				
Chloris virgata	1%	6	7.0	4.5	0.95	0.0008
	50%	7	6.3	9.4	0.96	0.0001
Echinochloa crusgalli	1%	7	4.4	22.9	0.95	0.0002
	50%	5	5.8	34.4	0.95	0.0042
Eriochloa villosa	1%	7	6.4	23.3	0.97	< 0.0001
	50%	2				
Portulaca oleracea	1%	7	9.4	11.3	0.96	< 0.0001
	50%	5	14.9	15.7	0.93	0.008
Euphorbia humifusa	1%	6	5.4	18.9	0.98	0.0001
	50%	5	8.1	27.3	0.98	0.0016
Artemisia anethifolia	1%	7	0.3	15.6	0.96	0.0001
	50%	4	2.0	22.2	0.99	0.0002

N, number of values; T_b, base temperature; θ_1, thermal time constant; R^2 and p represent the coefficient of determination and probability for the fitting.

Table 3. Estimated parameters of thermal time model for 1% and 50% germination percentiles (G) of C_3 species at suboptimal temperatures.

Species	G	N	T_b (°C)	θ_1 (°C·d)	R^2	p
Lappula echinata	1%	4	2.2	13.2	0.95	0.0241
	50%	3	1.0	25	0.90	0.20
Incarvillea sinensis	1%	5	3.2	27.5	0.95	0.0041
	50%	5	4.6	36.1	0.94	0.0072
Dracocephalum moldavica	1%	3	3.1	18.2	0.99	0.0231
	50%	3	2.3	34.4	0.99	0.0045
Bidens parviflora	1%	4	3.4	8.9	0.93	0.0352
	50%	3	3.6	12.2	0.93	0.17
Saussurea glomerata	1%	4	−0.2	31.1	0.84	0.08
	50%	3	2.0	44.6	0.96	0.11
Lactuca indica	1%	4	3.3	17.9	0.99	0.0057
	50%	4	3.6	22.7	0.99	0.0037
Achillea mongolica	1%	6	1.1	33.8	0.96	0.0006
	50%	3	5.7	41.8	0.98	0.09
Allium odorum	1%	4	5.5	16.4	0.90	0.0494
	50%	2				
Convolvulus arvensis	1%	5	2.9	13.2	0.92	0.0099
	50%	4	6.7	89.3	0.61	0.21
Pharbitis purpurea	1%	5	5.0	8.9	0.97	0.0026
	50%	3	7.7	13.3	0.90	0.20
Cynanchum sibiricum	1%	5	3.2	25.5	0.96	0.0037
	50%	3	7.6	28.2	0.99	0.0196
Cynanchum chinense	1%	6	4.6	26.6	0.97	0.0004
	50%	5	5.1	42.0	0.93	0.0076
Potentilla chinensis	1%	5	4.3	63.3	0.94	0.0058
	50%	2				
Stipa baicalensis	1%	6	3.8	43.1	0.94	0.0013
	50%					
Plantago asiatica	1%	4	7.9	28.6	0.96	0.0001
	50%	3	0.8	70.9	0.88	0.22

N, number of values; T_b, base temperature; θ_1, thermal time constant; R^2 and p represent the coefficient of determination and probability for the fitting.

Compared to tropical and subtropical legumes [18], the temperate grassland species in our study had lower base temperature and thermal time constants. This is not completely coincident with other studies. Trudgill [28] found the base temperature of tropical plants was higher than that of temperate plants, but they also demonstrated that the thermal time constant of tropical plants was lower than that for temperate plants [12]. Therefore, tropical plants germinate faster than temperate plants. They suggested that temperate plants will suffer frost injury if they germinate too early, while tropical plants might suffer high temperature or drought if they germinate too late. The germination response to temperature is also related to phylogeny. Plants in the Poaceae and Cyperaceae have been noted to have lower and higher base temperature, respectively [29].

3.3. Relationship between Seed Mass and Germination Parameters

The relationship between germination rate and seed mass depended on plant functional type and temperature. For C_4 species, big seeds had higher germination rates only at 5 °C ($p < 0.05$, Figure 5a). For C_3 species however, germination rate increased with seed mass significantly over the temperature range 15–40 °C ($p < 0.05$, Figure 5b). Neither the base temperature nor thermal time constant of either C_3 or C_4 species had a significant relationship with seed mass (Figure 5c,d). The T_b of C_3 species were more clustered around 5 °C, while C_4 species were scattered from −3.6 °C to 11.6 °C. The opposite was noted for the thermal time constant. θ_1 of C_4 species was confined to 5–23 °C·d, but that of C_3 species distributed from 9 °C·d to 63 °C·d.

To our knowledge, this is the first study to test the relationship between germination parameters and seed mass. It is interesting that larger seeds germinated faster for C_4 species at low temperature, while seed mass was positively related to germination rate for C_3 species at high temperatures. We speculate that larger seeds have an advantage under unfavorable conditions, although this hypothesis needs further study.

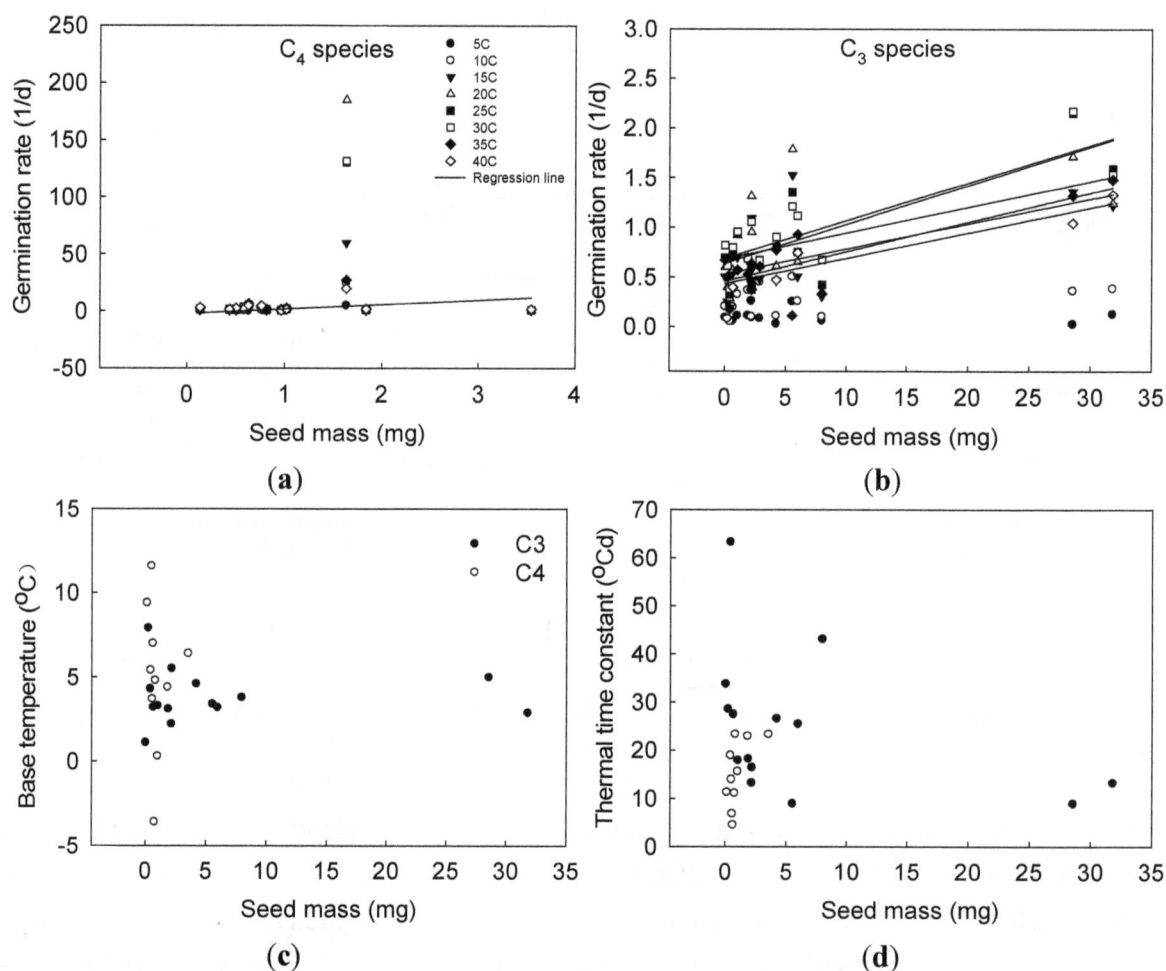

Figure 5. The relationship between seed mass and germination parameters of C_3 and C_4 species (germination rate, (**a**), (**b**); base temperature, (**c**); thermal time constant, (**d**); significant linear regressions were given in figures (**a**), 5 °C; (**b**), 15 °C, 20 °C, 25 °C, 30 °C, 35 °C, 40 °C).

4. Conclusions

The germination response to temperature was species-dependent. Significant differences in the thermal time constants were noted between C_3 and C_4, and between annual and perennial species. Although seed mass significantly influenced germination rate at certain temperatures for C_3 and C_4, base temperature and thermal time constant were not related to seed mass.

Acknowledgments

We thank Louis Irving in University of Tsukuba for improving the English and the anonymous reviewers for their insightful comments. This work was funded by the State Key Basic Research Development Program (973 Program) (2015CB150800).

Author Contributions

Conceived and designed the experiments: Hongxiang Zhang and Daowei Zhou. Performed the experiments: Hongxiang Zhang and Yu Tian. Analyzed the data and manuscript writing: Hongxiang Zhang.

Conflicts of Interest

The authors declare no conflict of interest.

References

1. Thompson, P.A. Characterisation of the germination response to temperature of species and ecotypes. *Nature* **1970**, *225*, 827–831.
2. Mott, J.J.; McKeon, G.M., Moore, C.M. The effect of seed bed conditions on the germination of four *Stylosanthes* species in the Northern Territory. *Aust. J. Agric. Res.* **1976**, *27*, 811–823.
3. Alvarado, V.; Bradford, K.J. A hydrothermal time model explains the cardinal temperatures for seed germination. *Plant Cell Environ.* **2002**, *25*, 1061–1069.
4. Zhang, H.; McGill, C.R.; Irving, L.J.; Kemp, P.D.; Zhou, D. A modified thermal time model to predict germination rate of ryegrass and tall fescue at constant temperature. *Crop Sci.* **2013**, *53*, 240–249.
5. Mott, J.J.; Groves, R.H. Germination strategies. In *The Biology of Australian Plants*; Pate, J.S., McComb, A., Eds.; University of Western Australia Press: Perth, Australia, 1981; pp. 307–341.
6. Murtagh, G.J. Effect of temperature on the germination of *Glycine javanica*. In *Proceedings of the International Grassland Congress*; Norman, M.J.T., Ed.; University of Queensland Press: Brisbane, Australia, 1970; pp. 574–578.
7. Mayer, A.M.; Poljakoff-Mayber, A. *The Germination of Seeds*, 3rd ed.; Pergammon Press: Oxford, UK, 1982.
8. Hegarty, T.W. Germination and other biological processes. *Nature* **1973**, *243*, 305–306.
9. Washitani, I.; Takenaka, A. Mathematical description of the seed germination dependency on time and temperature. *Plant Cell Environ.* **1984**, *7*, 359–362.

10. Covell, S.; Ellis, R.H.; Roberts, E.H.; Summerfield, R.J. The influence of temperature on seed germination rate in grain legumes. 1. A comparison of chickpea, lentil, soyabean and cowpea at constant temperatures. *J. Exp. Bot.* **1986**, *37*, 705–715.

11. Garcia-Huidobro, J.; Monteith, J.L.; Squire, G.R. Time, temperature and germination of pearl millet (*Pennisetum. typhoides* S. and H.). 1. Constant temperature. *J. Exp. Bot.* **1982**, *33*, 288–296.

12. Trudgill, D.L.; Perry, J.N. Thermal time and ecological strategies—A unifying hypothesis. *Ann. Appl. Biol.* **1994**, *125*, 521–532.

13. Steinmaus, S.J.; Prather, T.S.; Holt, J.S. Estimation of base temperatures for nine weed species. *J. Exp. Bot.* **2000**, *51*, 275–286.

14. Ellis, R.H.; Covell, S.; Roberts, E.H.; Summerfield, R.J. The influence of temperature on seed germination rate in grain legumes. 2. Intraspecific variation in chickpea (*Cicer. arietinum* L.) at constant temperatures. *J. Exp. Bot.* **1986**, *37*, 1503–1515.

15. Larsen, S.U.; Bibby, B.M. Differences in thermal time requirement for germination of three turfgrass species. *Crop. Sci.* **2005**, *45*, 2030–2037.

16. Nori, H.; Moot, D.J.; Black, A.D. Thermal time requirements for germination of four annual clover species. *N. Z. J. Agric. Res.* **2014**, *57*, 30–37.

17. Trudgill, D.L.; Squire, G.R.; Thompson, K. A thermal time basis for comparing the germination requirements of some British herbaceous plants. *New Phytol.* **2000**, *145*, 107–114.

18. McDonald, C.K. Germination response to temperature in tropical and subtropical pasture legumes. 1. Constant temperature. *Aust. J. Exp. Agric.* **2002**, *42*, 407–419.

19. Murphy, B.P.; Bowman, D.M.J.S. Seasonal water availability predicts the relative abundance of C_3 and C_4 grasses in Australia. *Global Ecol. Biogeogr.* **2007**, *16*, 160–169.

20. Moles, A.T.; Ackerly, D.D.; Tweddle, J.C.; Dickie, J.B.; Smith, R.; Leishman, M.R.; Mayfield, M.M.; Pitman, A.; Wood, J.T.; Westoby, M. Global patterns in seed size. *Global Ecol. Biogeogr.* **2007**, *16*, 109–116.

21. Westoby, M.; Falster, D.S.; Moles, A.T.; Vest, P.A.; Wright, J.J. Plant ecological strategies: Some leading dimensions of variation between species. *Ann. Rev. Ecol. Syst.* **2002**, *33*, 125–159.

22. Foster, S.A.; Janson, C.H. The relationship between seed size and establishment conditions in tropical woody plants. *Ecology* **1985**, *66*, 773–780.

23. Zhang, H.; Zhang, G.; Lü, X.; Zhou, D.; Han, X. Salt tolerance during seed germination and early seedling stages of 12 halophytes. *Plant Soil* **2015**, *388*, 229–241.

24. Yin, L.; Li, M. A study on the geographic distribution and ecology of C_4 plants in China. 1. C_4 plant distribution in China and their relation with regional climatic condition. *Acta Ecol. Sin.* **1997**, *17*, 350–363.

25. Tang, H.P.; Liu, S.R. The list of C_4 plants in Nei Mongol area. *Acta Sci. Nat. Univ. Nei Mong.* **2001**, *32*, 431–438.

26. Yin, L.; Zhu, L. C_3 and C_4 plants of forage resources in northeast grassland region of China. *Inn. Mong. Pratac.* **1990**, *3*, 32–40.

27. Zhang, H.; Yu, Q.; Huang, Y.; Zheng, W.; Tian, Y.; Song, Y.; Li, G.; Zhou, D. Germination shifts of C_3 and C_4 species under simulated global warming scenario. *PLoS ONE* **2014**, *9*, e105139.

28. Trudgill, D.L. Why do tropical poikilothermic organisms tend to have higher threshold temperatures for development than temperate ones? *Funct. Ecol.* **1995**, *9*, 136–137.

29. Grime, J.P.; Mason, G.; Curtis, A.V.; Rodman, J.; Band, S.R.; Mowforth, M.A.G.; Neal, A.M.; Shaw, S. A comparative study of germination characteristics in a local flora. *J. Ecol.* **1981**, *64*, 1017–1059.

7

Extension of Small-Scale Postharvest Horticulture Technologies—A Model Training and Services Center

Lisa Kitinoja [1,†,*] **and Diane M. Barrett** [2,†]

[1] World Food Logistics Organization, 1500 King Street, Alexandria, VA 22314, USA
[2] Department of Food Science and Technology, University of California, Davis, One Shields Ave, Davis, CA 95616, USA; E-Mail: dmbarrett@ucdavis.edu

[†] These authors contributed equally to this work.

[*] Author to whom correspondence should be addressed; E-Mail: kitinoja@postharvest.org

Academic Editor: Michael Blanke

Abstract: A pilot Postharvest Training and Services Center (PTSC) was launched in October 2012 in Arusha, Tanzania as part of a United States Agency for International Development (USAID) funded project. The five key components of the PTSC are (1) training of postharvest trainers, (2) postharvest training and demonstrations for local small-scale clientele, (3) adaptive research, (4) postharvest services, and (5) retail sales of postharvest tools and supplies. During the years of 2011–2012, a one year e-learning program was provided to 36 young horticultural professionals from seven Sub-Saharan African countries. These postharvest specialists went on to train more than 13,000 local farmers, extension workers, food processors, and marketers in their home countries in the year following completion of their course. Evaluators found that these specialists had trained an additional 9300 people by November 2014. When asked about adoption by their local trainees, 79% reported examples of their trainees using improved postharvest practices. From 2012–2013, the project supported 30 multi-day training programs, and the evaluation found that many of the improved practices being promoted were adopted by the trainees and led to increased earnings. Three PTSC components still require attention. Research activities initiated during the project are incomplete, and successful sales of postharvest goods and services will require commitment and improved partnering.

Keywords: postharvest technologies; e-learning; small-scale; impact evaluation; Sub-Saharan Africa

1. Introduction

A USAID-funded project in East Africa of a new model for extension of postharvest technologies was piloted during 2010–2014 under the Horticulture CRSP (Collaborative Research Support Program, now known as the Horticulture Innovation Lab, Davis, CA, USA) at the University of California (UC) at Davis. Project partners included the Department of Food Science and Technology at UC Davis, the International Programs office at the World Food Logistics Organization (WFLO, Alexandria, VA, USA), the University of Georgia, and the Asian Vegetable Research Development Center (AVRDC)—The World Vegetable Center (Arusha, Tanzania), and additional financial support and in-kind technical support was provided by postharvest specialists from The Postharvest Education Foundation (PEF), in which the authors are founding board members.

The pilot Postharvest Training and Services Center (PTSC) was launched in October 2012 on the campus of AVRDC—The World Vegetable Center in Arusha, Tanzania. The project operated as planned in Northern Tanzania (2012–2014), training young horticultural professionals from seven countries (Benin, Ethiopia, Ghana, Kenya, Rwanda, Tanzania, and Uganda) as "Postharvest Specialists" and was followed six months later by USAID Save the Children/Technical and Operational Performance Support (TOPS)-funded evaluation studies (October 2014–May 2015). Simultaneously, during 2012–2015, three years of mentoring were being provided by PEF for more than a dozen Tanzanian extension workers and postharvest trainers who are currently working within the local community in and near Arusha to improve postharvest handling practices and reduce losses in horticultural crops.

1.1. Background Information

Less than 5% of funding for horticultural research and extension (R&E) has been allocated to postharvest issues over the past 25 years [1,2] as the historical focus has been on increasing production. In the 1990s, the focus in the horticultural research sector moved to the field of marketing and more recently it has focused upon value chain development. Internet database searches show that less than one in 2000 agricultural projects undertaken globally have focused on fresh produce handling and marketing, according to advanced searches undertaken in five major online databases during 2010 (AidData [3]; USAID Documents [4]; World Bank [5]; United Nations Food and Agricultural Organization (UN FAO) Information Network for Postharvest Operations (INPHo) [6]; Devex [7]). While thousands of development projects have been launched in developing countries between 1990 and the present time by dozens of donor agencies and governmental bodies, few have focused on agriculture (less than 6% according to the AiDA database [3]; 25% according to the World Bank [5]), very few have focused on horticulture (approximately 1% of the agricultural projects), and only one-third of this 1% of horticultural projects included any kind of postharvest component [8].

Of the 1.3 billion tons of food losses and waste reported by the UN FAO [9], an external analysis shows that 44% is made up of fruits and vegetable crops, and 20% is roots and tuber crops [10]. In terms of the percent of kilocalories (kcal) lost or wasted for each type of food commodity, roots and tubers experience the greatest amount of loss and waste (63% on a caloric basis) while the rate of kcal losses for fruits and vegetables is 42%. In comparison, about 25% of the kcal of cereals and of seafood produced are lost or wasted.

A UN FAO–commissioned report published in 2011 advocated for "strengthening the supply chain through training and support for farmers, making investments in infrastructure and transportation, as well as investments in an expansion of the food processing and packaging industries", which experts believe could help to reduce the amount of food loss and waste [9]. In most countries where high levels of fresh produce wastage is occurring, the local farmers, traders, small-scale processors, and marketers of fruits and vegetable crops have little or no access to postharvest training, technical information, guidance on use of new technologies, or local access to the tools and supplies needed to utilize new technical knowledge or improved postharvest handling practices.

In developing countries, food losses and food waste occur mainly at earlier stages of the value chain and can be traced back to financial, managerial, and technical constraints in harvesting techniques and postharvest handling as well as a lack of storage and cooling facilities. While postharvest losses for fruits and vegetable crops are reported to be 44% of the total global production by weight [10], these losses have been measured in Sub-Saharan Africa (SSA) for many crops to be even higher levels of 40% to 80% [11].

1.2. Description of the Model

The five key components of the postharvest extension model known as the PTSC are (1) training of postharvest trainers, (2) postharvest training demonstrations for local clientele, (3) adaptive research, (4) postharvest services, and (5) retail sales of tools and postharvest supplies. The concept of the PTSC was developed during 2007–2008 by the lead author while she was serving as an independent consultant for a Millennium Challenge Corporation (MCC) project in Cape Verde. Several PTSCs were constructed in Cape Verde during 2008–2010, but the designs were "enhanced" by the local authorities to include beautiful architectural features, millipede inspection stations, and/or large-scale cold storage facilities, so by the time all the modifications were made, the costs were very high and the design was not promoted beyond the MCC project for the Cape Verde islands. The original concept and model were further developed as part of a research project for the Bill and Melinda Gates Foundation which the lead author developed and led for WFLO during 2009–2010 on the investigation of appropriate postharvest technologies that could be disseminated in a sustainable manner in SSA and South Asia.

Postharvest advocacy needs in developing countries are many and include enhanced funding for R&E, updating laws governing markets, access to micro-credit, reduced interest rates, support for associations, extension programs for women, hiring and training more women as horticultural extension agents, and access to high quality planting materials/seeds, simple postharvest tools, supplies (especially improved packages), equipment, and market information [8]. Training of trainers, building local postharvest extension capacity, and training local populations in the specialized skills needed for reducing food losses have been advocated and proposed for many years, but very few studies have been funded or implemented. This pilot project was the first attempt to bring all the needed components together in one location and provide potential users with a complete package of postharvest training, supplies, services, and support activities.

The PTSC was designed to serve as:

- a site for extension workers and local postharvest trainers to meet with growers and others working along the value chains to provide training to improve local capacity and knowledge on improved produce handling, harvesting, sorting/grading, packing, cooling, storage, food safety, processing, and marketing practices
- a training venue with permanent demonstrations for observing improved, cost-effective small-scale postharvest handling practices, facilities, and equipment
- a site where local private companies can demonstrate and explain the benefits of their goods and services related to improved postharvest handling, processing, or storage
- a retail shop with postharvest tools and supplies, packages, plastic crates, and other goods that can be purchased locally at reasonable prices
- a place where people can come to ask questions or get advice on how to use improved postharvest practices, learn about costs and benefits and marketing options
- a place where growers or traders can pay a small fee for services such as having their produce packed in improved containers, cooled and/or stored for a few days before marketing, leasing of a small insulated transport vehicle, using a solar dryer to produce dried fruits or vegetable snack products, *etc.*

2. The Evaluation Plan and Objectives

The final project report for the Horticulture CRSP project was submitted by WFLO in May 2014. The evaluation team reviewed the final report, and the objectives for an *ex post facto* evaluation of a completed postharvest extension project were set in cooperation with USAID/TOPS, and the data collection and analysis methods were planned during the proposal phase of the project.

2.1. Data Collection Methods

Data collection instruments (written surveys and structured interview schedules) were developed and field-tested in SSA by WFLO's local evaluation team, and data was collected over a six-month period via email surveys, phone, and face-to-face interviews to characterize the implementation, outcomes, and impacts of the five project components of the PTSC pilot project. Written surveys and questionnaires were developed using traditional evaluation methods, in collaboration with stakeholders, the local evaluation team, and clientele in order to ensure the questions were suitable and easy to understand, and that responses would be reliable [12]. Table 1 provides a summary of the data collection plan, target groups, sources of data, and response rate.

2.2. Objectives of the Ex Post Facto Evaluation

The evaluation project had three major objectives.

Objective 1: To determine the major capacity building outcomes and impact of the Postharvest Training and Services Center (PTSC) and Training the Trainers (ToT) program

Objective 2: To identify best practices in the management of the PTSC and its extension services

Objective 3: To identify problems, concerns, and obstacles to making the PTSC a sustainable and replicable model

Data was collected using a variety of methods (surveys, face-to-face or phone interviews, site visits) as shown in Table 1, where x indicates which method(s) were used for which target groups.

Table 1. Data collection plan and implementation.

Target Groups	Size of Target Group	Email Surveys	Face-to-Face Interviews	Phone Interviews	Site Visits for Observations	Response Rate
Training of trainers (ToT) participants	36 in total	x				92%
Local trainees (farmers and food processors)	50 people, random cluster sample selected from a population of 500 trainees		x		x	100%
Postharvest trainers	14 in total	x		x		100%
PTSC administrators and managers	7 in total		x		x	100%

3. Results and Discussion

Data analyses undertaken for the *ex post facto* evaluation project included simple counts and percentages for quantitative data and descriptive information (categories, lists, case studies) for qualitative data. This communication draws on both the final project report for Hort CRSP and preliminary evaluation reports for USAID/TOPS, none of which have been published. Project leaders at WFLO plan to formally present the results of the TOPS-funded postharvest extension project evaluation once the project has been completed.

3.1. Training of Postharvest Trainers

One of the major objectives of the Hort CRSP–funded project was to educate 30 trainers as postharvest specialists and provide them with the knowledge, tools, and motivation to train 5000 local clientele in their own home locations by the end of the project. The project outcomes exceeded these targets, as 36 young horticultural professionals from seven countries in SSA were provided with 18 months of e-learning–based training during 2011–2012. All 36 completed the postharvest e-learning program and were given certificates as postharvest specialists. Topics included postharvest loss assessment, commodity systems assessment, technical information on small-scale practices, designing postharvest demonstrations (on harvesting, handling, sorting/grading, improved packages, cooling practices, storage, food processing methods, and more), planning extension programs, and designing their own PTSCs for their home countries.

From October 2012 through October 2013, without any project-provided funding but with access to mentoring and technical support, these 36 trainers planned and implemented postharvest horticultural training programs for more than 13,000 local farmers, extension workers, food processors, and marketers in their home countries. The TOPS evaluation reached 33 of these trainers, and their reports revealed that from November 2013 to November 2014, an additional 9300 people were trained by 28 of the trainers. The five persons who did not report providing any postharvest training for local populations were either actively pursuing graduate studies (in Germany, South Africa, and the USA) or had taken new jobs outside the agricultural sector (in Ghana and in Tanzania).

When asked about adoption of postharvest practices by their local trainees, 79% of the respondents reported on examples of trainees who are now using improved postharvest practices. Among those practices were improved harvesting, packing practices, and postharvest handling practices, Zero Energy Cool Chamber (ZECC) storage, solar drying, use of a cool room (via second-hand reefer container), postharvest loss assessment, and use of a postharvest tool kit and materials. However, solar drying and improved packaging practices were the most commonly used practices. Some farmers and traders of fresh horticultural crops started to dry and package their surplus produce or products which are not sold at the daily market with improved (vented) plastic bags, in order to avoid postharvest losses. These simple practices provide value in addition to the commodity and allow trainees to increase their incomes.

3.2. Postharvest Training and Demonstrations for Local Farmers in Tanzania

During 2012–2013, 14 instructors provided more than 30 postharvest training programs in Tanzania for local farmers, food processors, and marketers. Nine of the 14 postharvest instructors were women (from the USA, Lebanon, New Zealand, and Tanzania), and five were men (from Uganda, the USA, Zimbabwe, and Tanzania). Most of the instructors were independent consultants working with local organizations in Tanzania. Four of the instructors (two men and two women) were recent graduate "Postharvest Specialists" of the Hort CRSP ToT program. All 14 instructors were evaluated during the TOPS project. They were each contacted via email by one of WFLO's local consultants, and she then followed up via phone calls to probe for details and any missing information and to ensure that each written survey was fully completed.

A few of the training programs were open to the public ("open house days" where local people were welcome to attend), but most were designed for a specific audience, based on requests or locally assessed training needs. Most of the training programs were offered over a period of several days, and all of the programs covered three or more postharvest topics. The sites for training included the PTSC at AVRDC, the Ministry of Agriculture, Food Security and Cooperatives (MAFC) Njiro training venue, and the OIKOS Mkuru Training Camp. Individuals or groups were invited by telephone or a group would make inquiries at AVRDC or the Selian Agricultural Research Institute (SARI) in Arusha about having a postharvest training program developed for them. According to these 14 instructors, the topics that women asked about most often were the use of shade, gentle handling, home storage, food processing, and marketing. The topics asked about most often by men were cooling, cold storage, transport, Good Agricultural Practices (GAPs), food safety, food processing, and marketing.

Two WFLO consultants, based locally in Arusha, conducted face-to-face interviews with 50 participants from the training programs. This sample was comprised of five persons who had been randomly selected from 10 randomly selected participant groups via cluster sampling. The survey results included responses from 42 women and eight men who had participated in PTSC training programs, with an average age of 46 years. The topics of the training programs were improved postharvest handling of fresh produce, ZECC storage, food processing methods (jam making, solar drying), cooling/cold storage, or marketing of horticultural crops. Most of the respondents identified themselves as farmers or food processors, and many were also marketers of their own fresh produce or processed products. Two of the respondents were also working as extension workers.

A wide range of demonstrations were mounted during the Hort CRSP project and have continued since the project ended. Specifications for demonstrations were based on published research and review articles [11,13–15].

These include:

- Use of shade
- Hand-washing/hygiene practices
- Improved containers (plastic crates, half-size wooden crates)
- Zero energy cool chamber (for storage of fresh fruits and vegetables)
- Solar drying
- Cold room equipped with CoolBot

Several new demonstrations have been added during 2014–2015 by the postharvest staff at AVRDC. These include:

- A simple hydro-cooler with recirculation system
- Improved wooden crates (smaller, smooth on the inside)
- Liners for use in traditional containers
- Wakati (a high relative humidity % storage container; under study with Arne Pauwels of Belgium)

Because the trainings took place at the PTSC sites where these demonstrations have been established, the 50 respondents reported having seen many of the following demonstrations (see Table 2 for a count of the number indicating they had seen each of the demonstrations). In addition, when they were asked to rate the usefulness (as most or least useful), most of the respondents did not select one demonstration as was anticipated, but each rated many of the demonstrations they had seen as "most useful".

Only three of the 50 respondents reported that they had not used one or more of the practices that they had first seen in a postharvest demonstration. Many reported using maturity indices (25), shade (27), sorting/grading (40), and/or improved containers (34) to help reduce fresh produce wastage.

The demonstrations designated as "least useful" were those with a high perceived cost (a cold room equipped with the CoolBot™, at a cost of approximately $2000 for specialized equipment and materials for self-construction, had not been adopted by any of the respondents, and the equipment required for jam-making was deemed too expensive by one respondent). Of the practices rated as "most useful", the least-cited was hand-washing and hygiene, which is already being practiced by those doing food processing.

All of the 50 persons interviewed for the evaluation indicated that they had adopted new postharvest handling practices or technologies for reducing losses and increasing the value of their crops after receiving training via the PTSC, and 42 people were able to provide details on local costs and benefits.

Table 2. Training participant ratings of postharvest demonstrations.

Demonstrations	No. that have seen it	Rated the demo as most useful	Rated the demo as least useful	No. that have been using the new or improved practice
Shade	44	31		27
Gentle handling	33	24		17
Maturity indices	33	24		25
Improved containers	45	32		34
Sorting/grading	47	36		40
Hand-washing/hygiene	45	10		35
ZECC	45	33		20
CoolBot™	37	3	24	0
Solar drying	44	34		29
Jam making	43	32	1	17

3.3. Adaptive Research

Adaptive research was planned on pest control, low-cost cool chambers, improved solar dryers, cool transport in insulated containers, food safety and/or other topics, but the research studies were not successfully carried out by AVRDC. Several meetings were held on the planning process during 2012 and 2013, and visits were made to Africa by the Principal Investigators and the Hort Innovation Lab management entity team in order to kick-start the process. In June 2013, a no-cost extension was granted to AVRDC in order to allow them more time to carry out some of these research studies.

In February 2014, AVRDC hired independent consultants to complete as much of the work as possible. Research studies were conducted on improved traditional containers (liners in wooden crates, use of plastic crates), consumer packages (recyclable clamshells and very thin plastic produce bags), insulated pallet covers for use during shipping (in order to measure effects on temperature change), and solar dryer modifications for enhancing drying during overcast weather (adding a black plastic wrap around the legs at the bottom of the dryer). AVRDC managed the work and provided the needed funds. A few of the studies have since been completed, and the results of a study on the use of low-cost insulated pallet covers and consumer packages for reducing losses of amaranth has been submitted to a major journal.

3.4. Postharvest Services at the PTSC in Arusha

Advisory services for those who are interested in adopting new practices and technologies were one of the only services being provided by the PTSC since the close of the project in 2014. The TOPS evaluation project revealed that since the end of the Hort CRSP pilot project, several local communities and training groups near Arusha have been receiving advice on constructing ZECCs for storing fresh foods on their farms or for food service. The ToT participants and AVRDC postharvest staff in Tanzania are actively making advisory field visits, meeting interested groups, and providing consulting for local clientele. Several local non-governmental organizations (NGOs) (Istituto OIKOS, The Mesula Project, ECHO-Impact Center, Arusha, Tanzania) have requested assistance and received postharvest training and advice on setting up their own demonstrations and training programs. None of the other planned "fee for services" activities (sorting/grading, packing, pre-cooling, cold storage, *etc.*) were being provided by the PTSC for local clientele.

The model included a variety of these postharvest services which were intended to serve as a ready source of income, generating funds to pay for utilities, management, and maintenance staff for the facilities. Without these sources of revenue, the PTSC will remain dependent upon the host organization for funding its management and training programs.

3.5. Retail Sales of Tools, Packages, and Postharvest Supplies

The evaluation revealed that the retail shop for the PTSC was never fully implemented, and much of the original inventory of tools and supplies provided by the Hort CRSP project is still on the shelves. The PTSC shop mainly functions for one week per year during the August "Nane-Nane" agricultural show at Njiro, Tanzania, when the shop is restocked with the most desired products and opened for business during a period of eight days so people can visit to make their purchases.

Both customers and the administrators of the PTSC project considered the retail shop to be one of the services provided for the local population, and expected the prices of goods and supplies to be kept lower than those at any competing vendor. The PTSC shop managers, on the other hand, did not have any wholesale buying expertise, and did not have the budget to be able to buy items in large enough quantities to make bulk purchases for the shop at lower wholesale prices.

When the 50 training program participants were queried regarding where they would obtain tools, supplies, and training if the PTSC did not exist, more than 50% said "nowhere" or said they "didn't know" or "were not sure" (Table 3).

Table 3. Respondent (*n* = 50) answers to queries on where to obtain postharvest tools, supplies, and training.

If the PTSC did not exist:	Where would you go for postharvest training?	Where would you go for postharvest demos/advice?	Where would you buy postharvest goods and services?
Nowhere	20	8	0
I don't know	7	15	3
Not sure	0	8	2
Other sources/charity organizations	21	16	16

The other sources listed by respondents included several local NGOs, churches, and international charities, but these sources were thought to be unreliable since they seldom offered training and did not always have the needed supplies. For example, OIKOS was included as a possible source for glass jars, but the evaluators learned that OIKOS buys its jars from the PTSC. The evaluation results make it clear that the PTSC shop has an important role to play in local postharvest loss reduction and small business development in Northern Tanzania, if it could be set up, stocked, operated, and marketed with a more business-like approach.

4. Timelines and Budgets

The PTSC set-up process began in early 2012 on the AVRDC campus in Arusha, Tanzania. AVRDC hired a postharvest specialist in June 2012, and assigned one of their campus staff as the PTSC manager in July. The PTSC designs, demonstration protocols, procurement lists, and equipment/supplies specifications were all provided by the project leaders. Renovations and procurements went over budget, but AVRDC paid the additional costs since they intended to use the PTSC for their own projects and programs as well as for Hort CRSP–funded project activities.

The total budget for the Hort CRSP project was $500,000, with approximately $100,000 for implementing the ToT program, and approximately $100,000 used for the PTSC renovations, set-up, and one year of locally offered postharvest training programs by guest instructors. A variety of adaptive research studies were funded but not completed as planned.

The postharvest demonstrations were of relatively low cost, and many could be used over and over again for training on site. When the 14 instructors were asked to estimate the costs for the postharvest demonstrations they had utilized during their training programs, they provided the following information:

- Use of shade—low cost ($50 to $100 for materials and labor), could use any type of local materials to make thatch or a woven roofing/poles structure
- Gentle handling—very low cost, mostly show and tell, $30 for a commercially purchased harvesting bag, much less ($5 to $6) to make one locally
- Use of maturity indices (color charts, sizes)—very low cost (for making color copies, lamination, strong wire to make sizing rings)
- Improved containers (liners, cushions, crates, *etc.*)—very low cost (a few cents for a paper liner, $5 to $7 for a plastic crate)
- ZECC—$400 to $500 for bricks, sand, shade covering, labor, and water tank to construct a new Zero Energy Cool Chamber
- Solar drying practices—$300 to $400 for materials and labor to construct a new direct-style solar drier with six large trays

5. Conclusions and Recommendations

5.1. Description of Outcomes and Benefits of PTSC Extension/Outreach Results

Local capacity building in postharvest knowledge, skills, and training expertise in seven countries in SSA was one of the positive outcomes of the project. The follow-on training activities reported on by the 36 ToT participants were able to reach a wide range of local clientele in their home countries, who gained enough knowledge to make changes in their postharvest practices. In Tanzania, the random cluster sampling of 50 farmers, food processors, and marketers who had participated in more than one dozen local training programs offered by the PTSC reported having many positive outcomes. 100% of those surveyed reported that they reduced produce wastage, and 42 of the 50 persons were able to provide details on the costs and positive financial benefits of their changes in postharvest practices. Table 4 provides four examples, and many more examples will be fully documented and published upon completion of the project evaluation for USAID/TOPS.

Table 4. Examples of relative costs and financial benefits of improved postharvest practices for fresh produce as reported in Tanzania.

Interviewee No., Sex, Age, Job Type, Site	Crop and Quantity, Traditional vs. New Practices	Relative Cost, % Losses, Market Value Using Traditional Practice	Relative Cost, % Losses, Market Value Using New Practice	Changes in Income per Load	ROI
#12, male, 52, farmer/processor, Nshupu	Tomatoes, 7600 kg. Selling without grading vs. sorting/grading before selling	0 Tsh, 40% 2,850,000 Tsh	160,000 Tsh 10% 3,600,000 Tsh	+590,000 Tsh (US $327) per 7600 kg	Immediate
#20, male, 54, farmer/marketer, Kindi	Cucumbers, 150 kg. Selling without sorting/grading vs. gentle harvest, sorting and grading before selling	0 Tsh 20% 16,000 Tsh	3000 Tsh 5% 30,000 Tsh	+11,000 Tsh (US$ 6.11) per 150 kg	Immediate
#48, female, 31, farmer/processor/marketer, Poli-Ndatu	Chinese cabbage, 100 kg. Selling without grading vs. grading before selling	0 Tsh 20% 20,000 Tsh	2000 Tsh 5% 35,000 Tsh	+13,000 Tsh (US$ 7.22) per 100 kg	Immediate
#49, female, 45, farmer/processor, Nshupu	African nightshade, 10 kg. Harvesting under full sun vs. harvesting in morning when temperature is lower	0 Tsh (did not consider her labor to be a cost) 50% 10,000 Tsh	No added cost (her labor only) 5% 20,000 Tsh	+10,000 Tsh (US$ 5.56) per 10 kg	Immediate

Each of the cost-benefit examples in Table 4 were individual case studies based upon recall information on key aspects of the specific technologies being adopted (*i.e.*, on cost of materials and supplies, market prices, percentage of postharvest losses). This detailed information was provided verbally to the interviewers during site visits undertaken for the project evaluation, and the PI performed the calculations based upon the information provided, in order to determine whether there was any relative gain in earnings. A simplified cost-benefit analyses method, developed for the Bill and Melinda Gates Foundation's Appropriate Postharvest Technologies Planning Project, which focuses on measuring relative changes in costs and relative benefits such as changes in percentage of losses and market value per kg was used to make the calculations [13]. All the examples provided by the 42 respondents showed a positive and relatively rapid or immediate return on investment (ROI), since the increase in their earnings using the improved postharvest practice was higher than their initial monetary investment. The change in percentage of losses shown in Table 4 is the key reason for these positive results, as losses using traditional practices ranged from 20% to 50%, and losses using the new practices were reported to be 5% to 10%.

The eight males who were interviewed tended to be involved with handling relatively larger quantities of produce (40 to 7600 kg) and in marketing fresh produce. The 42 females who were interviewed for this evaluation study tended to be involved in food processing of fruits and vegetables, which were sometimes purchased in the wholesale market, and in the handling, processing, and marketing of relatively smaller quantities of produce (5 to 150 kg).

The PTSC model has been observed by many international visitors to the AVRDC site, and was included as a case study on reducing postharvest food losses in Africa in a recent publication by the Global Knowledge Initiative (Rockefeller Fdn/GKI, 2014 [16]). Modified versions of these postharvest training programs are currently being implemented in rural India [17].

5.2. Identification of Constraints and Implementation Issues/Concerns

Key administrators and PTSC managers were interviewed by a WFLO consultant, who met them in their offices and spent a few hours with each person in Arusha during 23–27 February 2015. Seven people were interviewed using a written questionnaire as a guide, followed by probing for specifics and details.

Each person provided information on the constraints they had encountered during the initial PTSC set-up, the implementation of training programs and planned services, and in general management. Most of the constraints had to do with the lack of a budget for operating the PTSC once the Hort CRSP project funding ended in 2013. Several constraints had to do with the limited scope of the mission and operating rules for both AVRDC (which, as a registered NGO, is not allowed to sell goods in Tanzania) and the Ministry of Agriculture, Food Security and Cooperatives (MAFC) (which can offer services and goods for sale at cost, but is generally not allowed to price goods and services in order to make a profit).

In general, the administrators and managers believe the PTSC project to be a success, since they believe that having the PTSC enabled them to increase awareness (among farmers, food processors, marketers, visiting scientists, extension workers, and policy makers) of the role of postharvest technology in reducing food losses. Several people mentioned that the model PTSC was already being copied by other organizations, and that the SARI Agriculture Technology Transfer Centre site in Njiro had been selected as one of the sites for a "value addition/postharvest training center" under the Market Infrastructure, Value Addition and Rural Finance (MIVARF) project where the retail shop could be managed by a private firm and upgraded to better serve its intended role.

Feedback from participants and formal evaluation results provide guidance on general best practices for management, overcoming difficulties, and making improvements for future postharvest training and services centers.

The PTSC model is already being emulated by several Tanzanian organizations, and a variety of "Value addition centers", "Farmer services centers", "Postharvest training centers", and "Packinghouses/ postharvest training venues" are currently being developed in 16 districts under programs being implemented by MAFC and the Prime Minister's Office. These include:

- a packinghouse for 3500 members of a vegetable cooperative in Lushoto named LUKOVEG (MAFC and the local governmental authority)
- a large citrus/mango packinghouse/training center near Dar es Salaam for a farmers' association of 2000 members (MAFC)
- the Prime Minister's Office/African Development Bank's Market Infrastructure, Value Addition and Rural Finance (MIVARF) project with postharvest training and value addition centers in 12 districts in 2013–2014, with many more planned for 2014–2015, and
- MAFC/Tanzania Horticultural Association (TAHA) Farmer Services Centers (FSCs) under construction in four districts in southern Tanzania and Zanzibar, plus a plan for a large packinghouse to be located near the coast north of Dar es Salaam.

The MIVARF project followed a similar design plan as the Hort CRSP PTSC pilot project, starting with local needs assessments and commodity systems assessments, then working with stakeholders to select key crops and design-appropriate training programs, select and procure equipment, and provide local training for farmers, food processors, and marketers. The MIVARF project has already provided one year of capacity building via ToT programs, and identified local "service providers" in each district who will serve as private sector partners. These established partners are better able to operate the needed postharvest retail shops, provide maintenance, marketing support, and other postharvest services that will help to add value, reduce produce wastage, improve incomes, and create new local businesses and new local jobs. The MIVARF project's postharvest training and services centers will therefore include all five of the key components of the PTSC model.

Acknowledgments

The authors thank Lizanne Wheeler for her many contributions during the implementation of both the HORT CRSP–funded pilot PTSC project and the TOPS-funded project evaluation. We thank our colleagues at the WFLO International Programs Office for their administrative support.

Author Contributions

Lisa Kitinoja conceived and designed the Hort CRSP pilot PTSC project; Diane M. Barrett and Lisa Kitinoja implemented the project as co-PIs; Kitinoja designed the evaluation project for WFLO and

Save the Children/TOPS, implemented the evaluation project, and analyzed the data; both Kitinoja and Barrett wrote the paper.

Conflicts of Interest

The authors declare no conflict of interest.

References

1. Kader, A.A.; Rolle, R.S. *The Role of Post-Harvest Management in Assuring the Quality and Safety of Horticultural Crops*; FAO Agricultural Services Bulletin No.152; Food and Agriculture Organization of the United Nations: Rome, Italy, 2004.
2. Weinberger, K.; Lumpkin, A.T. Diversification into horticulture and poverty reduction: A research agenda. *World Dev.* **2007**, *35*, 1464–1480.
3. AiDA. AidData. Available online: http://aida.developmentgateway.org/aida/Search.do (accessed on 15 January 2010).
4. USAID Documents. Available online: http://dec.usaid.gov/index.cfm?p=search.sqlSearch&CFID =9413138&CFTOKEN=51188616 (accessed on 10 January 2010).
5. World Bank, Projects and Operations. Available online: http://web.worldbank.org/ (accessed on 4 January 2010).
6. UN FAO. INPhO. Available online: http://www.fao.org/inpho/isma?i=INPhO&p=index.jsp& lang=en (accessed on 10 January 2010).
7. DEVEX. Available online: http://www.devex.com/ (accessed on 15 January 2010).
8. Kitinoja, L.; Saran, S.; Roy, S.K.; Kader, A.A. Postharvest technology for developing countries: Challenges and opportunities in research, outreach and advocacy. *J. Sci. Food Agric.* **2011**, *91*, 597–603.
9. Gustavsson, J.; Cederberg, C.; Sonesson, U.; van Otterdijk, R. *Global Food Loss and Food Waste—Extent, Causes and Prevention*; Food and Agriculture Organization of the United Nations: Rome, Italy, 2011.
10. Lipinski, B.; Hanson, C.; Lomax, J.; Kitinoja, L.; Waite, R.; Searchinger, T. *Creating a Sustainable Food Future—Reducing Food Loss and Waste*; WRI Working Paper 39p; World Resources Institute: Washington, DC, USA, 2013.
11. Kitinoja, L.; AlHassan, H.A. Identification of appropriate postharvest technologies for improving market access and incomes for small horticultural farmers in Sub-Saharan Africa and South Asia. Part 1: Postharvest losses and quality assessments. *Acta Hortic.* **2012**, *934*, 31–40.
12. Diem, K.G. Using Research Methods to Evaluate Your Extension Program. Available online: http://www.joe.org/joe/2002december/index.php (accessed on 15 August 2014).
13. Saran, S.; Roy, S.K.; Kitinoja, L. Appropriate postharvest technologies for improving market access and incomes for small horticultural farmers in Sub-Saharan Africa and South Asia. Part 2: Field trial results and identification of research needs for selected crops. *Acta Hortic.* **2012**, *934*, 41–52.
14. Kitinoja, L; Thompson, J.F. Pre-cooling systems for small-scale producers. *Stewart Postharvest Rev.* **2010**, *6*, 1–14.

15. Winrock International. *Empowering Agriculture, Energy Options for Horticulture*; United States Agency for International Development: Washington, DC, USA, 2009; p. 79. Available online: http://pdf.usaid.gov/pdf_docs/PNADO634.pdf (accessed on 15 March 2015).

16. Rockefeller Foundation Global Knowledge Initiative. Reducing Food Waste and Spoilage: Assessing Resources Needed and Available to Reduce Post Harvest Food Loss in Africa, 2014. Available online: http://postharvest.org/Rockefeller%20Foundation%20Food%20Waste%20and%20Spoilage%20initiative%20Resource%20Assessment_GKI.pdf (accessed on 2 April 2015).

17. Raman, N.L.M.; Dubey, N. Panchayat and economic empowerment of rural women by hands on training. *Am. Int. J. Res. Humanity Arts Soc. Sci.* **2014**, *5*, 249–252.

8

Transferability of Models for Estimating Paddy Rice Biomass from Spatial Plant Height Data

Nora Tilly [1,*], Dirk Hoffmeister [1], Qiang Cao [2], Victoria Lenz-Wiedemann [1], Yuxin Miao [2] and Georg Bareth [1]

[1]   ICASD-International Center for Agro-Informatics and Sustainable Development, Institute of Geography (GIS & Remote Sensing Group), University of Cologne, 50923 Cologne, Germany; E-Mails: dirk.hoffmeister@uni-koeln.de (D.H.); victoria.lenz@uni-koeln.de (V.L.-W.); g.bareth@uni-koeln.de (G.B.)

[2]   ICASD-International Center for Agro-Informatics and Sustainable Development, Department of Plant Nutrition, China Agricultural University, 100193 Beijing, China; E-Mails: qiangcao@cau.edu.cn (Q.C.); ymiao@cau.edu.cn (Y.M.)

*   Author to whom correspondence should be addressed; E-Mail: nora.tilly@uni-koeln.de

Academic Editor: Yanbo Huang

Abstract: It is known that plant height is a suitable parameter for estimating crop biomass. The aim of this study was to confirm the validity of spatial plant height data, which is derived from terrestrial laser scanning (TLS), as a non-destructive estimator for biomass of paddy rice on the field scale. Beyond that, the spatial and temporal transferability of established biomass regression models were investigated to prove the robustness of the method and evaluate the suitability of linear and exponential functions. In each growing season of two years, three campaigns were carried out on a field experiment and on a farmer's conventionally managed field. Crop surface models (CSMs) were generated from the TLS-derived point clouds for calculating plant height with a very high spatial resolution of 1 cm. High coefficients of determination between CSM-derived and manually measured plant heights (R^2: 0.72 to 0.91) confirm the applicability of the approach. Yearly averaged differences between the measurements were ~7% and ~9%. Biomass regression models were established from the field experiment data sets, based on strong coefficients of determination between plant height and dry biomass (R^2: 0.66 to 0.86 and 0.65 to 0.84 for linear and exponential models, respectively). The spatial and temporal transferability of the models to the farmer's conventionally managed fields is supported by strong coefficients of determination between estimated and measured values (R^2: 0.60 to 0.90 and 0.56 to 0.85 for linear and exponential models, respectively). Hence, the suitability of TLS-derived spatial plant height as a non-destructive estimator for biomass of paddy rice on the field scale was verified and the transferability demonstrated.

Keywords: terrestrial laser scanning; plant height; biomass; rice; precision agriculture; field level

1. Introduction

Solutions to ensure the world's food security are required due to the growing world population. Focusing on the supply with staple food, the cultivation of rice is essential. This is in particular for the Asian world important, where 2011 and 2012 about 90% of the estimated world rice production was cultivated, each year about 650 million tons [1]. Miao *et al.* [2] reviewed long-term experiments on sustainable field management and highlighted the required increase in cereal production to ensure food security in China. The authors emphasized the combination of traditional practices and modern sensor-based management approaches for addressing this challenge.

In this context, precision agriculture (PA) rises in importance, which focuses on spatial and temporal variabilities of natural conditions and an adequate dealing with resources [3]. PA-improved management methods support farmers in closing the gap between potential and current yield [4]. Based on analyses of long-term field experiments, Roelcke *et al.* [5] concluded that there is a great need for on-farm experiments. Therefore accurate crop monitoring based on remote and proximal sensing has become increasingly important within PA in recent years [6,7]. A widely used indicator for quantifying the actual status of plants is the nitrogen nutrition index (NNI) [8–10]. The index shows the ratio between measured and critical N content. The latter is determined by the crop-specific N dilution curve, showing the relation between N concentration and biomass. Consequently, the accurate and non-destructive determination of biomass is a precondition for calculating the NNI.

For rice, it has been shown that grain yield is positively correlated to biomass and nitrogen (N) translocation efficiency [11], but over-fertilization affects the nutrient balance in soil and groundwater. Consequently, the NNI should be used for optimizing rice production with PA-improved management methods. Therefore, non-invasive approaches for biomass estimation are of key importance as rice paddies should be entered with machinery as little as possible during the growing season. Satellite remote sensing images serve for estimating the actual biomass and yield of large paddy rice fields [12–16]. However, for monitoring within-field variability and more accurately estimating biomass, a higher spatial resolution is required. The potential of ground-based plant parameter measurements as input for biomass estimation models was recently demonstrated for rice, maize, cotton, and alfalfa [7]. However, therein, plant height was manually measured, which is prone to selection bias. A ground-based multi-sensor approach showed good results for predicting biomass of grassland [17]. For biomass estimation of paddy rice, *in-situ* approaches with hand-held sensors for measuring canopy reflectance provided good results [18–20]. Moreover, Confalonieri *et al.* [21] emphasized rice plant height as a key factor for predicting yield potential and developed a model for estimating plant height increase, but accurate *in-situ* measurements of plant height on field level are rare. Although virtual geometric models of single rice plants in a high resolution exist [22,23], uncertainties remain about the transferability to the field, due to varying patterns of plant growth. Hence, accurate methods for determining plant height on field level are desirable.

Light detection and ranging (LIDAR) sensors have been increasingly used in vegetational studies since the 1980s [24]. *In-situ* studies confirmed the potential of ground-based LIDAR methods, also known as terrestrial laser scanning (TLS), for the assessment of plant parameters in agricultural applications. Previous studies focused on the acquisition of plant height [25], post-harvest growth [26], leaf area index [27], crop density [28,29], nitrogen status [30], or the detection of individual plants [31,32]. Moreover, the potential of TLS for estimating the biomass of small-grain cereals was emphasized [33–36]. Regarding the accuracy, Lumme *et al.* [33] found that estimated heights of cereal plants correlated with tape measurements. The high precision for mapping of maize plants was shown by Höfle [31]. Little research has been done so far on TLS *in-situ* measurements of paddy rice. Hosoi and Omasa [37] examined vertical plant area density as an estimator for biomass, achieved with a portable scanner in combination with a mirror. Besides, biomass estimations based on TLS-derived spatial plant height was evaluated for some of the fields considered in the presented study [38,39]. But as stated above, multi-annual on-farm experiments are necessary for achieving a comprehensive understanding of plant growth and developing objective sensor-based measuring methods and models for biomass estimations [2,5].

Based on the promising results of the single year analyses [38,39], this study focused on (I) the robustness of the method, (II) the spatial and temporal transferability of the models, and (III) a model improvement. For the latter, in addition to partially existing linear models, exponential models were established, as a better suitability of these models is denoted in other studies of biomass estimations over different growth stages [20,40,41]. In two consecutive growing seasons, rice fields were monitored during the pre-anthesis period. Based on the data sets of a field experiment, estimation models for biomass were established and then applied on a farmer's conventionally managed fields.

2. Data and Methods

2.1. Study Area

Heilongjiang Province in the northeast of China is an important region for agricultural production [42]. Almost 25% of the total area is covered by the Sanjiang Plain (~120,000 km²). The regional climate with cold and dry winters and short but warm, humid summers is marked by the East Asian summer monsoon [43,44]. Three field sites around the city of Jiansanjiang (N 47°15'21″ E 132°37'43″) were considered in this study.

At the Keyansuo experimental station (Jiansanjiang, Heilongjiang Province, China) the same field experiment was monitored in 2011 and 2012 (Figure 1). For the experiment, nine N fertilizer treatments were repeated three times for the rice varieties *Kongyu 131* and *Longjing 21*. Hence, the field with a spatial extent of 60 m by 63 m consisted of 54 plots, each about 10 m by 7 m in size. A detailed description of the experimental set-up is given by Cao *et al.* [45]. Related to the amount of N input, variations in plant height and biomass were expected. These differences were useful for the TLS monitoring approach to capture varying patterns of plant growth at one growing stage.

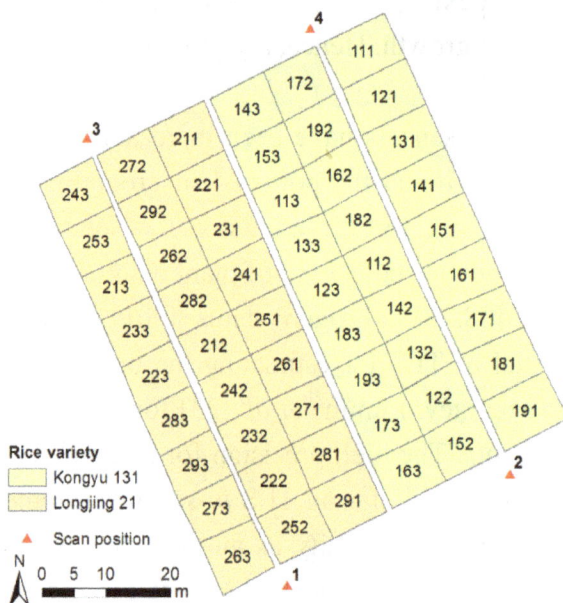

Figure 1. Design of the field experiment and scan positions. Three-digit number in the plot represents rice variety (1 = *Kongyu 131*; 2 = *Longjing 21*); treatment (1 to 9); and repetition (1 to 3). Modified from Tilly *et al*. [39].

In addition, one farmer's conventionally managed field was investigated each year (hereafter referred to as farmer's field). The aim was to provide independent validation data sets for checking the spatial and temporal transferability of the findings from the field experiment data. For the following, they are termed *village 69* (year 2011) and *village 36* (year 2012). In both years, it was not possible to find a field with one of the field experiment rice varieties, where destructive sampling was possible several times during the growing season. In *village 69* the variety *Kenjiandao 6* was cultivated, in *village 36* the variety *Longjing 31*. Moreover, in *village 36* management units with very heterogeneous development were chosen, including parts without any plants (Figure 2). On each field two management units were investigated. In *village 69* and *village 36* each unit was about 60 m by 40 m and 50 m by 70 m in size, respectively.

Figure 2. One management unit with very heterogeneous plant growth in *village 36*.

2.2. Field Measurements

On each site, three TLS campaigns were carried out in June and July of the respective year to capture the key vegetative stages of the rice plants. During this pre-anthesis period, differences in plant development occur mainly due to the increase of tillers and plant height. This period is important for fertilizer management decisions. In both years, the campaigns on the field experiment and the farmer's field were carried out on two consecutive days to reach a best possible comparison regarding the plant development. For quantifying the phenological stages of plants and steps in plant development the BBCH-scale was used [46,47]. The abbreviation BBCH is derived from the funding organizations: Biologische Bundesanstalt (German Federal Biological Research Centre for Agriculture and Forestry), Bundessortenamt (German Federal Office of Plant Varieties), and Chemical industry. The campaign dates and BBCH-values for all sites are given in Table 1.

Table 1. Dates of the terrestrial laser scanning (TLS) campaigns and corresponding phenological stages.

Date/	2011		2012	
BBCH-scale [a]	Field experiment	Village 69	Field experiment	Village 36
1. Campaign	21 June 2011/ 13	22 June 2011/ 13	1 July 2012/ 37	30 June 2012/ 37
2. Campaign	4 July 2011/ 13–15; 22–23	5 July 2011/ 13; 21	9 July 2012/ 42	8 July 2012/ 37; 39
3. Campaign	18 July 2011/ 19; 29; 32	19 July 2011/ 19; 29; 34	17 July 2012/ 50	16 July 2012/ 19; 29; 34

[a] Multiple values due to several samples.

Terrestrial laser scanners operating with the time-of-flight technique were used for all campaigns. The relative positions of survey points are calculated from the distances, as well as the horizontal and vertical angles between sensor and targets. For this, the time between transmitting and receiving a pulsed laser signal and its angles are measured. In 2011 and 2012, the Riegl VZ-1000 and Riegl LMS-Z420i, respectively, were provided by the company Five Star Electronic Technologies (Beijing, China) [48,49]. Both devices operate with a near-infrared laser beam and have a beam divergence of 0.3 mrad (VZ-1000) and 0.25 mrad (LMS-Z420i). The angular resolution was set to 0.04 deg. All scans were conducted from the dikes between the paddies to avoid entering them, resulting in an oblique perspective. More detailed descriptions are given in Tilly *et al.* [39].

The set-up for the campaigns on the field experiment was similar in both years. Each time, nine scan positions were established for covering all fields of the Keyansuo experimental station and minimizing shadowing effects. For this analysis, the scans from all positions were used, but four positions were of major importance, as they were located close to the investigated field experiment. Following, the largest number of points was acquired from these positions. Point clouds from other positions were used to avoid gaps in the final point cover due to information signs close to the field. As shown in Figure 1, two positions respectively were set up at the north and south edges. At each position the scanner was mounted on a tripod which raised the sensor up to 1.5 m above ground. Additionally, a small tractor-trailer system was used for the positions at the south edge of the field for achieving a greater height of about 3 m. The narrow dikes along the other edges made it impossible to reach those positions with the tractor-trailer system.

Due to a limited access on the dikes between the management units of both farmer's fields, it was also impossible to use a trailer. Hence, the sensor height of the scanner on the tripod was about 1.5 m above ground. In *village 69* the scan positions were established close to the four corners of the management units (Figure 3). As the investigated units in *village 36* were located at the edge of the whole field, this set-up was slightly changed. Two positions in the north were established on a small hill close to the field for reaching a higher position and an additional position was placed at the center of the edge (scan position 5 in Figure 3). Further two positions were set up close to the south corners. In both fields, twelve thin, long bamboo sticks per management unit were stuck in the ground. These bamboo sticks can be easily detected in the TLS point clouds and located in the field to ensure the spatial linkage to other plant parameter measurements.

Figure 3. Scan positions and bamboo stick positions on the farmer's fields.

Furthermore, ranging poles with high-reflective cylinders [50] were built upon the dikes between the fields, homogeneously distributed around the field. These can be detected by the laser scanner and act as tie points for merging the scan data in post-processing. In the first campaigns, the position of each pole was marked in the fields. By re-establishing the ranging poles at exactly the same position for the following campaigns, all scans of one site can be merged. In the data sets from 2011, alignment errors occurred due to imprecise re-establishing of the ranging poles or where an exact marking of the positions was difficult, particularly on the farmer's fields. These errors could be rectified with software options but caused time-consuming post-processing. In 2012, additional tie points were used to avoid this. As shown in Figure 4 for *village 36*, five small, round reflectors were permanently attached to trees close to the fields and remained there during the observation period. A homogeneous distribution around the field was not possible, as no other stationary objects were available.

Figure 4. Small, round reflectors were permanently attached to trees in *village 36*.

At all sites, manual measurements of plant height and biomass were performed during the whole vegetation period. Corresponding to each TLS campaign on the field experiment, the heights of eight to ten and four hills per plot were measured in 2011 and 2012, respectively. Each hill consisted of four to six rice plants.

Regarding the measurement of biomass, differences between the sites and years must be pointed out. As part of the field experiment, destructive sampling was performed several times during the vegetation period. Samples were taken from both varieties, but only from the three repetitions of five treatments ($n = 30$). The dates of sampling differed from the TLS campaign dates in 2011, but due to the small plot size, it was not feasible to take additional samples. Thus, the biomass values were linearly interpolated. In 2012, the measurements could be carried out on the same day.

On the farmer's fields, four hills around each bamboo stick were destructively taken after the TLS measurements (each $n = 24$). For the following campaign, the bamboo sticks were moved in a defined direction to the center of four other hills. In each management unit of *village 36*, one bamboo stick was placed in the part without any plant and left at its position for all campaigns (no. 12 in Figure 3).

The cleaned above ground biomass was weighed after drying. All samples were oven dried at 105 °C for 30 min and dried to constant weight at 75 °C. The dry biomass per m^2 was calculated, considering the specific number of hills per m^2.

2.3. Post-Processing of the TLS Data

The post-processing of the scan data was similar for all sites. A detailed description is given for the data sets from 2011 in Tilly *et al.* [39]. Riegl's software RiSCAN Pro, also applied for the data acquisition, was used for the first steps of the data handling. The scans from all campaigns were imported into one RiSCAN Pro project file for each site. Following, a co-registration of all scan positions was carried out, based on the reflectors acting as tie points. As mentioned above, the data sets of 2011 showed alignment errors, due to non-optimal positioning or imprecise re-establishing of the ranging poles. The iterative closest point (ICP) algorithm [51], implemented in RiSCAN Pro as Multi Station Adjustment, was used to modify the position and orientation of each scan position in multiple iterations for getting the best fitting result. For the campaigns in 2012, additional small reflectors were permanently established. By first registering one scan position of each campaign based on these permanent tie points and aligning all other positions to these, an accurate alignment was possible. After optimizing the alignment with the ICP algorithm the error, measured as standard deviation between the used point-pairs, was 0.06 m and 0.01 m on average for both sites of 2011 and 2012, respectively.

Following, the point clouds were merged to one data set per campaign and the area of interest (AOI) was manually extracted. Clearly identifiable noise in the point clouds far below and above the field, caused by reflections on water in the field or on small particles in the air, was previously removed. The crop surface was then determined from the point clouds with a filtering scheme for selecting maximum points. A common reference surface is required for the calculation of plant heights. Therefore, the AOI is usually scanned without any vegetation. As it was not possible to obtain such data on the rice fields, the lowest parts in the point clouds from the first campaigns were selected. At this stage, the rice plants were small enough for clearly identifying points at the bottom of the hills, as shown in Tilly *et al.* [39]. The point clouds of the field experiment data sets were subdivided plot-wise to attain a common spatial base. Each management unit of the farmer's fields was regarded as one data set. All data sets were exported as ASCII files, which contained the XYZ coordinates of each point for spatial and statistical analyses.

2.4. Calculation of Plant Height and Visualization as Maps of Plant Height

For the spatial analyses, crop surface models (CSMs) were constructed from the TLS-derived point clouds. CSMs were introduced by Hoffmeister *et al.* [50] for an objective and non-invasive deriving of spatial crop height and crop growth patterns. A CSM represents the crop surface at a specific date with a high spatial resolution. Therefore, the exported point clouds were interpolated to raster data sets with a consistent spatial resolution of 1 cm with the inverse distance weighting (IDW) algorithm in ArcGIS Desktop 10 (Esri, Redlands, CA, USA). IDW is suitable for preserving the accuracy of measurements with a high density, as it is a deterministic, exact interpolation and retains a measured value at its location [52]. Likewise, a digital terrain model (DTM) was generated from the manually selected ground points as common reference surface. Next, the DTM was subtracted from the CSM for calculating the plant heights. In the same way, plant growth between two dates can be spatially measured by calculating the difference between two CSMs. Herein, growth is defined as spatio-temporal difference in height. Finally, maps of plant height were created for visualizing the pixel-wise calculated values.

For the following analyses, one plant height value per campaign for comparable spatial units was necessary. Therefore, the CSM-derived plant heights were averaged plot-wise for the field experiment ($n = 54$). Previously, each plot was clipped with an inner buffer of 60 cm for preventing border effects. As the manual measurements were used for validating the laser scanning results, these plant height values were also averaged plot-wise ($n = 54$). Around each bamboo stick on the farmer's fields, a circular buffer with a radius of 1 m was generated to attain a common spatial base, for which the CSM-derived plant heights were averaged (each $n = 24$).

2.5. Estimation of Biomass

The field experiment analyses were taken to express the correlation between plant height and dry above ground biomass (hereafter referred to as biomass) in a biomass regression model (BRM). Since only the above ground plant height is determinable from the TLS data, statements about the subsurface cannot be done. As mentioned above, other studies showed that exponential models performed better for biomass estimations over different growth stages. For establishing exponential models in addition to the linear ones, the biomass values were natural log-transformed. The models were used for estimating the biomass on the farmer's fields based on the TLS-derived spatial plant height data. Previously, linear and exponential biomass regression models (BRMs) were established, only regarding the field experiment for checking the general concept and evaluating differences between the results for 2011 and 2012 (hereinafter referred to as trial BRMs). Afterwards, the transferability of the model to the farmer's fields was evaluated. The workflow can be structured as following:

I.   Examination of concept with trial BRMs: Each linear and exponential model was derived from the measurements of two field experiment repetitions from one year. The biomass of the remaining third repetition was estimated and validated against the destructive measurements.

II. Generation of BRM: Overall six models were established based on the measurements of all field experiment repetitions, separately for each year and as a combination of both years, each as linear and exponential model.

III. Application of the BRMs: Each model was used for estimating the biomass at all campaign dates on both farmer's fields based on the CSM-derived plant height of the buffer areas around the bamboo sticks.

IV. Validation of the BRMs: By comparing estimated and destructively measured biomass values the general validity, robustness, and suitability of the linear and exponential BRMs were evaluated.

The accuracy of each BRM was evaluated based on the coefficient of determination, index of agreement and root mean square error, calculated for each estimated value in comparison with the destructively measured biomass. The coefficient of determination (R^2) is widely used as measure of the dependence between two variables, but often unrelated to the size of the difference between them. For validating models, Willmott's index of agreement (d) shows to which degree a measured value can be estimated [53,54]. The index ranges between 0 and 1, from total disagreement to entire agreement. In addition, the root mean square error (RMSE) indicates how well the estimated values fit to the measured values [55].

3. Results

3.1. Maps of CSM-Derived Plant Height

The TLS-derived CSMs and the DTM were used to calculate plant height pixel-wise for all plots of the field experiment and each management unit of both farmer's fields. The resulting raster data sets have a high resolution of 1 cm. Maps of plant height were created for visualizing spatial and temporal patterns and variations. In Figure 5, maps of plant height are shown for two field experiment plots for all campaigns of both years. The respective first repetition of two fertilizer treatments for the rice variety *Kongyu 131* are selected as an example, whereby the plot numbers, 111 and 151, refer to the lower and higher amount of applied N fertilizer, respectively. In particular in the maps of plot 111, the linear structure of the rice plant rows is detectable in both years. In 2012, Plot 151 shows a discernible pattern with higher plant height values in the north corner, which is visible in all campaigns. Moreover, differences in plant height occur between the different fertilizer treatments. The mean plant heights are higher for plots with a higher amount of applied N fertilizer, with a difference ranging from ~7 cm to ~13 cm and ~4 cm to ~16 cm for 2011 and 2012, respectively.

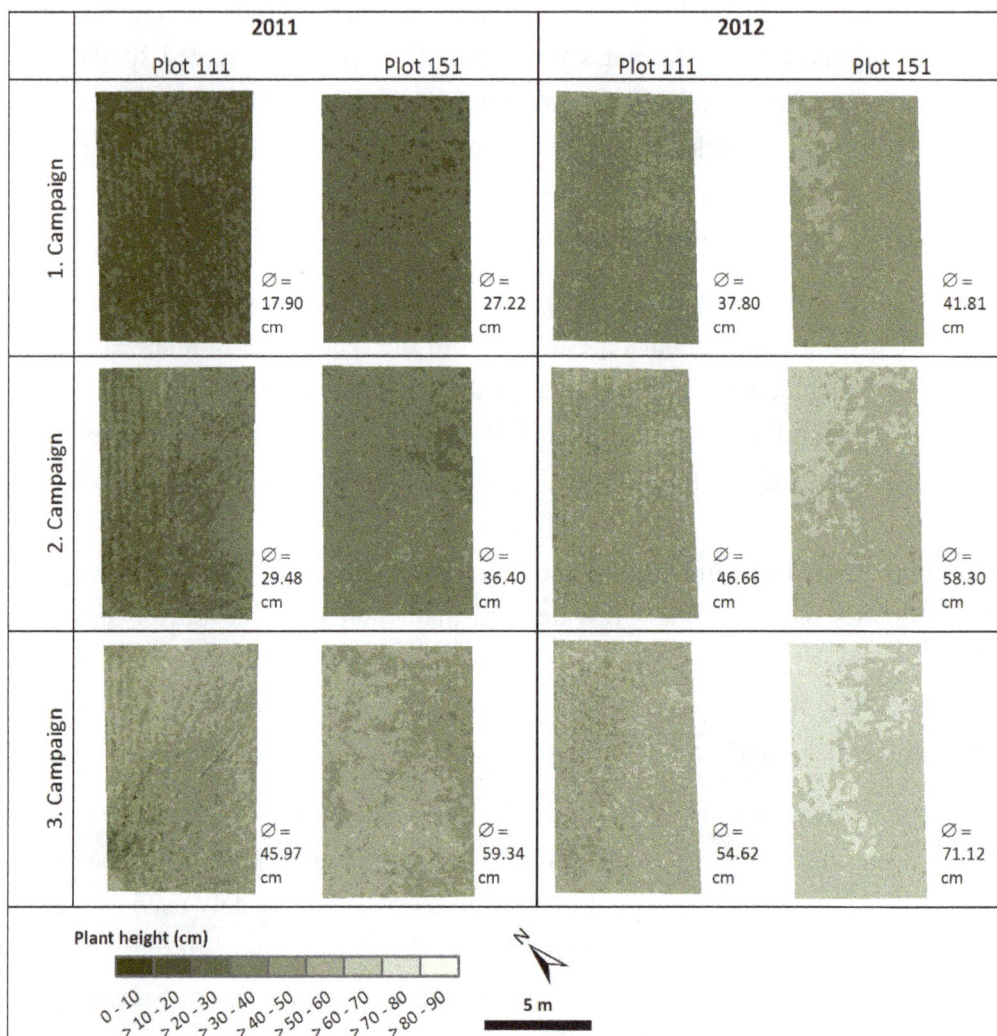

Figure 5. Crop surface model (CSM)-derived maps of plant height for two field experiment plots of both years, given with mean plant height per plot.

3.2. Analysis of Plant Height Data

Regarding the field experiment, averaged CSM-derived and manually measured plant heights were used for validating the accuracy of the scan data (Table 2). The mean heights are quite similar for both years, with an average difference of ~7% and ~9% for 2011 and 2012, respectively. The standard deviation within each campaign increases over time. All values and the resulting regression lines are shown in Figure 6. The coefficients of determination are high for 2011 and 2012 with $R^2 = 0.91$ and $R^2 = 0.72$, respectively.

Table 2. Mean crop surface model (CSM)-derived and manually measured plant heights of the field experiment (*n*, number of samples; x̄, mean value; SD, standard deviation; min, minimum; max, maximum).

Date	*n*	Plant height from CSM (cm)				Measured plant height (cm)				Difference
		x̄	SD	min	max	x̄	SD	min	max	%
21 June 11	54	**24.84**	3.63	17.90	32.99	**24.37**	2.06	19.13	28.88	**1.89**
04 July 11	54	**34.62**	4.36	24.59	42.71	**37.94**	2.42	32.38	44.13	**9.59**
18 July 11	54	**55.38**	7.22	44.28	70.30	**63.56**	4.25	53.10	70.70	**14.77**
01 July 12	54	**44.72**	3.08	37.80	53.25	**40.85**	4.87	31.00	49.50	**8.64**
09 July 12	54	**57.09**	3.61	48.87	64.64	**46.84**	4.30	37.50	56.50	**17.95**

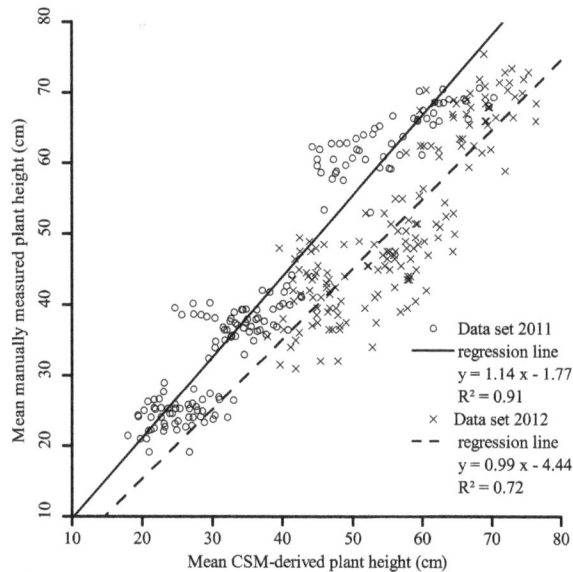

Figure 6. Regression of the mean CSM-derived and manually measured plant heights of the field experiment of both years (each *n* = 162).

3.3. Analysis of Estimated Biomass

Following the set-up of the field experiment, only five treatments were considered for the destructive biomass sampling (*n* = 30). Thus, the number of samples and averaged plant height values differ from the comparison shown in Table 2. On both farmer's fields, biomass was taken around all bamboo sticks (each *n* = 24). Mean value, standard deviation, minimum, and maximum were calculated for the plant height and dry biomass of all campaigns on each site (Table 3). The analysis of the mean plant heights can be summarized to: (I) the differences between the field experiment 2011 and *village 69* are less than ~5 cm, (II) the data sets from the field experiment 2012 and *village 36* show considerably larger differences with ~25 cm, (III) the difference between the data sets of the field experiment lies between ~10 cm and ~20 cm, (IV) comparing the farmer's fields, the difference increases over the growing season from ~2 cm to ~20 cm, and (V) the standard deviations within each campaign are almost similar and below ~5 cm, despite the results from *village 36* with values between ~6 cm and ~8 cm.

Regarding the biomass measurements, comparative statements have to be limited, due to the interpolated values for the field experiment 2011. Nevertheless, the results can be summed up as following: (I) all mean values are considerable higher for 2012, (II) the difference between the values of the field experiment 2011 and *village 69* increases over time from less than 5% for the first campaign to ~40% and ~30% for the second and third campaign, respectively, (III) the difference between the values of the field experiment 2012 and *village 36* is constantly less than 5% during the whole observation period, and (IV) the standard deviation is much higher for all measurements in 2012, ranging from ~75 g/m^2 to ~145 g/m^2, in contrast to ~15 g/m^2 to ~80 g/m^2 for the measurements in 2011.

Table 3. Mean CSM-derived plant heights and destructively measured biomass values of all sites (*n*, number of samples; $\bar{x}$, mean value; SD, standard deviation; min, minimum; max, maximum).

Site/		Plant height from CSM (cm)				Biomass (g/m^2) [a]			
Date	*n*	$\bar{x}$	SD	min	max	$\bar{x}$	SD	min	max
Field experiment									
21.06.11	30	**24.93**	2.85	20.59	30.33	**59.51**	18.86	24.04	100.70
04.07.11	30	**33.80**	3.74	27.25	40.75	**131.72**	30.03	66.71	199.41
18.07.11	30	**56.69**	5.49	44.91	63.03	**422.27**	80.90	274.74	599.53
01.07.12	30	**43.81**	2.95	37.80	48.14	**231.42**	74.48	104.47	421.35
09.07.12	30	**56.08**	3.73	46.66	62.28	**449.92**	105.62	225.40	673.79
17.07.12	30	**66.63**	5.05	54.62	75.24	**636.10**	127.87	372.06	946.15
Village 69									
22.06.11	24	**20.80**	4.82	13.39	31.44	**57.58**	13.02	25.64	80.01
05.07.11	24	**34.09**	4.52	27.13	44.60	**217.43**	29.44	146.54	278.12
19.07.11	24	**59.49**	4.87	51.79	72.58	**589.71**	73.01	482.33	723.32
Village 36									
30.06.12	24	**18.13**	7.59	1.96	45.00	**251.67**	91.46	123.00	479.88
08.07.12	24	**30.23**	6.22	19.25	41.73	**469.93**	104.00	171.90	639.00
16.07.12	24	**40.36**	8.28	21.54	52.82	**717.61**	143.73	399.36	966.42

[a] values for the field experiment 2011 are linearly interpolated from other dates.

The regression equations from the field experiment data were used to establish linear and exponential BRMs. Previously, the general concept was examined with trial BRMs, each achieved from two field experiment repetitions of one year, validated against the third repetition. Table 4 shows the equations of the linear and exponential trial BRMs with the estimated and measured biomass values. In both years over- and underestimations occur, depending on the repetition combination and linear or exponential model. However, for the linear models the mean deviations of the estimated values from the actual measured values are small for 2011, less than 19% and very small for 2012, less than 1%. On the contrary, for 2011 the coefficients of determination (R^2) as well as the indices of agreement (d) between estimated and measured biomass values are higher and the root mean square error (RMSE) is lower. Similar R^2 and d values were achieved with the exponential models. Due to the log-transferred biomass values, the RMSE values cannot be directly compared. However, whereas the differences between estimated and measured values are much lower for 2011 (below 5%), they are slightly higher for 2012 (up to ~2.5%).

Table 4. Trial biomass regression models (BRMs) and validation of estimated against measured biomass (R^2, coefficient of determination; d, index of agreement; RMSE, root mean square error).

	Year/ Repetition	Trial BRMs [a]	Estimated Repetition	Mean biomass (g/m²) estimated	measured	Difference (%)	R^2	d	RMSE
Linear	**2011**								
	1 & 2	y = 11.06x − 211.23	3	249.79	210.61	−18.60	0.92	0.96	61.54
	1 & 3	y = 11.12x − 237.97	2	174.05	208.32	16.45	0.81	0.93	79.90
	2 & 3	y = 11.15x − 229.41	1	189.38	194.56	2.66	0.88	0.97	52.90
	2012								
	1 & 2	y = 14.33x − 379.96	3	427.12	426.06	−0.25	0.72	0.91	93.27
	1 & 3	y = 14.87x − 413.65	2	404.44	402.35	−0.52	0.55	0.85	125.13
	2 & 3	y = 14.36x − 379.12	1	413.28	417.20	0.94	0.71	0.91	92.77
Exponential [b]	**2011**								
	1 & 2	y = 0.06x + 2.76	3	4.99	5.22	4.58	0.88	0.95	0.38
	1 & 3	y = 0.06x + 2.64	2	5.01	4.83	−3.64	0.80	0.93	0.41
	2 & 3	y = 0.06x + 2.80	1	4.91	5.05	2.91	0.91	0.97	0.30
	2012								
	1 & 2	y = 0.04x + 3.79	3	5.95	5.96	0.22	0.68	0.89	0.28
	1 & 3	y = 0.04x + 3.82	2	5.88	6.02	2.44	0.58	0.82	0.36
	2 & 3	y = 0.04x + 3.67	1	5.94	5.88	−1.03	0.72	0.91	0.25

[a] x = plant height (cm); y = biomass (g/m²); [b] biomass values are natural log-transformed.

The final linear and exponential BRMs were established from the field experiment data sets for each year separately and for both years combined (Table 5). All values and the resulting regression lines are plotted in Figure 7 for the linear and exponential models, the corresponding equations are given in Table 5. Strong coefficients of determination for all data sets prove the dependency of biomass on plant height during the regarded pre-anthesis period. Comparable results were achieved for linear (2011: $R^2 = 0.86$; 2012: $R^2 = 0.66$; combination: $R^2 = 0.81$) and exponential models (2011: $R^2 = 0.84$; 2012: $R^2 = 0.65$; combination: $R^2 = 0.84$). Each model was used for estimating the biomass of the buffer areas around the bamboo sticks on both farmer's fields based on the CSM-derived plant height. The reliability of the estimated values was validated against the measured biomass values. In Table 5 the mean differences are given, averaged for each campaign and over all campaigns on each farmer's field. Further, the coefficient of determination (R^2), index of agreement (d), and root mean square error (RMSE) are given for each BRM. Generally, the estimations for *village 69* are better overall, verifiable through smaller percentage deviations, higher R^2 and d as well as lower RMSE values for linear and exponential models. The differences between linear and exponential models for each site are small with slightly better R^2 values for the linear BRMs. Within each site, the three models yielded almost similar results. Regarding the BRMs of the single years, the linear function showed slightly lower percentage deviations with the data set from 2011, whereas the exponential with the one from 2012. For the combined data set, the linear model functioned slightly better than both single year BRMs, whereas with the exponential models it performed weaker.

Table 5. Biomass regression models (BRMs), derived from field experiment and validation of estimated against measured biomass for the farmer's fields (R^2, coefficient of determination; d, index of agreement; RMSE, root mean square error).

	Site/	BRM [a]	Mean difference					R^2	d	RMSE
			per campaign (g/m²)			all campaigns				
	Data set		1.	2.	3.	(g/m²)	%			
Linear	**Village 69**									
	2011	y = 11.06x − 224.18	51.69	64.56	110.79	90.73	31.48	0.90	0.92	119.70
	2012	y = 14.51x − 390.58	146.33	113.35	115.10	125.59	43.57	0.90	0.91	146.90
	combination	y = 12.37x − 273.19	73.47	68.95	98.30	89.83	31.16	0.90	0.93	115.22
	Village 36									
	2011	y = 11.06x − 224.18	254.34	320.62	380.60	336.87	74.48	0.60	0.53	377.04
	2012	y = 14.51x − 390.58	281.90	382.73	425.57	375.82	83.09	0.60	0.51	429.33
	combination	y = 12.37x − 273.19	175.02	330.06	383.54	312.30	69.04	0.60	0.53	383.62
Exponential [b]	**Village 69**									
	2011	y = 0.06x + 2.74	0.04	0.59	0.32	0.23	4.35	0.85	0.95	0.46
	2012	y = 0.04x + 3.76	−0.58	0.25	0.24	−0.03	−0.65	0.85	0.92	0.45
	combination	y = 0.05x + 2.95	0.07	0.72	0.58	0.41	7.81	0.85	0.91	0.56
	Village 36									
	2011	y = 0.06x + 2.74	1.58	1.52	1.42	1.47	24.31	0.56	0.44	1.47
	2012	y = 0.04x + 3.76	0.65	1.12	1.13	0.97	15.97	0.56	0.51	1.06
	combination	y = 0.05x + 2.95	1.38	1.62	1.58	1.51	24.92	0.56	0.42	1.51

[a] x = plant height (cm); y = biomass (g/m²); [b] biomass values are natural log-transformed.

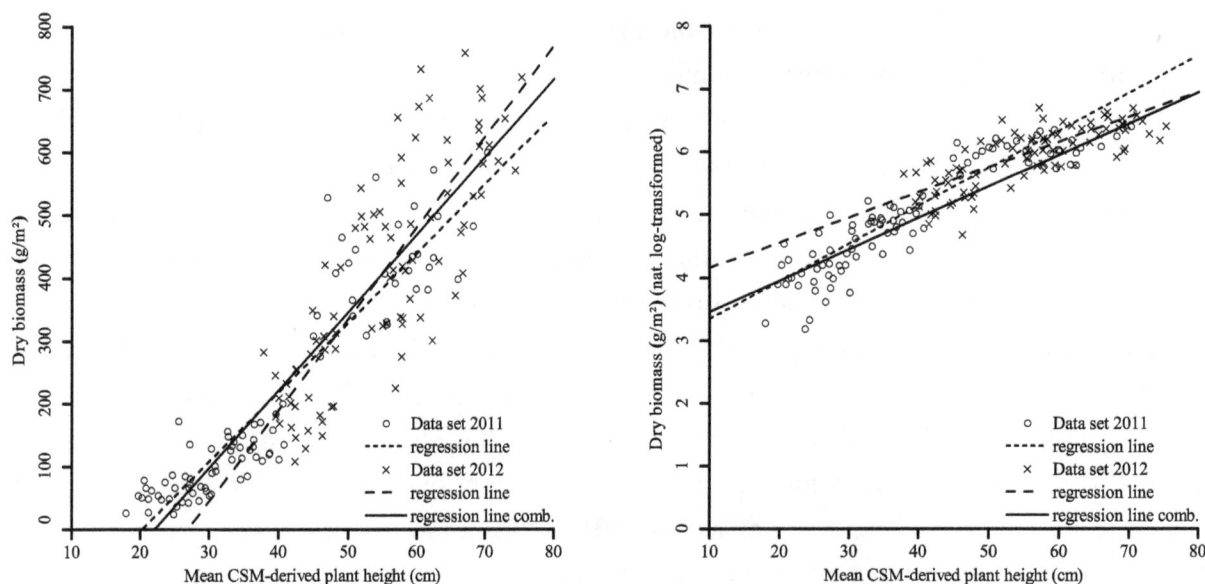

Figure 7. Linear (**left**) and exponential (**right**) regression between mean CSM-derived plant height and dry biomass for the field experiment of both years (each n = 90); regression equations are given in Table 5. Biomass values for the exponential regression are natural log-transformed.

4. Discussion

Overall, the acquisition with both laser scanners worked very well. The reliability of the devices was shown in earlier studies [38,39,50]. Due to the lightweight build-up and higher measurement rate the Riegl VZ-1000 is preferable to the Riegl LMS-Z420i, but was not available in 2012. As mentioned, alignment errors in the data sets from 2011 caused time-consuming post-processing. The positioning of additional reflectors was helpful for aligning the data sets from 2012 and led to better results, reflected by the lower error after the whole alignment process. A further source of error in TLS measurements is noise in the point cloud, caused by reflections on rain, insects, or other small particles in the air. Due to the small size of the measured crops and uneven surfaces, this issue has to be regarded in particular for applications in agriculture, as also reported from other studies [33,35]. The measuring speed of the used time-of-flight scanners reduced the noise already and filter options in RiSCAN Pro simplified its removal, but further developments are desirable. In this context, intensity values should be investigated for establishing filtering schemes. So far, they are used for separating laser returns on canopy from ground returns [56] or for detecting single plants [31,32].

Regarding the practical implementation, this approach indicates advantages towards similar studies. Good results were achieved for estimating biomass of rice plants based on the vertical plant area density, measured with a portable scanner in combination with a mirror [37]. However, for the application on larger-scale fields their set-up might be less practical. Through the non-invasive TLS acquisition from the edges of the field, undisturbed plant growth can be ensured and the scan positions with the tractor-trailer system profited from the greater height. As the linear structure of the rice plant rows is observable, a more precise acquisition of the crop surface can be assumed. Thus, lightweight scanners are desirable, which can easier be brought to a lifted position. Moreover, cost-effective systems like the Velodyne HDL-64E LiDAR sensor [57] and mobile laser scanning systems like the ibeo ALASCA XT [58] should be considered for realizing practical applications of the presented approach for farmers.

Further, the oblique perspective of the scanner must be taken into account, which is unavoidable from a ground-based system without entering the field. Studies indicate that the height of reflection points might be overestimated through the influence of the scanning angle [59]. As the measured signal is influenced by the scanning geometry and beam divergence [31,60], a radiometric calibration is supported for stationary TLS by other studies [26,61]. In this study, the merged and cleaned point clouds were filtered with a scheme for selecting maximum points. Hence, the crop surface was determined from an evenly distributed coverage of the field and overestimations should be precluded.

Manual measurements of plant height were conducted for validating the TLS data. However, therein differences between the measurement methods must be denoted. Whereas with less than ten hills per field experiment plot, only a small and mostly the highest part of the entire crop surface was considered for the manual measurements, the scanner captures the whole plot, including the lower parts. Hence, only plot-wise averaged values could be compared but the high R^2 values up to 0.91 between both measurements confirmed the accuracy of the TLS data. However, the approach of using the 90th percentile [36] instead of the maximum values for the CSM-based plant height calculation should be considered for achieving values which are more robust against low scanning resolutions. Generally, the precision of the TLS-derived CSMs is difficult to determine by the manual measurements due to these differences. The good performance of TLS measurements for agricultural applications is presumed from other studies [31,33] and performance tests by the manufacturer validate the high accuracy and precision of the Riegl scanners [48,49]. Nevertheless, a main advantage is the objective assessment of plant height by CSMs, which avoids the selection bias of manual measurements. The non-invasive acquisition of the whole area in a high spatial resolution is one of the main benefits of the presented approach. In the context of PA, this is required for accurate crop monitoring [6].

Considering the upscaling of known plant information, the transferability of the virtually modelled geometry of single rice plants to field level might be evaluated with the high resolution CSMs [22,23]. Referring to the model of predicting yield potential for rice [21], the CSM-derived plant heights can be used as input data. Border effects cause problems in estimating rice yield, due to differences between internal and external rice plants in a plot [62]. In this study, an inner buffer was used to avoid border effects. For further studies, the high resolution of the TLS-derived CSMs might be useful for determining the differences between internal and external rows.

The pixel-wise calculated plant heights were visualized in maps of plant height for discovering spatial or temporal patterns and variations. As shown in Figure 5 the high resolution of 1 cm allowed an exact representation. In contrast, rice field mapping based on space-borne data has not been carried out with resolutions finer than 1 m so far [12,15,16]. However, new satellites like the WorldView-3 [63], providing a panchromatic resolution of ~0.3 m, should enable a more detailed acquisition. The high resolution is one of the major advantages of TLS data and enables the usability as *in-situ* validation for space-borne data. Although, the spatial extent of air- or space-borne methods cannot be reached with ground-based methods and the data acquisition effort is high, they are more flexible for the application in the field. Consequently, the presented approach may offer a tool for comparative analyses between TLS and airborne laser scanning (ALS). As shown by Bendig *et al.* [64] good results were achieved for the creation of CSMs from unmanned aerial vehicle (UAV)-based imaging for barley (R^2 up to 0.82 between CSM-derived and manually measured plant heights). Furthermore, promising results for the assessment of trees have already been achieved with UAV-based laser scanning systems [58,65]. However, the influence of the oblique and nadir scanning perspectives of ground- and air-borne measurements, respectively, have been less investigated so far. A comparative study on TLS and common plane-based ALS showed that the scanning angle and possible resolution influences the results [66]. Therefore, multiple sensors and acquisition levels should be combined for comprehensive analyses.

For confirming the general validity of spatial plant height data as a non-destructive estimator for biomass of paddy rice and proving the robustness as well as the spatial and temporal transferability of all established models, destructive biomass sampling was performed on all sites, revealing differences between the fields (Table 3). Basic differences were a lower human impact and larger size of the management units on the farmer's fields as well as the presence of different rice varieties and fertilizer treatments on all sites.

The three repetitions of each fertilizer treatment on the field experiment were useful to set up trial BRMs for proving the general concept (Table 4). High coefficients of determination and indices of agreement between the estimated and measured biomass values for each repetition of both years support linear and exponential models with comparable results. Nevertheless, further research is necessary for defining the differences between rice varieties and the influence of varying fertilizer treatments.

In addition to the final BRMs of each year, a model based on the combined data set of both years was established, each as a linear and an exponential model. The transferability of the BRMs from the small scale field experiment for estimating biomass on larger scale farmer's fields was shown (Table 5). Besides the transferability of existing models, a model improvement through the combined data set and through additional exponential models was investigated. As shown in Figure 7 for the data sets of the field experiment, the dependency of biomass on plant height can be described by linear and exponential regressions with similar high coefficients of determination. However, herein, only the pre-anthesis period was regarded. After anthesis, increasing biomass is mostly related to the development of grains while plant height remains almost constant. Thus, further studies are necessary for investigating the performance of linear and exponential BRMs for the estimation of rice biomass during the later stages.

The results of the linear and exponential models are almost similar for each site, with overall better values for *village 69*. As stated above the linear and exponential BRM yielded better results with the data sets from 2011 and 2012, respectively. A possible explanation might be the slightly different captured growth stages or the interpolated biomass values for 2011. Moreover, analyses are necessary, concerning the influence of different rice varieties, fertilizer treatments, or soil conditions. Additionally, the lower human impact on the farmer's fields might influence the plant development. For *village 36* the heterogeneous plant development in the management units has to be stated as a source for the differences between estimated and measured values. The varying performance of the combined model might be caused by these differences. Of most importance might be the fact that the relation between plant height and biomass in the two regarded periods seems to be best represented by different models. Overall, the results support the applicability of BRMs for biomass estimations based on TLS-derived spatial plant height data and substantiate the potential of ground-based plant parameter measurements as input for biomass estimation models [7,17].

5. Conclusions

The applicability and high suitability of terrestrial laser scanning for monitoring plant height of paddy rice based on multi-temporal CSMs were confirmed. An outstanding feature is the objective assessment of the whole field in a very high spatial resolution. Moreover, as the scans are non-invasively acquired from the field edges, entering the rice paddies is avoided. By investigating a repeated field experiment and two farmer's conventionally managed fields in two years, varying patterns of plant development and growth were covered.

For PA, monitoring of plant parameters for adjusting site-specific fertilization is a major topic. Strong coefficients of determination between plant height and biomass show the applicability of spatial plant height data as a non-destructive estimator for biomass of rice plants. Based on the promising results of single year analyses [38,39], in this contribution, the annual transferability of the BRMs and the applicability on different fields were regarded. Moreover, a model improvement through exponential models was examined. During the regarded pre-anthesis period, the linear and exponential models performed equally well. Further studies are necessary regarding a presumed differing performance during the later stages. However, the spatial and temporal transferability of the BRMs to a larger scale is supported by estimations of biomass on farmer's fields based on TLS-derived CSMs. High coefficients of determination and indices of agreement between estimated and measured values demonstrate the coherence of the results and prove the robustness of the method. Regarding the accuracy of the estimation, best results were achieved with different models, depending on the used data. Overall, higher R^2 values were achieved with the linear models, whereas the exponential models yielded smaller percentage deviations.

To summarize, the novelty in this contribution is the comparative analysis of linear and exponential models based on objectively assessed plant height as a reliable estimator for the biomass of paddy rice over different growing seasons and different fields. Further long-term experiments and comprehensive monitoring approaches are required for investigating the performance of linear and exponential models for the pre-anthesis and for later growing stages.

In the future, combined approaches involving plant height and spectral measurements should be developed for accurately determining the actual biomass and N content of plants. Following, spatially resolved NNI calculations could be executed for improving N management strategies [67]. Thereby, over-fertilization could be reduced while keeping or enhancing the yield.

Acknowledgments

The accomplished field work was affiliated to the activities of the International Center for Agro-Informatics and Sustainable Development (ICASD). Founded in 2009 by the Department of Plant Nutrition of the China Agricultural University in Beijing and the Institute of Geography at the University of Cologne in Germany, ICASD is an open, international, and multidisciplinary cooperative research center (www.ICASD.org).

We would like to thank all colleagues and student assistants for their great effort during the field work (Juliane Bendig, Simon Bennertz, Jonas Brands, Erik Boger, Martin Gnyp, Anne Henneken, and Maximilian Willkomm). Further we extend our thanks to Five Star Electronic Technologies (Beijing, China), the Qixing Research and Development Centre and the Jiansanjiang Agricultural Research Station (both located in Heilongjiang Province, China) for good cooperation as well as RIEGL GmbH (Horn, Austria) for continuous support. This work was financially supported by the International Bureau of the German Federal Ministry of Education and Research (BMBF, project number 01DO12013) and the German Research Foundation (DFG, project number BA 2062/8-1).

Author Contributions

Nora Tilly with advice from Dirk Hoffmeister was responsible for the TLS measurements. The field experiment at the Keyansuo experimental station was carried out by Qiang Cao supervised by Yuxin Miao, who is scientific coordinator of the ICASD project, together with Victoria Lenz-Wiedemann. Nora Tilly carried out the post-processing and wrote this article, greatly supported by Dirk Hoffmeister and Victoria Lenz-Wiedemann. Georg Bareth directed the research and supervised the whole acquiring, analyzing, and writing process.

Conflicts of Interest

The authors declare no conflict of interest.

References

1. FAO FAOSTAT. Available online: http://faostat3.fao.org/faostat-gateway/go/to/home/E (accessed on 1 July 2014).

2. Miao, Y.; Stewart, B.A.; Zhang, F. Long-term experiments for sustainable nutrient management in China. A review. *Agron. Sustain. Dev.* **2011**, *31*, 397–414.

3. Oliver, M.; Bishop, T.; Marchant, B. An overview of precision agriculture. In *Precision Agriculture for Sustainability and Environmental Protection*; Oliver, M., Bishop, T., Marchant, B., Eds.; Routledge: London, UK, 2013.

4. Van Wart, J.; Kersebaum, K.C.; Peng, S.; Milner, M.; Cassman, K.G. Estimating crop yield potential at regional to national scales. *Field Crops Res.* **2013**, *143*, 34–43.

5. Roelcke, M.; Han, Y.; Schleef, K.H.; Zhu, J.-G.; Liu, G.; Cai, Z.-C.; Richter, J. Recent trends and recommendations for nitrogen fertilization in intensive agriculture in eastern China. *Pedosphere* **2004**, *14*, 449–460.

6. Mulla, D.J. Twenty five years of remote sensing in precision agriculture: Key advances and remaining knowledge gaps. *Biosyst. Eng.* **2012**, *114*, 358–371.

7. Marshall, M.; Thenkabail, P. Developing *in situ* Non-Destructive Estimates of Crop Biomass to Address Issues of Scale in Remote Sensing. *Remote Sens.* **2015**, *7*, 808–835.

8. Greenwood, D.J.; Gastal, F.; Lemaire, G.; Draycott, A.; Millard, P.; Neeteson, J.J. Growth rate and %N of field grown crops: Theory and experiments. *Ann. Bot.* **1991**, *67*, 181–190.

9. Lemaire, G.; Jeuffroy, M.-H.; Gastal, F. Diagnosis tool for plant and crop N status in vegetative stage. *Eur. J. Agron.* **2008**, *28*, 614–624.

10. Elia, A.; Conversa, G. Agronomic and physiological responses of a tomato crop to nitrogen input. *Eur. J. Agron.* **2012**, *40*, 64–74.

11. Ntanos, D.A.; Koutroubas, S.D. Dry matter and N accumulation and translocation for Indica and Japonica rice under Mediterranean conditions. *Field Crops Res.* **2002**, *74*, 93–101.

12. Ribbes, F.; Le Toan, T. Rice field mapping and monitoring with RADARSAT data. *Int. J. Remote Sens.* **1999**, *20*, 745–765.

13. Yang, X.; Huang, J.; Wu, Y.; Wang, J.; Wang, P.; Wang, X.; Huete, A.R. Estimating biophysical parameters of rice with remote sensing data using support vector machines. *Sci. China. Life Sci.* **2011**, *54*, 272–281.

14. Li, W.; Li, H.; Zhao, L. Estimating Rice Yield by HJ-1A Satellite Images. *Rice Sci.* **2011**, *18*, 142–147.

15. Lopez-Sanchez, J.M.; Ballester-Berman, J.D.; Hajnsek, I. First Results of Rice Monitoring Practices in Spain by Means of Time Series of TerraSAR-X Dual-Pol Images. *IEEE J. Sel. Top. Appl. Earth Obs. Remote Sens.* **2011**, *4*, 412–422.

16. Koppe, W.; Gnyp, M.L.; Hütt, C.; Yao, Y.; Miao, Y.; Chen, X.; Bareth, G. Rice monitoring with multi-temporal and dual-polarimetric TerraSAR-X data. *Int. J. Appl. Earth Obs. Geoinf.* **2012**, *21*, 568–576.

17. Reddersen, B.; Fricke, T.; Wachendorf, M. A multi-sensor approach for predicting biomass of extensively managed grassland. *Comput. Electron. Agric.* **2014**, *109*, 247–260.

18. Casanova, D.; Epema, G.F.; Goudriaan, J. Monitoring rice reflectance at field level for estimating biomass and LAI. *Field Crops Res.* **1998**, *55*, 83–92.

19. Gnyp, M.L.; Yu, K.; Aasen, H.; Yao, Y.; Huang, S.; Miao, Y.; Bareth, G. Analysis of crop reflectance for estimating biomass in rice canopies at different phenological stages. *Photogramm. Fernerkund. Geoinf.* **2013**, *4*, 351–365.

20. Aasen, H.; Gnyp, M.L.; Miao, Y.; Bareth, G. Automated hyperspectral vegetation index retrieval from multiple correlation matrices with HyperCor. *Photogramm. Eng. Remote Sens.* **2014**, *80*, 785–796.

21. Confalonieri, R.; Bregaglio, S.; Rosenmund, A.S.; Acutis, M.; Savin, I. A model for simulating the height of rice plants. *Eur. J. Agron.* **2011**, *34*, 20–25.

22. Watanabe, T.; Hanan, J.S.; Room, P.M.; Hasegawa, T.; Nakagawa, H.; Takahashi, W. Rice morphogenesis and plant architecture: Measurement, specification and the reconstruction of structural development by 3D architectural modelling. *Ann. Bot.* **2005**, *95*, 1131–1143.

23. Ding, W.; Zhang, Y.; Zhang, Q.; Zhu, D.; Chen, Q. Realistic Simulation of Rice Plant. *Rice Sci.* **2011**, *18*, 224–230.

24. Lee, W.S.; Alchanatis, V.; Yang, C.; Hirafuji, M.; Moshou, D.; Li, C. Sensing technologies for precision specialty crop production. *Comput. Electron. Agric.* **2010**, *74*, 2–33.

25. Zhang, L.; Grift, T.E. A LIDAR-based crop height measurement system for *Miscanthus giganteus*. *Comput. Electron. Agric.* **2012**, *85*, 70–76.

26. Koenig, K.; Höfle, B.; Hämmerle, M.; Jarmer, T.; Siegmann, B.; Lilienthal, H. Comparative classification analysis of post-harvest growth detection from terrestrial LiDAR point clouds in precision agriculture. *ISPRS J. Photogramm. Remote Sens.* **2015**, *104*, 112–125.

27. Gebbers, R.; Ehlert, D.; Adamek, R. Rapid mapping of the leaf area index in agricultural crops. *Agron. J.* **2011**, *103*, 1532–1541.

28. Hosoi, F.; Omasa, K. Estimating vertical plant area density profile and growth parameters of a wheat canopy at different growth stages using three-dimensional portable lidar imaging. *ISPRS J. Photogramm. Remote Sens.* **2009**, *64*, 151–158.

29. Saeys, W.; Lenaerts, B.; Craessaerts, G.; De Baerdemaeker, J. Estimation of the crop density of small grains using LiDAR sensors. *Biosyst. Eng.* **2009**, *102*, 22–30.

30. Eitel, J.U.H.; Vierling, L.A.; Long, D.S.; Raymond Hunt, E. Early season remote sensing of wheat nitrogen status using a green scanning laser. *Agric. For. Meteorol.* **2011**, *151*, 1338–1345.

31. Höfle, B. Radiometric correction of terrestrial LiDAR point cloud data for individual maize plant detection. *Geosci. Remote Sens. Lett. IEEE* **2014**, *11*, 94–98.

32. Hoffmeister, D.; Tilly, N.; Bendig, J.; Curdt, C.; Bareth, G. Detektion von Wachstumsvariabilität in vier Zuckerrübensorten durch multi-temporales terrestrisches Laserscanning. In Proceedings of the 32. GIL-Jahrestagung: Informationstechnologie für eine Nachhaltige Landbewirtschaftung, Freising, Germany, 29 February–1 March 2012; Clasen, M., Fröhlich, G., Bernhardt, H., Hildebrand, K., Theuvsen, B., Eds.; Köllen Verlag: Bonn, Germany, 2012; pp. 135–138.

33. Lumme, J.; Karjalainen, M.; Kaartinen, H.; Kukko, A.; Hyyppä, J.; Hyyppä, H.; Jaakkola, A.; Kleemola, J. Terrestrial laser scanning of agricultural crops. In Proceedings of the International Archives of the Photogrammetry, Remote Sensing and Spatial Information Sciences 37 (Part B5), Beijing, China, 3–11 July 2008; Chen, J.; Jiang, J.; Maas, H.-G., Eds.; Organising Committee of the XXIst International Congress for Photogrammetry and Remote Sensing: Beijing, China, 2008; pp. 563–566.

34. Ehlert, D.; Horn, H.-J.; Adamek, R. Measuring crop biomass density by laser triangulation. *Comput. Electron. Agric.* **2008**, *61*, 117–125.

35. Ehlert, D.; Adamek, R.; Horn, H.-J. Laser rangefinder-based measuring of crop biomass under field conditions. *Precis. Agric.* **2009**, *10*, 395–408.

36. Hämmerle, M.; Höfle, B. Effects of Reduced Terrestrial LiDAR Point Density on High-Resolution Grain Crop Surface Models in Precision Agriculture. *Sensors* **2014**, *14*, 24212–24230.

37. Hosoi, F.; Omasa, K. Estimation of vertical plant area density profiles in a rice canopy at different growth stages by high-resolution portable scanning LiDAR with a lightweight mirror. *ISPRS J. Photogramm. Remote Sens.* **2012**, *74*, 11–19.

38. Tilly, N.; Hoffmeister, D.; Cao, Q.; Lenz-Wiedemann, V.; Miao, Y.; Bareth, G. Precise plant height monitoring and biomass estimation with Terrestrial Laser Scanning in paddy rice. In Proceedings of the ISPRS Annals of the Photogrammetry, Remote Sensing and Spatial Information Sciences Conference, Antalya, Turkey, 11–13 November 2013; Scaioni, M., Lindenbergh, R.C., Oude Elberink, S., Schneider, D., Pirotti, F., Eds.; Volume II-5/W2, pp. 295–300.

39. Tilly, N.; Hoffmeister, D.; Cao, Q.; Huang, S.; Lenz-Wiedemann, V.; Miao, Y.; Bareth, G. Multitemporal crop surface models: Accurate plant height measurement and biomass estimation with terrestrial laser scanning in paddy rice. *J. Appl. Remote Sens.* **2014**, *8*, 083671.

40. Thenkabail, P.S.; Smith, R.B.; De Pauw, E. Hyperspectral vegetation indices and their relationships with agricultural crop characteristics. *Remote Sens. Environ.* **2000**, *71*, 158–182.

41. Hansen, P.M.; Schjoerring, J.K. Reflectance measurement of canopy biomass and nitrogen status in wheat crops using normalized difference vegetation indices and partial least squares regression. *Remote Sens. Environ.* **2003**, *86*, 542–553.

42. Gao, J.; Liu, Y. Climate warming and land use change in Heilongjiang Province, Northeast China. *Appl. Geogr.* **2011**, *31*, 476–482.

43. Domrös, M.; Gongbing, P. *The Climate of China*; Springer-Verlag: Berlin, Germany, 1988.

44. Ding, Y.; Chan, J.C.L. The East Asian summer monsoon: An overview. *Meteorol. Atmos. Phys.* **2005**, *89*, 117–142.

45. Cao, Q.; Miao, Y.; Wang, H.; Huang, S.; Cheng, S.; Khosla, R.; Jiang, R. Non-destructive estimation of rice plant nitrogen status with Crop Circle multispectral active canopy sensor. *Filed Crops Res.* **2013**, *154*, 133–144.

46. Meier, U. *Growth Stages of Mono- and Dicotyledonous Plants*, 2nd ed.; Blackwell: Berlin, Germany, 2001.

47. Lancashire, P.D.; Bleiholder, H.; van den Boom, T.; Langelüddeke, P.; Strauss, R.; Weber, E.; Witzenberger, A. A uniform decimal code for growth stages of crops and weeds. *Ann. Appl. Biol.* **1991**, *119*, 561–601.

48. Riegl LMS GmbH Datasheet Riegl LMS-Z420i. Available online: http://www.riegl.com/uploads/tx_pxprieldownloads/10_DataSheet_Z420i_03-05-2010.pdf (accessed on 1 July 2014).

49. Riegl LMS GmbH Datasheet Riegl VZ-1000. Available online: http://www.riegl.com/uploads/tx_pxprieldownloads/DataSheet_VZ-1000_18-09-2013.pdf (accessed on 1 July 2014).

50. Hoffmeister, D.; Bolten, A.; Curdt, C.; Waldhoff, G.; Bareth, G. High resolution Crop Surface Models (CSM) and Crop Volume Models (CVM) on field level by terrestrial laser scanning. In Proceedings of the SPIE, 6th International Symposium on Digital Earth, Beijing, China, 4 November 2010; Guo, H., Wang, C., Eds.; Volume 7840.

51. Besl, P.J.; McKay, N.D. A Method for Registration of 3D Shapes. *IEEE Trans. Pattern Anal. Mach. Intell.* **1992**, *14*, 239–256.

52. Johnston, K.; Ver Hoef, J.M.; Krivoruchko, K.; Lucas, N. *Using ArcGIS Geostatistical Analyst*; ESRI: Redlands, CA, USA, 2001.

53. Willmott, C.J.; Wicks, D.E. An empirical method for the spatial interpolation of monthly precipitation within California. *Phys. Geogr.* **1980**, *1*, 59–73.

54. Willmott, C.J. On the validation of models. *Phys. Geogr.* **1981**, *2*, 184–194.

55. Hair, J.F.; Black, W.C.; Babin, B.J.; Anderson, R.E. *Multivariate Data Analysis*, 7th ed.; Pearson: Upper Saddle River, NJ, USA, 2010.

56. Guarnieri, A.; Pirotti, F.; Vettore, A. Comparison of discrete return and waveform terrestrial laser scanning for dense vegetation filtering. In Proceedings of the International Archives of Photogrammetry, Remote Sensing and Spatial Information Sciences, Melbourne, Australia, 25 August–1 September 2012; Shortis, M., Mills, J., Eds.; Volume 39, pp. 511–516.

57. Velodyne Velodyne HDL-64E User's Manual. Available online: http://www.velodynelidar.com/lidar/products/manual/63-HDL64E S2 Manual_Rev D_2011_web.pdf (accessed on 1 July 2014).

58. Jaakkola, A.; Hyyppä, J.; Kukko, A.; Yu, X.; Kaartinen, H.; Lehtomäki, M.; Lin, Y. A low-cost multi-sensoral mobile mapping system and its feasibility for tree measurements. *ISPRS J. Photogramm. Remote Sens.* **2010**, *65*, 514–522.

59. Ehlert, D.; Heisig, M. Sources of angle-dependent errors in terrestrial laser scanner-based crop stand measurement. *Comput. Electron. Agric.* **2013**, *93*, 10–16.

60. Kaasalainen, S.; Jaakkola, A.; Kaasalainen, M.; Krooks, A.; Kukko, A. Analysis of incidence angle and distance effects on terrestrial laser scanner intensity: Search for correction methods. *Remote Sens.* **2011**, *3*, 2207–2221.

61. Kaasalainen, S.; Krooks, A.; Kukko, A.; Kaartinen, H. Radiometric calibration of terrestrial laser scanners with external reference targets. *Remote Sens.* **2009**, *1*, 144–158.

62. Wang, K.; Zhou, H.; Wang, B.; Jian, Z.; Wang, F.; Huang, J.; Nie, L.; Cui, K.; Peng, S. Quantification of border effect on grain yield measurement of hybrid rice. *Field Crops Res.* **2013**, *141*, 47–54.

63. DigitalGlobe Datasheet WorldView-3. Available online: https://www.digitalglobe.com/sites/default/files/DG_WorldView3_DS_forWeb_0.pdf (accessed on 26 June 2015).

64. Bendig, J.; Yu, K.; Aasen, H.; Bolten, A.; Bennertz, S.; Broscheit, J.; Gnyp, M.L.; Bareth, G. Combining UAV-based plant height from crop surface models, visible, and near infrared vegetation indices for biomass monitoring in barley. *Int. J. Appl. Earth Obs. Geoinf.* **2015**, *39*, 79–87.

65. Wallace, L.; Watson, C.; Lucieer, A. Detecting pruning of individual stems using airborne laser scanning data captured from an Unmanned Aerial Vehicle. *Int. J. Appl. Earth Obs. Geoinf.* **2014**, *30*, 76–85.

66. Luscombe, D.J.; Anderson, K.; Gatis, N.; Wetherelt, A.; Grand-Clement, E.; Brazier, R.E. What does airborne LiDAR really measure in upland ecosystems? *Ecohydrology* **2014**, doi:10.1002/eco.1527.

67. Yao, Y.; Miao, Y.; Huang, S.; Gao, L.; Ma, X.; Zhao, G.; Jiang, R.; Chen, X.; Zhang, F.; Yu, K.; *et al.* Active canopy sensor-based precision N management strategy for rice. *Agron. Sustain. Dev.* **2012**, *32*, 925–933.

9

Aflatoxin Accumulation in a Maize Diallel Cross

W. Paul Williams * and Gary L. Windham

United States Department of Agriculture, Agricultural Research Service, Corn Host Plant Resistance Research Unit, P.O. Box 9555, MS 39762, USA;
E-Mail: gary.windham@ars.usda.gov

*   Author to whom correspondence should be addressed; E-Mail: paul.williams@ars.usda.gov

Academic Editor: Anna Andolfi

Abstract: Aflatoxins, produced by the fungus *Aspergillus flavus*, occur naturally in maize. Contamination of maize grain with aflatoxin is a major food and feed safety problem and greatly reduces the value of the grain. Plant resistance is generally considered a highly desirable approach to reduction or elimination of aflatoxin in maize grain. In this investigation, a diallel cross was produced by crossing 10 inbred lines with varying degrees of resistance to aflatoxin accumulation in all possible combinations. Three lines that previously developed and released as sources of resistance to aflatoxin accumulation were included as parents. The 10 parental inbred lines and the 45 single crosses making up the diallel cross were evaluated for aflatoxin accumulation in field tests conducted in 2013 and 2014. Plants were inoculated with an *A. flavus* spore suspension seven days after silk emergence. Ears were harvested approximately 60 days later and concentration of aflatoxin in the grain determined. Parental inbred lines Mp717, Mp313E, and Mp719 exhibited low levels (3–12 ng/g) of aflatoxin accumulation. In the diallel analysis, both general and specific combining ability were significant sources of variation in the inheritance of resistance to aflatoxin accumulation. General combining ability effects for reduced aflatoxin accumulation were greatest for Mp494, Mp719, and Mp717. These lines should be especially useful in breeding for resistance to aflatoxin accumulation. Breeding strategies, such as reciprocal recurrent selection, would be appropriate.

Keywords: *Aspergillus flavus*; aflatoxin; maize; plant resistance

1. Introduction

Aflatoxins, produced by the fungus *Aspergillus flavus*, occur naturally in maize. The most potent carcinogens found in nature, aflatoxins are toxic not only to humans, but also to livestock, pets, and wildlife [1–4]. Dietary exposure to aflatoxins is one of the major causes of hepatocellular carcinoma, the fifth most common cancer in humans worldwide [5]. Their acute and chronic toxicity poses a major threat to humans in developing countries where maize is a dietary staple. The US Food and Drug Administration restricts the sale of grain with aflatoxin levels exceeding 20 ng/g [6]. In the United States aflatoxin contamination, first recognized as a major problem associated with maize production in the Southeast in the 1970s [7,8], continues to be a chronic problem in the Southeast and a sporadic problem in the Midwest [9,10]. Drought, high temperatures, and insect damage are often associated with aflatoxin contamination [8,9,11].

Plant resistance is generally considered a highly desirable approach toward reduction or elimination of aflatoxin accumulation in maize grain. Awareness of the serious consequences associated with pre-harvest aflatoxin contamination and the potential value of plant resistance in reducing losses to aflatoxin led to the establishment of several federally and state supported maize breeding programs in the southeastern USA to address the problem [12]. One such program was initiated by the U.S. Department of Agriculture's Agricultural Research Service at Mississippi State, Mississippi. The first line released from this program for resistance to *A.flavus* infection and aflatoxin accumulation was Mp313E [13]. Five additional lines were released subsequently: Mp420, Mp715, Mp717, Mp718, and Mp719 [14–17]. In addition, other potential useful sources of resistance have been identified [12,18].

This investigation was undertaken to compare aflatoxin accumulation in 10 maize inbred lines with varying levels of resistance to aflatoxin accumulation and in crosses among the 10 lines. The second objective was to determine the importance of general combining ability (GCA) and specific combining ability (SCA) in the inheritance to resistance to aflatoxin accumulation for this set of crosses.

2. Results

Among the 10 parental inbred lines, mean aflatoxin accumulation over years ranged from 742 ng/g for PHW79 to a low of 3 ng/g for Mp717 (Table 1). Aflatoxin concentration was lowest in Mp717, Mp313E, and Mp719, three lines that were developed and released as sources of resistance to *A. flavus* infection and aflatoxin accumulation by USDA-ARS [13,16,17]. NC388 also accumulated a low level of aflatoxin and did not differ significantly from Mp717 in mean aflatoxin concentration.

Differences in aflatoxin accumulation among single cross hybrids ranged from a high of 525 ng/g for Va35 × NC388 to a low of 2 ng/g for Mp313E × Mp494 and Mp313E × NC388 (Table 2). Interestingly, NC388 was a parent of the hybrids at both the high and low limits of aflatoxin accumulation. Mean aflatoxin accumulation was also high in B73 × T173 and Va35 × T173, crosses between two known susceptible lines. In addition to Mp313E × NC388 and Mp313E × Mp494, other hybrids that exhibited very low levels of aflatoxin accumulation (2–4 ng/g) were Mp719 × CML322, Mp719 × NC388, PHW79 × NC388, Mp717 × CML222, Mp313E × CML322, and Mp313E × Mp719.

Table 1. Mean aflatoxin accumulation in 10 parental inbred lines grown at Mississippi State in 2013 and 2014.

Inbred Line	Aflatoxin [†]	
	Logarithmic Mean [ln (y + 1)] ng/g	Geometric Mean [‡] ng/g
PHW79	6.61	742
B73	6.24	513
T173	5.96	387
Va35	5.04	154
CML322	5.04	154
Mp494	4.29	72
NC388	3.40	28
Mp719	2.60	12
Mp313E	2.49	11
Mp717	1.50	3
LSD(0.05)	2.28	

[†] Data were transformed (ln(y + 1), where y = concentration of aflatoxin in a sample) before statistical analysis; [‡] Geometric means were calculated by converting logarithmic means to the original scale.

Table 2. Mean aflatoxin accumulation in 45 F1 hybrids constituting a diallel cross grown at Mississippi State in 2013 and 2014.

F$_1$ Hybrid	Aflatoxin [†]	
	Logarithmic mean [ln (y + 1)] ng/g	Geometric Mean [‡] ng/g
Va35 × NC 388	6.27	525
B73 × T173	5.79	327
Va35 × T173	5.33	206
B73 × CML322	5.31	201
B73 × PHW79	5.07	158
PHW79 × Mp313E	4.83	125
Va35 × CML322	4.78	118
T173 × Mp719	4.76	116
T173 × CML322	4.69	108
B73 × Va35	4.48	87
Mp494 × CML322	4.41	81
Mp494 × Mp719	4.20	66
Mp717 × Mp719	4.00	54
T173 × Mp717	3.93	50
PHW79 × Mp494	3.81	44
Va35 × Mp494	3.76	42
B73 × Mp717	3.71	40
B73 × Mp313E	3.51	32
B73 × Mp719	3.45	31
B73 × Mp494	3.33	27
T173 × NC388	3.28	25
Mp494 × NC388	3.21	24

Table 2. *Cont.*

F₁ Hybrid	Aflatoxin [†]	
	Logarithmic mean [ln (y + 1)] ng/g	Geometric Mean [‡] ng/g
B73 × NC388	3.16	23
T173 × PHW79	3.18	23
Va35 × PHW79	3.13	22
T173 × Mp313E	3.00	19
Mp717 × NC388	2.63	13
Va35 × Mp717	2.46	11
NC388 × CML322	2.38	10
Va35 × Mp719	2.31	9
Mp313E × Mp717	22.5	9
Mp494 × Mp717	2.32	9
Va35 × Mp313E	2.22	8
T173 × Mp494	2.07	7
PHW79 × Mp717	1.79	5
PHW79 × Mp719	1.75	5
PHW79 × CML322	1.85	5
Mp313E × Mp719	1.54	4
Mp313E × CML322	1.69	4
Mp717 × CML322	1.53	4
PHW79 × NC388	1.36	3
Mp719 × NC388	1.30	3
Mp719 × CML322	1.30	3
Mp313E × Mp494	0.92	2
Mp313E × NC388	0.94	2
LSD (0.05)	2.02	

[†] Data were transformed (ln(y + 1), where y = concentration of aflatoxin in a sample) before statistical analysis; [‡] Geometric means were calculated by converting logarithmic means to the original scale.

The analysis of variance indicated that variation associated with years was not significant ($P < 0.05$) although the interaction of hybrids × years was a highly significant ($P < 0.01$) source of variation (Table 3). GCA was a highly significant ($P < 0.01$) source of variation in both 2013 and 2014 and in the combined analysis over years. SCA was a significant ($P < 0.01$) source of variation only in 2013. The interaction of GCA × years was not significant. Although SCA was not significant in the combined analysis, the interaction SCA × years was highly significant ($P < 0.01$).

GCA effects were calculated for 2013, 2014, and for the two years combined (Table 4). GCA effects were highly significant ($P < 0.01$) and negative for Mp494, Mp717, Mp719, and NC388 each year and over years. In an earlier investigation of a diallel cross produced from a different set of lines, Mp717 and Mp494 exhibited highly significant GCA effects for reduced aflatoxin accumulation [18]. GCA effects for aflatoxin accumulation were highly significant ($P < 0.01$) and positive for PHW79 in 2013, 2014, and over years. GCA effects for T173 and B73 were significant ($P < 0.05$) and positive in the combined analysis, but highly significant ($P < 0.01$) within each year. PHW79, B73, and T173 contributed to increased susceptibility when used in production of hybrids.

Table 3. Analysis of variance for aflatoxin concentration in grain harvested from a diallel cross grown in 2013 and 2014 at Mississippi State.

Source	df	Mean squares [ln(y+1)] ng/g [†]		
		2013	2014	Over Years
Years	1	-	-	0.67
Reps (year)	6	-	-	1.18
Hybrids	44	12.24 **	7.19 **	15.50 **
GCA [‡]	9	43.23 **	25.27 **	65.59 **
SCA [‡]	35	4.21 **	2.61	2.61
Hybrids × Years	44	-	-	3.98 **
GCA × Years	9	-	-	3.01
SCA × Years	35	-	-	4.23 **
Error	132,264 [§]	1.88	1.73	1.81

** Significant at $P < 0.01$; [†] Data were transformed (ln(y + 1), where y = concentration of aflatoxin in a sample) before statistical analysis; [‡] GCA, general combining ability; SCA = specific combining ability; [§] df = 132 for individual years and 264 for combined years.

Table 4. Estimates of general combining ability (GCA) effects for 10 inbred parental lines of a diallel cross of maize grown at Mississippi State in 2013 and 2014.

Inbred Line	GCA effects [ln(y + 1)] ng/g [†]		
	2013	2014	Over Years
PHW79	2.10 **	1.26 **	1.68 **
T173	1.46 **	1.07 **	1.27 *
B73	1.16 **	0.87 **	1.01 *
Va35	0.09	0.81 **	0.45
Mp313E	−0.39	−0.37	−0.38 *
CML322	−0.60 **	−0.31	−0.46 **
NC388	−0.68 **	−0.72 **	−0.70 **
Mp717	−0.84 **	−0.87 **	−0.85 **
Mp719	−1.13 **	−0.77 **	−0.95 **
Mp494	−1.17 **	−0.96 **	−1.07 **

* Significantly different from 0 at $P < 0.05$; ** Significantly different from 0 at $P < 0.05$; [†] Data were transformed (ln(y + 1), where y = concentration of aflatoxin in a sample) before analysis.

3. Discussion

Aflatoxin accumulation in the inbred lines per se (Table 1) was generally consistent with their performance in hybrid combinations as indicated by the GCA effects (Table 4). The susceptible lines, PHW79, B73, T173, and Va35, had the greatest accumulation of aflatoxin as inbred lines and the highest positive GCA effects. Although aflatoxin accumulation was only 11 ng/g for Mp313E, the second lowest among the inbred lines, the associated GCA effect for Mp313E indicated the least reduction in aflatoxin accumulation among the resistant lines. This is consistent with the results of a previous investigation in which Mp313E was included in a diallel cross produced from a different set of lines [18]. The negative GCA effects associated with Mp494, Mp717, and Mp719, lines that were developed specifically as sources of resistance to accumulation of aflatoxin, indicate that these lines could be useful in breeding programs. Although NC388 was not developed and released as a source of resistance, it appears to have potential for use in breeding for resistance as well. Breeding methods that maximize the use of GCA should be most effective in using these lines to develop germplasm with resistance to aflatoxin accumulation.

Although conventional breeding strategies based on phenotypic selection should be useful in the production of improved germplasm lines and hybrids, molecular marker-assisted selection should be effective in increasing resistance to aflatoxin accumulation as well. USDA-ARS at Mississippi State has invested considerable effort in mapping quantitative trait loci (QTL) associated with resistance to aflatoxin accumulation [19] and using that information to develop near-isogenic lines. Mp313E was the primary source of resistance in much of this effort. Mp494, Mp717, Mp719, and NC388, the lines that exhibited highly significant GCA effects for resistance in this investigation, may prove to be a better source of QTL, and eventually gene-based markers, for resistance to aflatoxin accumulation. For a trait such as resistance to aflatoxin accumulation that is highly sensitive to environmental effects and variations, molecular markers may well be the most effective approach to efficiently producing resistant hybrids. Investigations such as this one can be useful in choosing the best sources of germplasm for inclusion in gene-mapping studies. Efforts to identify gene-based molecular markers for resistance in Mp494, Mp719, and other resistant lines are currently underway.

4. Materials and Methods

A diallel cross was produced by crossing 10 inbred lines of maize in all possible combinations. Three of the parental inbred lines were developed and released by USDA-ARS at Mississippi State as sources of resistance to *A. flavus* infection and aflatoxin accumulation: Mp313E, Mp717, and Mp719 [13,16,17]. Mp494, NC388, and CML322 had exhibited resistance, and Va35, T173, B73, and PHW79, susceptibility in field trials conducted at Mississippi State [12,18,20]. The 10 inbred lines were planted on 14 May 2013 and 21 April 2014 in a Leeper silty clay loam (fine, smectitic, non-acid, thermic Vertic Epiaquepts) soil at Mississippi State, MS. In separate experiments, the 45 hybrids constituting the diallel cross were planted on 15 May 2013 and 21 April 2014. Both inbred lines and hybrids were planted in single row plots that were 4 m long, spaced 0.97 m apart, and arranged in a randomized complete block design with four replications. Standard maize production practices for the area were followed.

Seven days after silks had emerged from 50% of the plants in a plot, the primary ear of each plant was inoculated with *A. flavus* isolate NRRL 3357, which is known to produce aflatoxin in maize, using the side-needle technique [21]. Using a tree-marking gun fitted with a 14-gage needle, a 3.4 mL suspension containing 3×10^8 *A. flavus* conidia was injected underneath the husks into the side of the ear. Inoculum was prepared as described by [22].

The inoculated ears were harvested approximately 60 d after inoculation and dried at 38° for 7 day. Ears from each plot were bulked and shelled. The grain was thoroughly mixed before grinding with a Romer mill (Union, MO, USA). The concentration of aflatoxin in a 50-g sample was determined by the Vicam AflaTest procedure (Watertown, MA, USA). This procedure detects aflatoxin at levels as low a 1 ng/g.

The values for aflatoxin concentration were transformed before statistical analysis as ($\ln(y + 1)$, where y is the concentration of aflatoxin in a sample). The transformation was performed to provide a more nearly normal distribution of the data. Data were analyzed using the SAS General Linear Models procedure [23]. In the analysis of the diallel cross, the variance was partitioned using DIALLEL-SAS [24,25] based on Griffing's [26] Method 4, Model I, into GCA and SCA components and their interactions with years. To test the significance of GCA and SCA components in the 2-year analysis, the interactions between years and the corresponding component as the error term for *F*-tests. Estimates of GCA and SCA were calculated and their significance determined by *t*-tests.

5. Conclusions

The identification of maize germplasm with heritable resistance to *A. flavus* infection, and the subsequent accumulation, is important in reducing aflatoxin contamination of maize. A diallel cross, produced by crossing 10 parental lines with varying degrees of resistance, was evaluated for resistance to aflatoxin accumulation in field tests. Both GCA and SCA were significant sources of variation in the inheritance of resistance to aflatoxin accumulation. GCA effects for reduced aflatoxin accumulation were greatest for Mp494, Mp717, and Mp719 indicating that these lines should be especially useful in breeding for resistance to aflatoxin accumulation in maize. Conventional breeding strategies such as reciprocal recurrent selection should be effective in increasing levels of resistance. Mp494, Mp717, and Mp719 are also potentially useful sources of gene-based markers for resistance to aflatoxin accumulation.

Acknowledgments

Funding for the research was provided by the United States Department of Agriculture's Agricultural Research Service. The authors express their appreciation to Paul M. Buckley, Gerald A. Matthews, Ladonna T. Owens, Stephanie S. Pitts, and Patrick L. Tranum for technical assistance in conducting the research and preparing the manuscript. This paper is a joint contribution of USDA-ARS and the Mississippi Agricultural and Forestry Experiment Station and is published as journal No. J-12663 of the Missississippi Agricultural and Forestry Experiment Station. Mention of trade names or commercial products in this publication is solely for the purpose of providing specific information and does not imply recommendation or endorsement by USDA.

Author Contributions

W.W. conceived and designed the study, produced the seed, analyzed the data, and wrote the manuscript. G.W. inoculated developing ears with *A. flavus*, analyzed grain samples for aflatoxin content, and summarized the data.

Conflicts of Interest

The authors declare no conflicts of interest.

References

1. Bokhari, F.M. Implications of fungal infections and mycotoxins in camel diseases in Saudi Arabia. *Saudi J. Biol. Sci.* **2010**, *17*, 73–81.

2. Castegnaro, M.; McGregor, D. Carcinogenic risk assessment of mycotoxins. *Revue Med. Vet.* **1998**, *149*, 671–678.

3. Gourama, H.; Bullerman, L.B. *Aspergillus flavus* and *Aspergillus parasiticus* aflatoxigenic fungi of concern in foods and feeds. *J. Food Prot.* **1995**, *58*, 1395–1404.

4. Leung, M.C.K.; Diaz-Liano, G.; Smith. T.K. Mycotoxins in pet foods: A review on worldwide prevalence and preventative strategies. *J. Agric. Food Chem.* **2006**, *54*, 9623–9635.

5. Wild, C.P.; Hall, A.J. Primary prevention of hepatocellular carcinoma in developing countries. *Mutat. Res.* **2000**, *462*, 381–393.

6. Park, D.L.; Liang, B. Perspectives on aflatoxin control for human food and animal feed. *Trends Food Sci. Tech.* **1993**, *4*, 334–342.

7. McMillian, W.W.; Wilson, D.M.; Widstrom, N.W. Insect damage, *Aspergillus flavus* ear mold, and aflatoxin contamination in south Georgia corn fields in 1977. *J. Environ. Qual.* **1978**, *7*, 564–566.

8. McMillian, W.W.; Wilson, D.M.; Widstrom, N.W. Aflatoxin contamination of pre-harvest corn in Georgia: A six-year study of insect damage and visible *Aspergillus flavus*. *J. Environ. Qual.* **1985**, *14*, 200–202.

9. Payne, G.A. Aflatoxin in maize. *Crit. Rev. Plant Sci.* **1992**, *10*, 423–440.

10. Widstrom, N.W. The aflatoxin problem with corn grain. *Adv. Agron.* **1996**, *56*, 219–280.

11. Dowd, P. Insect management to facilitate preharvest mycotoxins management. *Toxin Rev.* **2003**, *22*, 327–350.

12. Williams, W.P.; Krakowsky, M.D.; Scully, B.T.; Brown, R.L.; Menkir, A.; Warburton, M.L.; Windham, G.L. Identifying and developing maize germplasm with resistance to accumulation of aflatoxins. *World Mycotoxin J.* **2015**, *8*, 193–209.

13. Scott, G.E.; Zummo, N. Registration of Mp313E parental line of maize. *Crop Sci.* **1990**, *30*, 1378.

14. Scott, G.E.; Zummo, N. Registration of Mp420 germplasm line of maize. *Crop Sci.* **1992**, *32*, 1296.

15. Williams, W.P.; Windham, G.L. Registration of dent corn germplasm line Mp715. *Crop Sci.* **2001**, *41*, 1374–1375.

16. Williams, W.P.; Windham, G.L. Registration of maize germplasm line Mp717. *Crop Sci.* **2006**, *46*, 1407.

17. Williams, W.P.; Windham, G.L. Registration of Mp718 and Mp719 germplasm lines of maize. *J. Plant Reg.* **2012**, *6*, 1–3.

18. Williams, W.P.; Windham, G.L.; Buckley, P.M. Diallel analysis of aflatoxin accumulation in maize. *Crop Sci.* **2008**, *48*, 134–138.

19. Brooks, T.D.; Williams, W.P.; Windham, G.L.; Willcox, M.C.; Abbas, H.K. Quantitative trait loci contributing resistance to aflatoxin accumulation in the maize inbred Mp313E. *Crop Sci.* **2005**, *45*, 171–174.

20. Williams, W.P. Breeding for resistance to aflatoxin accumulation in maize. *Mycotoxin Res.* **2006**, *22*, 27–32.

21. Zummo, N.; Scott, G.E. Evaluation of field inoculation techniques for screening maize genotypes against kernel infection by *Aspergillus flavus* in Mississippi. *Plant Dis.* **1998**, *72*, 313–316.

22. Windham, G.L.; Williams, W.P. Evaluation of corn inbreds and advanced breeding lines for resistance to aflatoxin contamination in the field. *Plant Dis.* **2002**, *86*, 232–234.

23. SAS Institute. *SAS System (version 9.3) for Windows*; SAS Institute: Cary, NC, USA, 2010.

24. Zhang, Y.; Kang, M.S. DIALLEL-SAS: A SAS program for Griffing's diallel analyses. *Agron. J.* **1997**, *89*, 176–182.

25. Zhang, Y.; Kang, M.S. DIALLEL-SAS: A program for Griffing's diallel methods. In *Handbook of Formulas and Software for Plant Geneticists and Breeders*; Kang, M.S., Ed.; Haworth Inc.: New York, NY, USA, 2003; pp. 1–19.

26. Griffing, B. Concept of general and specific combining ability in relation to diallel crossing systems. *Aust. J. Biol. Sci.* **1956**, *9*, 463–495.

10

Testing of Eight Medicinal Plant Extracts in Combination with Kresoxim-Methyl for Integrated Control of *Botrytis cinerea* in Apples

Burtram C. Fielding [1], Cindy-Lee Knowles [2,†], Filicity A. Vries [3,†] and Jeremy A. Klaasen [2,*]

[1] Molecular Biology and Virology Laboratory, Department of Medical BioSciences, Faculty of Natural Sciences, University of the Western Cape, Bellville 7535, South Africa;
E-Mail: bfielding@uwc.ac.za

[2] Plant Pathology Laboratory, Department of Medical BioSciences, Faculty of Natural Sciences, University of the Western Cape, Bellville 7535, South Africa; E-Mail: clknowles@uwc.ac.za

[3] Fruit, Vine and Wine Institute of the Agricultural Research Council, ARC Infruitec-Nietvoorbij, Private Bag X5026, Stellenbosch 7599, South Africa; E-Mail: VriesF@arc.agric.za

† These authors contributed equally to this work.

* Author to whom correspondence should be addressed; E-Mail: jklaasen@uwc.ac.za

Academic Editor: Nieves Goicoechea

Abstract: *Botrytis cinerea* is a fungus that causes gray mold on many fruit crops. Despite the availability of a large number of botryticides, the chemical control of gray mold has been hindered by the emergence of resistant strains. In this paper, tests were done to determine the botryticidal efficacy of selected plant extracts alone or combined with kresoxim-methyl. In total, eight South African medicinal plants *viz Artemisia afra, Elyptropappus rhinocerotis, Galenia africana, Hypoxis hemerocallidea, Siphonochilus aetheopicus, Sutherlandia frutescens, Tulbaghia violacea* and *Tulbaghia alliacea* were screened. *Allium sativum*, a plant species known to have antifungal activity, was included in the *in vivo* studies. For the *in vitro* studies, synergistic interactions between the plant extracts and the kresoxim-methyl fungicide were tested with radial growth assays. Data indicated synergistic inhibitory effects between the fungicide and the plant extracts. Next, different doses of plant extracts combined with kresoxim-methyl were used for decay inhibition studies on Granny Smith apples. Synergistic and additive effects were observed for many of the combinations. Even though this study was done using only one strain of *B. cinerea*, results showed that the tested indigenous South African plant species possess natural compounds that potentiate the activity of kresoxim-methyl.

Keywords: *Botrytis cinerea*; gray mold; kresoxim-methyl; medicinal plants; plant extracts; strobilurin; fungicides

1. Introduction

Growth of fungal pathogens leads to considerable economic losses during postharvest handling, transportation and storage of crops [1,2]. As a solution, synthetic fungicides have been used globally since the 1950s to protect major crops from damage by phytopathogenic fungi [3,4]. Many modern fungicides are effective and exhibit relatively low mammalian toxicity [5]. Disconcertingly, however, as a result of fungicide overuse, resistant strains of these pathogens are now appearing with alarming frequency [6,7]. This progression of resistance has now become a major problem globally, particularly where high resistance factors have been reported and the frequencies of mutant phenotypes in the population are high. This phenomenon greatly lowers the efficacy of active ingredients in the fungicides, resulting in an increase in the cost of chemical control and potentially resulting in environmental damage if repeated treatment is required [4].

Botrytis cinerea (Pers. ex. Fr) is a ubiquitous fungus causing gray mold on various crops, even when the most advanced postharvest technologies have been applied [8]. It is considered an important pathogen of vegetables, ornamental plants and fruits [4,6,9]. Together with blue mold, caused by *Penicillium* spp., gray mold is one of the main postharvest diseases of apples in storage. Gray mold can spread from fruit to fruit and whole bins of fruit can potentially become infected, often leading to severe economic losses [10]. In spite of the availability of numerous botryticides, the chemical control of gray mold has been encumbered by the emergence of resistant strains [4,9,11–14]. Of greatest concern, multiple drug resistant (MDR) strains have recently been isolated from French and German vineyards. MDR populations increase the risk of gray mold rot and, more worryingly, impede the effectiveness of current strategies for the management of fungicide resistance [6,15].

In recent years, researchers have focused on the discovery of plant-derived fungicides which are considered safer than synthetic fungicides. Plant extracts with fungistatic or fungicidal activities have shown potential as effective synthetic fungicide alternatives for postharvest control of various plant diseases of fruit [16–20]. Also, biofungicides applied in mixtures with synthetic fungicides, result in a reduction of the amounts of active ingredients applied and could possibly prevent the development of fungicide resistance [21,22]. In this study, tests were done to determine the botryticidal efficacy of selected plant extracts alone or in combination with kresoxim-methyl fungicide (Stroby®, BASF, Ludwigshaven, Germany). In total, eight South African medicinal plant species including *Artemisia afra*, *Elyptropappus rhinocerotis*, *Galenia africana*, *Hypoxis hemerocallidea*, *Siphonochilus aetheopicus*, *Sutherlandia frutescens*, *Tulbaghia violacea* and *Tulbaghia alliacea* were screened. *Allium sativum* (commercial garlic), a plant with known antifungal characteristics [23–25], was included in the *in vivo* studies. For the *in vitro* studies, synergistic interactions between the plant extracts and the kresoxim-methyl fungicide were tested with radial growth assays. Next, different doses of plant extracts combined with kresoxim-methyl were used for decay inhibition studies on Granny Smith apples. Taken together, results showed that the tested indigenous South African plants possess natural compounds that potentiate the activity of kresoxim-methyl. This report offers an attractive prospect for the development of alternative strategies for controlling *B. cinerea*.

2. Results

2.1. Effect of Plant Extracts Alone and in Combination with Kresoxim-Methyl on the Growth of B. Cinerea in vitro

Firstly, the *in vitro* inhibitory effects of eight medicinal plant extracts alone and in combination with kresoxim-methyl on *B. cinerea* growth were studied (Table 1). In the absence of kresoxim-methyl, most of the plant extracts tested produced weak inhibitory effects (>50% inhibition). Only *G. africana* and *E. rhinocerotis* extracts showed radial growth inhibition >50%, at all doses tested and 500 mg·mL^{-1}, respectively. In general, the mycelium growth of *B. cinerea* decreased with increasing plant extract doses. *In vitro* studies with methanol extracts of *A. afra*, *A. sativum*, *E. rhinocerotis*, *H. hemerocallidea*, *S. aethiopicus*, *S. frutescens*, *T. alliacea*, and *T. violacea* showed weak antifungal properties against *B. cinerea*. The strongest antifungal activity in the radial growth bioassay was observed for *G. africana* extracts.

Although kresoxim-methyl (2.5 mg·mL^{-1}) in combination with *G. africana* extract at doses of 125.0, 250.0 and 500.0 mg·mL^{-1} showed statistically significantly high inhibition levels compared to the kresoxim-methyl control, it only produced additive interactions (SR 0.5–1.5). The *E. rhinocerotis* extracts showed optimal botryticidal activity at 250.0 and 500.0 mg·mL^{-1}, while *A. sativum*, *S. aethiopicus*, *T. alliacea*, *H. hemerocallidea* and *T. violacea* showed optimal botryticidal activity at 125.0 and 250.0, but not at 500.0 mg·mL^{-1}. For the other combination treatments synergistic interactions (SR > 1.5; $p < 0.5$) were observed for *S. aethiopicus*, *S. frutescens* and *T. alliaceae* extracts at 62.5 mg·mL^{-1}, for *A. afra*, *A. sativum*, *H. hemerocallidea*, *S aethiopicus*, *T. alliacea* and *T. violacea* at 125.0 mg·mL^{-1}, and for *A. afra*, *A. sativum*, *H. hemerocallidea*, *T. alliacea* and *T. violacea* at 250.0 mg·mL^{-1}. Antagonistic effects (SR < 0.5) were primarily observed for the 500.0 mg·mL^{-1} plant extract and 2.5 mg·mL^{-1} Stroby combinations.

Total radial growth inhibition (100%) was only observed for the combinations of 5.0 mg·mL^{-1} kresoxim-methyl and 250 mg·mL^{-1} *T. alliacea* and *T. violacea* (Table 2). Most of the plant extracts and 5.0 mg·mL^{-1} kresoxim-methyl combinations showed additive interactions. Synergistic interactions for plant extracts and the 5.0 mg·mL^{-1} kresoxim-methyl combinations were observed for *T. alliacea* and *T. violacea* at 62.5 mg·mL^{-1}, for *A. afra*, *S. frutescens*, *T. alliacea* and *T. violacea* at 125.0 mg·mL^{-1}, and for *T. alliacea* and *T. violacea* at 250.0 mg·mL^{-1}. No synergistic effects were observed for the 500.0 mg·mL^{-1} plant extract and 5.0 mg·mL^{-1} kresoxim-methyl combinations. However, when the plant extracts were combined with kresoxim-methyl, significant reduction in mycelium growth of the fungus was observed for almost all of the combinations. The inhibitory effects in combinations with kresoxim-methyl were especially significant for extracts of *A. afra*, *E. rhinocerotis*, *G. africana*, *H. hemerocallidea*, *S. aethiopicus*, *S. frutescens*, *T. alliacea*, and

T. violacea. Using the Abbott method, we were able to calculate mathematical synergistic ratios for all the plant extracts at their respective doses and in combination with kresoxim-methyl. With the exception of *G. africana* and *E. rhinocerotis*, all the plant extracts showed synergism (>1.5) for the *in vitro* study. *Siphonochilus aethiopicus, S. frutescens, T. alliacea* and *T. violacea* showed the most potent synergistic interaction at the lowest extract dose (*i.e.*, 62.5 mg·mL^{-1}) used for the *in vitro* study.

2.2. In Storage: Effect of Plant Extracts Alone and in Combination with Kresoxim-Methyl on the Growth of B. cinerea on Apples

In this study, single extracts of the medicinal plant species *A. afra, E. rhinocerotis, G. africana, H. hemerocallidea, S. aethiopicus, S. frutescens, T. alliacea,* and *T. violacea* exhibited weak or no antifungal properties against *B. cinerea in vivo*. However, when low doses of the plant extracts were combined with a subinhibitory concentration of kresoxim-methyl, synergistic and additive interactions were observed. Synergistic effects were especially obvious for the lowest extract doses of *E. rhinocerotis, H. hemerocallidea, S. frutescens, T. alliacea* and *T. violacea* (Table 2). Significant differences in the inhibition of decay progression for the plant extract and kresoxim-methyl (0.005 mg·mL^{-1}) combinations compared to the control were observed for *A. afra* (15.63 mg·mL^{-1}), *G. africana* (62.5 mg·mL^{-1}), *E. rhinocerotis* (1.95, 3.91 mg·mL^{-1}), *H. hemerocallidea* (1.95, 3.91, 15.63, 31.25 mg·mL^{-1}), *S. frutescens* (1.95, 3.91, 7.81 mg·mL^{-1}), *T. alliacea* (1.95, 3.91 mg·mL^{-1}) and *T. violacea* (1.95, 3.91, 7.81, 31.25, 62.0 mg·mL^{-1}) ($p < 0.05$). Synergistic interactions were observed for the plant extract doses of *E. rhinocerotis* (7.81 mg·mL^{-1}), *H. hemerocallidea* (7.81, 15.63, 31.25 mg·mL^{-1}), *S. frutescens* (3.91 mg·mL^{-1}), *T. alliacea* (7.81 mg·mL^{-1}) and *T. violacea* (15.63 mg·mL^{-1}). Kresoxim-methyl, in combination with the higher plant extract dose of 62.5 mg·mL^{-1}, produced primarily additive interactions.

3. Experimental Secion

3.1. Pathogen Preparations

Non-resistant *Botrytis cinerea* was isolated from the surface of "Granny Smith" apples infected with gray mold disease and maintained on potato-dextrose agar (PDA) at 25 °C for 14 days (Disease Management Division, Agricultural Research Council (ARC) Infruitec—Nietvoorbij, South Africa). Fresh inoculum was prepared by transferring spores from stock cultures to PDA and incubating these at 25 °C in the dark under a white fluorescent light with a 12:12 light:dark photoperiod.

Table 1. Inhibition of radial growth and synergy ratio (SR) between the kresoxim methyl fungicide and different plant extract doses against *Botrytis cinerea in vitro*.

	Radial Growth Inhibition (%)													
	Plant Extract Dose (mg·mL⁻¹)				Plant Extract Dose + 2.5 mg·mL⁻¹ Stroby®					Plant Extract Dose + 5.0 mg·mL⁻¹ Stroby®				
	62.5	125	250	500	0	62.5	125	250	500	0	62.5	125	250	500
Stroby®					37					72.5				
A. afra	0.9	−0.5	2.6	15.3	-	22.6* (1.2)	38.3* (2.3)	34.6* (1.7)	17.2** (0.5)	-	76.7* (1.4)	98.5*/** (1.9)	80.9* (1.4)	39.3*/** (0.6)
A. sativum	−0.4	25.7	8.9	12.1		52.0* (3.1)	74.5*/** (1.8)	74.6*/** (2.9)	26.4* (0.9)		75.7* (0.8)	96.8*/** (1.2)	99.8*/** (2.3)	77.5* (1.2)
E. rhinocerotis	17.8	27.6	47.4	62		0.7** (0)	44.7 (0.1)	88.1*/** (1.4)	74.4** (0.9)		29.6** (0.4)	79.8* (1)	85.8* (0.9)	73.7 (0.6)
G. africana	67.1	71.4	76.9	74		61.7** (0.7)	81.8** (0.9)	87.0** (0.9)	84.3** (0.9)		58.1 (0.5)	73 (0.6)	83.8 (0.6)	90.8 (0.7)
H. hemerocallidea	7.2	11.4	28.6	24		13.7 (0.6)	52.6* (1.9)	79.3*/** (1.8)	49.3* (1.2)		87.3* (1.4)	89.8* (1.4)	94.7*/** (1.2)	61.6* (0.8)
S. aethiopicus	−4.6	22.6	−6.8	26.8		43.5** (3.5)	90.0*/** (2.3)	78.8*/** (7.9)	27.5 (0.6)		−2.4** (0.1)	19.6** (0.3)	28.4*/** (0.6)	0.7*/** (0)
S. frutescens	−3.4	3.4	−0.2	8.8		49.2* (3.6)	12.4** (0.6)	26.1* (1.5)	16.8** (0.6)		59.7* (1.2)	93.9*/** (1.7)	80.0* (1.5)	22.8*/** (0.4)
T. alliacea	−7.5	3.2	5	18.6		38.0* (4)	79.3*/** (4)	62.3*/** (2.8)	23.6 (0.7)		71.3* (1.6)	90.0* (1.6)	100.0*/** (1.7)	71.1* (1)
T. violacea (EC)	−0.9	−3.6	−10.1	32.2		−3.6*/** (0.2)	19.5* (1.4)	37.6* (1.4)	52.4* (1.1)		47.7* (0.9)	84.4*/** (1.7)	88.1* (1.4)	48.1** (0.6)
T. violacea	0.7	11.8	9.6	19.6		59.6*/** (3.4)	68.6*/** (2.4)	76.1*/** (2.9)	35.4 (1)		97.5*/** (1.8)	97.5*/** (1.5)	100.0*/** (1.6)	92.5*/** (1.3)

The synergism ratio for percentage inhibition (values in bold) was based on the Abbott formula as described by [26]. Radial growth inhibition values of the plant extract and Stroby® combinations with one asterisk (*) are significantly different ($p > 0.05$) from the respective values of the plant extract doses without Stroby® in the same row. Radial growth inhibition values of the plant extract and Stroby® combinations with two asterisks (**) are significantly different ($p > 0.05$) from the value of the Stroby® dose (2.5 or 5.0 mg·mL⁻¹) in the column.

Table 2. Percentage inhibition of grey mold decay on Granny Smith apples inoculated with conidia of *B. cinerea*, and the relative level of synergism (SR) of mixtures containing 0.0 and 0.005 mg·mL⁻¹ of kresoxim-methyl and different doses of medicinal plant extracts.

	Decay Inhibition (%)												
	Plant Extract Dose (mg·mL⁻¹)							Plant Extract Dose + 0.005 mg·mL⁻¹ Stroby®					
	1.95	3.91	7.81	15.63	31.25	62.5	0	1.95	3.91	7.81	15.63	31.25	62.5
Stroby®							57.4						
A. afra	−16.9	11.2	12.4	30	−13.8	19.2		53.9*	88.0*	72.6*	93.6*/**	36.6*	49.6*
								1.3	1.3	1.1	1.1	0.8	0.7
E. rhinocerotis	49.1	10.4	−9.6	86	−6.2	−13.9		100.0*/**	96.0*/**	77.5*	52.3*	26.9*	37.4*
								0.9	1.4	1.6	0.4	0.5	0.9
G. africana	10.6	19.7	1.6	37.6	59.2	74.1		86.7*	70.9*	70.3*	72.2*	82.4	96.3**
								1.3	0.9	1.2	0.8	0.7	0.7
H. hemerocallidea	8.4	28.5	−6.1	−0.1	−3.4	24.6		92.5*/**	95.9*/**	85.6*	94.0*/**	94.0*/**	70.0*
								1.4	1.1	1.7	1.7	1.8	0.9
S. aethiopicus	−3.9	4.1	27.9	23.8	32.4	16.8		47.8*	83.3*	72.8*	75.4*	81.6*	30.3
								0.9	1.4	0.9	0.9	0.9	0.3
S. frutescens	8.9	5.1	12.2	8.6	3.6	16.1		90.9*/**	100.0*/**	93.1*/**	69.1*	86.2*	35.3
								1.4	1.6	1.4	1.1	1.4	0.5
T. alliacea	16.6	24	−11.1	13.6	18.2	−8.7		100.0*/**	96.2*/**	76.6*	73.4*	60.2*	68.8*
								1.4	1.2	1.7	1	0.8	1.4
T. violacea	15.8	46.4	40.8	−8.24	11.2	20.5		100.0*/**	93.2*/**	100.0*/**	81.9*	100.0*/**	24.5**
								1.4	0.9	1	1.7	1.5	0.3

The synergism ratio for percentage inhibition (values in bold) was based on the Abbott formula as described by [26]. Decay inhibition mean values of the plant extract and Stroby® combinations with one asterisk (*) are significantly different ($p > 0.05$) from the respective values of the plant extract doses without Stroby® in the same row. Decay inhibition mean values of the plant extract and Stroby® combinations with two asterisks (**) are significantly different ($p > 0.05$) from the value of the Stroby® dose (0.005 mg·mL⁻¹) in the column.

3.2. Preparation of Plant Extracts

The plant parts of the following medicinal plant species used in traditional medicine practices in South Africa were obtained for the preparation of plant extracts: *Artemisia afra* Jacq. (fresh leaves), *Elytropappus rhinocerotis* (Lf) Less (fresh leaves), *Galenia africana* (L.) (dried leaves), *Hypoxis hemerocallidea* (Fisch. and C.A. Mey) (fresh corms), *Siphonochilus aethiopicus* (Schweif.) BL Burt (fresh rhizomes + fleshy roots), *Sutherlandia frutescens* (L.) R. Br. (dried leaves), *Tulbaghia alliacea* (fresh corms), *Tulbaghia violacea* Harv. (fresh leaves + rhizomes from Western Cape greenhouse plants), *Tulbaghia violacea* from the Eastern Cape (fresh leaves + rhizomes from "wild" plant). *Allium sativum* (bulbs) was included in the *in vivo* studies. Identity of the various plant species was authenticated and assigned a voucher number by Frans Weitz (Herbarium, Department of Botany, University of the Western Cape, Bellville, South Africa).

Where fresh plant organs were used, the organs were homogenized in a Waring blender and 50 g of the macerated plant was then extracted overnight in 100 ml methanol (MeOH; 99.8% Sigma Aldrich) in a closed conical flask at room temperature to obtain 50% (w/v) (500 mg·mL^{-1}) crude, viscous liquid extract. Where dried leaves were used, the leaves were powdered in a hammer mill and 50 g extracted overnight in a closed conical flask at room temperature in MeOH to obtain 50% viscous liquid extracts. Each crude methanol extract was filtered through Whatman No. 4 qualitative filter paper and stored at 4 °C until used. For the *in vitro* and apple fruit bioassays two-fold dilution series of the 50% crude plant extracts were prepared with deionized sterile water as described below.

3.3. In vitro Assay

A 1 ml suspension of the plant extracts and kresoxim-methyl (Stroby® WG fungicide, 500 g·kg^{-1}, BASF, Ludwigshaven, Germany), single or combinations of both products, were spread evenly onto a solidified PDA surface in 9 mm Petri dishes and allowed to air dry under sterile conditions in a laminar flow to evaporate the solvents. Controls consisted of sterile water and methanol. The 50% methanolic plant extracts were diluted in sterile water in 10 ml doses of 0.0, 62.5, 125.0, 250.0, 500.0 mg·mL^{-1}, with kresoxim-methyl (0.0, 2.5 and 5.0 mg·mL^{-1}), respectively. *In vitro* studies showed that assays with 2.5 and 5.0 mg·mL^{-1} kresoxim-methyl provided the highest inhibitory effects with plant extracts in the radial growth experiments. Each plate was inoculated with 3-mm mycelial plugs removed from the margins of actively growing 14-day-old *B. cinerea* cultures, and placed upside down on the PDA surface. Radial growth was assessed 5 days after incubation at 22–23 °C.

3.4. Postharvest Assay on Apples

The apple cultivar, Granny Smith, were removed from cold storage, surface-sterilized with 70% EtOH for 2 min and air-dried. Each apple was wounded (5 mm in diameter and 3 mm in depth) three times halfway between the calyx and the stem end. A 20-μL drop of each plant extract and kresoxim-methyl was placed in the wounds and allowed to air-dry for two hours before application of a 20-μL conidial suspension (1×10^4 spores mL^{-1}); the 20-μL drops had final plant extract doses of 0.0, 1.95, 3.91, 7.81, 15.63, 31.25 and 62.5 mg·mL^{-1}, with or without kresoxim-methyl at 0.0 and 0.005 mg·mL^{-1}. Controls consisted of sterile water and methanol without the compounds to be tested.

Fruits were stored in commercial cardboard boxes at 20 °C in a high humidity (95% RH) walk-in incubator. Diameter of *B. cinerea* decay lesions was determined after a 7-day incubation period.

3.5. Statistical Analysis

The *in vitro* experimental design was completely randomized. Each mycelial plug in a Petri dish constituted a replicate. To assess differences in the mycelial growth of *B. cinerea* among the treatments the percentage inhibition were calculated from the radial growth as:

$$\% = 100 - [(treatment/control) \times 100] \tag{1}$$

All analyses were carried out using SAS Version 8.2 [27]. For the apple bioassay each treatment was performed on three apples, each with triplicate wounds. Percentage inhibition was determined for each replicate and averaged for the three apples. Data for percentage decayed fruit were subjected to a standard analysis of variance. The analysis of variance was performed using SAS Version 8.2 [27]. Student *t*-Least Significant Difference was calculated at the 5% significant level to compare treatment means of percentage inhibition. The synergism ratio for percentage inhibition was based on the Abbott formula [28] as described by Gisi [26]:

$$C_{exp} = (A + B) - (AB/100) \tag{2}$$

In this study, C_{exp} was the expected efficacy of the mixture, with A and B the control levels given by kresoxim-methyl and the medicinal plant extract, respectively. The synergy ratio (SR) between the observed (C_{obs}) and expected (C_{exp}) efficacies of the mixture was calculated as:

$$SR = C_{ob}/C_{exp} \tag{3}$$

An SR >1.5 indicates a synergistic interaction between compounds; 0.5–1.5 indicates an additive interaction between compounds; <0.5 indicates an antagonistic interaction between compounds [29].

4. Discussion and Conclusions

In recent years there has been a drive to find alternatives to chemical fungicides considered as safe and with lower risk to human health and the environment. Satisfactory results have been reported using biocontrol or antagonistic microorganisms [30] and natural compounds including various plant extracts [31,32]. Natural plant protectants with antimicrobial activity have been studied since they generally tend to have low mammalian toxicity, low environmental impact and wider public acceptance [5,20]. Secondary plant metabolites are often active against a small number of specific target microbial species and are biodegradable to nontoxic products, making it potentially valuable in integrated pest management programs [2]. Also, identifying synergistic combinations of fungicides and natural plant compounds could result in control strategies with high biological activity, low dose rate application and a low risk of pathogen-resistance development [5,20]. In this study, the reason for the reduced *B. cinerea* sensitivity to the combinations of kresoxim-methyl and plant extracts at higher doses has not yet been elucidated. We speculate that this could be indicative of competitive inhibition resulting from constituents in the plant extracts competing for the mode of action sites of kresoxim-methyl. This would effectively impede the complete inhibition of mitochondrial respiration, since the binding site of kresoxim-methyl is the ubihydroquinone cytochrome-c oxidoreductase [33]. In fact, more recent studies with plant extracts showed that polyphenols could stimulate and/or enhance mitochondrial function [34–36]. Whether polyphenols act directly or indirectly to enhance mitochondrial functions and whether this requires binding of polyphenols to a specific receptor remain to be investigated. Studies with arbuscular mycorrhizal fungi showed that plant root-exuded factors cause changes in fungal mitochondrial morphology, orientation, and overall biomass and rapidly induces the expression of certain fungal genes and, in turn, respiratory activity before intense branching [37].

Interestingly, in this study the extract from *T. violacea* from the Western Cape showed higher inhibitory and synergistic activities in combination with Stroby® compared to the *T. violacea* extracts from the Eastern Cape and the *A. sativum* extracts. We do not know the reason for this observation as yet. However, due to the close relationship between *Allium* and *Tulbaghia* species within the *Alliaceae* family the biological and chemical characteristics of the two species seems to be comparable [38]. Interestingly, even though the *E. rhinocerotis* extract showed synergistic effects in the *in vivo* study, it did not produce a similar effect in the *in vitro* study. This could be a result of the type of biological formulation used, which could have affected the efficacy of the extract in the *in vitro* study [39,40].

As yet, we do not have an explanation for the lower ratios of synergistic reactions for the plant extract and kresoxim-methyl combinations *in vivo* compared to *in vitro*. However, this paper provided evidence that compounds in the tested plant extracts contributed additively and/or synergistically with the fungicide, kresoxim-methyl. In the *in vitro* radial growth assays, synergistic inhibitory effects were observed between kresoxim-methyl and plant extracts. Finally, in decay inhibition studies on Granny Smith apples synergistic and additive effects were observed for many of the combinations. Interestingly, the antifungal activities of the plant extracts in combination with the kresoxim-methyl did not appear to be dose-dependent. Although only one strain of *B. cinerea* was used in this study, we showed that sub-inhibitory concentrations of the kresoxim-methyl fungicide and very low dose rates of plant extracts can act in synergy for the potential control of gray mold in apples.

Acknowledgments

Jeremy Klaasen receives funding from the Technology Innovation Agency, South Africa. Burtram C. Fielding receives funding from the National Research Foundation, South Africa. Any opinion, findings and conclusions or recommendations expressed in this material are those of the author and therefore the National Research Foundation or Technology Innovation Agency do not accept any liability in regard thereto.

Author Contributions

J.A.K. conceptualized the experiments. C-L.K. and F.V. performed the experiments. J.A.K., F.V. and B.C.F. analyzed the results. B.C.F. and J.A.K. wrote the manuscript.

Conflicts of Interest

The authors declare no conflict of interest.

References

1. Zhang, H.; Li, R.; Liu, W. Effects of chitin and its derivative chitosan on postharvest decay of fruits: A review. *Int. J. Mol. Sci.* **2011**, *12*, 917–934.

2. Camele, I.; Altieri, L.; De Martino, L.; de Feo, V.; Mancini, E.; Rana, G.L. *In vitro* control of post-harvest fruit rot fungi by some plant essential oil components. *Int. J. Mol. Sci.* **2012**, *13*, 2290–2300.

3. Knight, S.C.; Anthony, V.M.; Brady, A.M.; Greenland, A.J.; Heaney, S.P.; Murray, D.C.; Powell, K.A.; Schulz, M.A.; Spinks, C.A.; Worthington, P.A.; *et al.* Rationale and perspectives on the development of fungicides. *Ann. Rev. Phytopathol.* **1997**, *35*, 349–372.

4. Leroux, P.; Gredt, M.; Leroch, M.; Walker, A.S. Exploring mechanisms of resistance to respiratory inhibitors in field strains of *Botrytis cinerea*, the causal agent of gray mold. *Appl. Environ. Microbiol.* **2010**, *76*, 6615–6630.

5. Shao, X.; Cheng, S.; Wang, H.; Yu, D.; Mungai, C. The possible mechanism of antifungal action of tea tree oil on *Botrytis cinerea. J. Appl. Microbiol.* **2013**, *114*, 1642–1649.

6. Kretschmer, M.; Leroch, M.; Mosbach, A.; Walker, A.S.; Fillinger, S.; Mernke, D.; Schoonbeek, H.J.; Pradier, J.M.; Leroux, P.; de Waard, M.A.; *et al.* Fungicide-driven evolution and molecular basis of multidrug resistance in field populations of the grey mould fungus *Botrytis cinerea. PLoS Pathog.* **2009**, *5*, e1000696, doi:10.1371/journal.ppat.1000696.

7. Gabriolotto, C.; Monchiero, M.; Negre, M.; Spadaro, D.; Gullino, M.L. Effectiveness of control strategies against *Botrytis cinerea* in vineyard and evaluation of the residual fungicide concentrations. *J. Environ. Sci. Health Part B* **2009**, *44*, 389–396.

8. Spadaro, D.; Gullino, M.L. State of the art and future prospects of the biological control of postharvest fruit diseases. *Int. J. Food Microbiol.* **2004**, *91*, 185–194.

9. Kim, Y.K.; Xiao, C.L. Stability and fitness of pyraclostrobin- and boscalid-resistant phenotypes in field isolates of *Botrytis cinerea* from apple. *Phytopathology* **2011**, *101*, 1385–1391.

10. Notes on Apple Diseases Blue Mould and Grey Mould in Stored Apples. Available online: http://www.omafra.gov.on.ca/english/crops/pub360/notes/applemould.htm (accessed on 5 November 2013).

11. Walker, A.S.; Micoud, A.; Remuson, F.; Grosman, J.; Gredt, M.; Leroux, P. French vineyards provide information that opens ways for effective resistance management of *Botrytis cinerea* (grey mould). *Pest Manag. Sci.* **2013**, *69*, 667–678.

12. Billard, A.; Fillinger, S.; Leroux, P.; Lachaise, H.; Beffa, R.; Debieu, D. Strong resistance to the fungicide fenhexamid entails a fitness cost in *Botrytis cinerea*, as shown by comparisons of isogenic strains. *Pest Manag. Sci.* **2012**, *68*, 684–691.

13. Samuel, S.; Papayiannis, L.C.; Leroch, M.; Veloukas, T.; Hahn, M.; Karaoglanidis, G.S. Evaluation of the incidence of the G143A mutation and cytb intron presence in the cytochrome bc-1 gene conferring QoI resistance in *Botrytis cinerea* populations from several hosts. *Pest Manag. Sci.* **2011**, *67*, 1029–1036.

14. Bardas, G.A.; Veloukas, T.; Koutita, O.; Karaoglanidis, G.S. Multiple resistance of *Botrytis cinerea* from kiwifruit to SDHIs, QoIs and fungicides of other chemical groups. *Pest Manag. Sci.* **2010**, *66*, 967–973.

15. Leroch, M.; Plesken, C.; Weber, R.W.; Kauff, F.; Scalliet, G.; Hahn, M. Gray mold populations in german strawberry fields are resistant to multiple fungicides and dominated by a novel clade closely related to *Botrytis cinerea. Appl. Environ. Microbiol.* **2013**, *79*, 159–167.

16. Zabka, M.; Pavela, R.; Gabrielova-Slezakova, L. Promising antifungal effect of some Euro-Asiatic plants against dangerous pathogenic and toxinogenic fungi. *J. Sci. Food Agric.* **2011**, *91*, 492–497.

17. Liu, X.; Wang, L.P.; Li, Y.C.; Li, H.Y.; Yu, T.; Zheng, X.D. Antifungal activity of thyme oil against *Geotrichum citri-aurantii in vitro* and *in vivo. J. Appl. Microbiol.* **2009**, *107*, 1450–1456.

18. Askarne, L.; Talibi, I.; Boubaker, H.; Boudyach, E.H.; Msanda, F.; Saadi, B.; Ait Ben Aoumar, A. Use of Moroccan medicinal plant extracts as botanical fungicide against citrus blue mould. *Lett. Appl. Microbiol.* **2013**, *56*, 37–43.

19. Talibi, I.; Askarne, L.; Boubaker, H.; Boudyach, E.H.; Msanda, F.; Saadi, B.; Ben Aoumar, A.A. Antifungal activity of Moroccan medicinal plants against citrus sour rot agent *Geotrichum candidum. Lett. Appl. Microbiol.* **2012**, *55*, 155–161.

20. Soylu, E.M.; Kurt, S.; Soylu, S. *In vitro* and *in vivo* antifungal activities of the essential oils of various plants against tomato grey mould disease agent *Botrytis cinerea. Int. J. Food Microbiol.* **2010**, *143*, 183–189.

21. De Lapeyre de Bellaire, L.; Essoh Ngando, J.; Abadie, C.; Chabrier, C.; Blanco, R.; Lescot, T.; Carlier, J.; Côte, F. Is chemical control of *Mycosphaerella* foliar diseases of bananas sustainable. In Proceedings of the International Symposium on Recent Advances in Banana Crop Protection for Sustainable Production and Improved Livelihoods, White River, South Africa, 10–14 September 2007; Jones, D., van den Bergh, I., Eds.; ISHS Acta Horticulturae: White River, South Africa, 2009; pp. 161–170.

22. Schmitt, A.; Seddon, B. Biocontrol of plant pathogens with microbial BCAs and plant extracts—Advantages and disadvantages of single and combined use. In Proceedings of the 14th International Reinhardsbrunn Symposium 2004, Friedrichroda, Thuringia, Germany, 25–29 April 2004; BCPC: Atlon, UK, 2005; pp. 205–225.

23. Goncagul, G.; Ayaz, E. Antimicrobial effect of garlic (*Allium sativum*). *Recent Pat. Anti-Infect. Drug Discov.* **2010**, *5*, 91–93.

24. Low, C.F.; Chong, P.P.; Yong, P.V.; Lim, C.S.; Ahmad, Z.; Othman, F. Inhibition of hyphae formation and SIR2 expression in *Candida albicans* treated with fresh *Allium sativum* (garlic) extract. *J. Appl. Microbiol.* **2008**, *105*, 2169–2177.

25. Iciek, M.; Kwiecien, I.; Wlodek, L. Biological properties of garlic and garlic-derived organosulfur compounds. *Environ. Mol. Mutagen.* **2009**, *50*, 247–265.

26. Gisi, U. Synergistic Interactions of fungicides in mixtures. *Phytopathology* **1996**, *86*, 1273–1279.

27. SAS Institute Inc. *SAS/STAT User's Guide*, Version 8; SAS Campus Drive: Cary, NC, USA, 1999; Volume 2.

28. Abbott, W.S. A method of computing the effectiveness of an insecticide. *J. Am. Mosq. Control Assoc.* **1987**, *3*, 302–303.

29. Gisi, U.; Binder, H.; Rimbach, E. Synergistic interactions of fungicides with different modes of action. *Trans. Br. Mycol. Soc.* **1985**, *84*, 299–306.

30. Fiume, F.; Fiume, G. Biological control of *Botrytis* gray mould and Sclerotinia drop in lettuce. *Commun. Agric. Appl. Biol. Sci.* **2005**, *70*, 157–168.

31. Martinez-Romero, D.; Serrano, M.; Bailen, G.; Guillen, F.; Zapata, P.J.; Valverde, J.M.; Castillo, S.; Fuentes, M.; Valero, D. The use of a natural fungicide as an alternative to preharvest synthetic fungicide treatments to control lettuce deterioration during postharvest storage. *Postharvest Biol. Technol.* **2008**, *47*, 54–60.

32. Rial-Otero, R.; Gonzalez-Rodriguez, R.M.; Cancho-Grande, B.; Simal-Gandara, J. Parameters affecting extraction of selected fungicides from vineyard soils. *J. Agric. Food Chem.* **2004**, *52*, 7227–7234.

33. Sauter, H.; Ammermann, E.; Benoit, R.; Brand, S.; Gold, R.E.; Grammenos, W.; Köhle, H.; Lorenz, G.; Müller, B.; Schirmer, U.; Speakman, J.B.; Wenderoth, B.; Wingert, H. Mitochodrial respiration as a target for antifungals: Lessons from research on strobilurins. In *Antifungal Agents—Discovery and Mode of Action*; Dixon, G.K., Copping, L.G., Holloman, D.W., Eds.; BIOS Scientific Publishers: Oxford, UK, 1995; pp. 173–191.

34. Davinelli, S.; Sapere, N.; Visentin, M.; Zella, D.; Scapagnini, G. Enhancement of mitochondrial biogenesis with polyphenols: Combined effects of resveratrol and equol in human endothelial cells. *Immun. Ageing* **2013**, *10*, 28, doi:10.1186/1742-4933-10-28.

35. Rehman, H.; Krishnasamy, Y.; Haque, K.; Thurman, R.G.; Lemasters, J.J.; Schnellmann, R.G.; Zhong, Z. Green tea polyphenols stimulate mitochondrial biogenesis and improve renal function after chronic cyclosporin a treatment in rats. *PLoS ONE* **2013**, *8*, e65029.

36. Rasbach, K.A.; Schnellmann, R.G. Isoflavones promote mitochondrial biogenesis. *J. Pharmacol. Exp. Ther.* **2008**, *325*, 536–543.

37. Tamasloukht, M.; Sejalon-Delmas, N.; Kluever, A.; Jauneau, A.; Roux, C.; Becard, G.; Franken, P. Root factors induce mitochondrial-related gene expression and fungal respiration during the developmental switch from asymbiosis to presymbiosis in the arbuscular mycorrhizal fungus *Gigaspora rosea*. *Plant Physiol.* **2003**, *131*, 1468–1478.

38. Lyantagaye, S.L. Ethnopharmacological and phytochemical review of *Allium* species (sweet garlic) and *Tulbaghia* species (wild garlic) from southern Africa. *Tanzan. J. Sci.* **2011**, *37*, 58–71.

39. McLean, K.L.; Swaminathan, J.; Frampton, C.M.; Hunt, J.S.; Ridgway, H.J.; Stewart, A. Effect of formulation on the rhizosphere competence and biocontrol ability of Trichoderma atroviride C52. *Plant Pathol.* **2005**, *54*, 212–218.

40. Adandonon, A.; Aveling, T.A.S.; Labuschagne, N.; Tamo, M. Biocontrol agents in combination with *Moringa oleifera* extract for integrated control of Sclerotium-caused cowpea damping-off and stem rot. *Eur. J. Plant Pathol.* **2006**, *115*, 409–418.

Nutrient Composition, Forage Parameters, and Antioxidant Capacity of Alfalfa (*Medicago sativa*, L.) in Response to Saline Irrigation Water

Jorge F. S. Ferreira *, Monica V. Cornacchione [†,‡], Xuan Liu [‡] and Donald L. Suarez [‡]

US Salinity Laboratory, 450 W. Big Springs Rd., Riverside, CA 92507, USA;
E-Mails: Xuan.Liu@ars.usda.gov (X.L.); Donald.Suarez@ars.usda.gov (D.L.S.)

[†] Currently at INTA- Estación Experimental Agropecuaria Santiago del Estero, Jujuy 850, Santiago del Estero 4200, Argentina; E-Mail: cornacchione.monica@inta.gob.ar.

[‡] These authors contributed equally to this work.

* Author to whom correspondence should be addressed; E-Mail: Jorge.Ferreira@ars.usda.gov

Academic Editor: Cory Matthew

Abstract: Although alfalfa is moderately tolerant of salinity, the effects of salinity on nutrient composition and forage parameters are poorly understood. In addition, there are no data on the effect of salinity on the antioxidant capacity of alfalfa. We evaluated four non-dormant, salinity-tolerant commercial cultivars, irrigated with saline water with electrical conductivities of 3.1, 7.2, 12.7, 18.4, 24.0, and 30.0 dS·m^{-1}, designed to simulate drainage waters from the California Central Valley. Alfalfa shoots were evaluated for nutrient composition, forage parameters, and antioxidant capacity. Salinity significantly increased shoot N, P, Mg, and S, but decreased Ca and K. Alfalfa micronutrients were also affected by salinity, but to a lesser extent. Na and Cl increased significantly with increasing salinity. Salinity slightly improved forage parameters by significantly increasing crude protein, the net energy of lactation, and the relative feed value. All cultivars maintained their antioxidant capacity regardless of salinity level. The results indicate that alfalfa can tolerate moderate to high salinity while maintaining nutrient composition, antioxidant capacity, and slightly improved forage parameters, thus meeting the standards required for dairy cattle feed.

Keywords: alfalfa; salinity; forage quality; nutrient composition; antioxidant capacity; total phenolics

---

1. Introduction

Alfalfa (*Medicago sativa*, L.) is the most cultivated legume worldwide and the fourth most cultivated crop in the United States. Alfalfa is cultivated in most continents and in more than 80 countries occupying more than 35 million ha [1]. In the USA, it is among the top three field crops cultivated in 26 states, thus contributing more than US $10 billion a year to the farm economy, primarily as an animal feed [2]. Alfalfa is considered to be the most important forage crop for providing protein to dairy and beef cattle, sheep, horses, birds, and other livestock [1]. Feeding of alfalfa hay to lactating dairy cows has decreased sharply in the past 10 years, primarily as a result of economic issues associated with high water use, the costs of multiple harvests, and storage [3]. These authors also mentioned the increased use of corn and cereal silages in animal diets to replace alfalfa. However, dry matter intake is significantly higher for cows fed alfalfa and barley silages than for cows fed oat and triticale silages [4]. According to these authors, alfalfa silage contains higher concentrations of all minerals analyzed compared with cereal silages, except for Na. Moreover, the cows also absorbed K better from alfalfa silage (89%) than from cereal silages (74% to 83%). Alfalfa is highly important to livestock considering its fast canopy recovery after each harvest, its relative tolerance of salinity, its capacity to endure temperature extremes (e.g., hot days and cold nights), its nutritional value, and palatability to livestock.

In arid lands, irrigation is necessary for high forage mass production. However, this irrigation is often associated with salinization. Among the approximately 270 million hectares of irrigated land worldwide, about 40% is located in arid/semiarid zones [5] where soil salinization generally occurs. Some of the typical agronomic parameters used to evaluate the salinity tolerance of crops include yield, survival, plant height, and relative growth rate or reduction [6–8]. Few researchers have evaluated alfalfa forage mass production, nutrient composition, and forage parameters for livestock under high salinity stress [9–12]. Further, we found no published reports on the effects of salinity on the antioxidant capacity of alfalfa. It has been reported that salinity stress imposed on a model legume (*Lotus japonicus*) increased antioxidant enzyme levels in leaves [13], and that the expression of genes associated with antioxidant enzymes increased in response to excessive levels of reactive oxygen species (ROS) generated by salinity stress [14]. These authors postulated that these enzymes protect plant tissues from ROS damage triggered by salinity stress, but there are no reports on the biosynthesis of non-enzymatic antioxidants, such as flavonoids and phenolic compounds, by alfalfa in response to salinity. Alfalfa shoots are a rich source of antioxidant flavonoids, mainly apigenin, tricin, luteolin, and chrysoeriol glycosides [15], and of phenolic compounds reported to have anti-inflammatory [16], antioxidant, and neuroprotective activity in mice [17]. The ratio of alfalfa antioxidant flavones acylated with hydroxycinnamic acid to non-acylated (lower antioxidant capacity) flavones increases in summer when plants are exposed to a higher amount of UV-B radiation [15]. Antioxidant flavonoids in *Ligustrum vulgare* were reported to increase under both UV-B and NaCl salinity stress [18]. Thus, although alfalfa is fed to livestock for its high protein content, digestibility, and palatability, there is a scarcity of information on the effects of salinity on alfalfa mineral composition and forage quality, while there is no information on its antioxidant capacity under salinity stress.

In this work, we evaluated four commercial alfalfa cultivars, tolerant to salinity, for their response to salinity when cultivated in outdoor sand tanks and irrigated at six salinity levels with water high in sodium, chloride, and sulfate. The goal of our work was to evaluate the effects of increasing salinity on the mineral nutritional composition, forage quality, and antioxidant capacity of alfalfa shoots.

2. Experimental Section

2.1. Plant Material and Growth Conditions

Four commercial non-dormant, salinity-tolerant, *Medicago sativa* L. cultivars "Salado", "SW8421S", "SW9215", and "SW9720" (S&W, Fresno, CA, USA, www.swseedco.com) were grown from seeds in 24 outdoor sand tanks from 23 June 2011 to 17 April 2012 at the Salinity Laboratory (USDA-ARS) in Riverside, California. Irrigation water at different levels of electrical conductivity (EC) was applied to four cultivars in a split-plot design. The irrigation water EC (measured in deciSiemens per meter) levels consisted of a control using Riverside tap water (EC = 0.6 dS·m^{-1}) plus fertilizers (EC = 3.1 dS·m^{-1}), and treatments of 7.2, 12.7, 18.4, 24.0 and 30.0 dS·m^{-1}, with four tanks (replicates) per treatment. The tanks measured 82 cm wide by 202 cm long by 85 cm deep. Further details on sowing density per cultivar and irrigation frequency are described elsewhere [19]. Salinity treatments and the irrigation water control (EC of 3.1 dS·m^{-1}) were designed to simulate the drainage water composition of the Central Valley, CA, with subsequent concentration of salts considering mineral precipitation (calcite and/or gypsum) using the UNSATCHEM model [20], which simulates typical soil water interactions. All reservoirs had modified Hoagland's solution, and added Na$^+$, SO$_4^{2-}$, and Cl$^-$ (including control water) to reach the target EC; the detailed composition is described elsewhere [19]. The composition of Riverside tap water (EC = 0.6 dS·m^{-1}) in mmolc·L^{-1} was: 3.4 Ca^{2+}, 0.8 Mg^{2+}, 1.6 Na$^+$, 0.1 K$^+$, 1.3 SO$_4^{2-}$, 0.8 Cl$^-$, and 0.49 NO$_3^-$. The water composition of all the treatment waters is shown in Table 1.

2.2. Plant Growth and Nutrient Composition

Growth and forage mass measurements were collected at seven harvest dates except for the plants that were irrigated with water with an EC = 24.0 dS·m^{-1}, which were harvested three times (4th, 6th, and 7th harvests) during the 299 days of cultivation and are presented elsewhere [19]. For this work, we present data on ionic and nutrient composition at 84 days after seeding (DAS) (2nd harvest, on 15 September 2011) and at 299 DAS (7th harvest, on 17 April 2012). The second harvest was conducted when the control plants were at the early flowering stage, corresponding to morphological stage 5 [21]. The seventh harvest was conducted when the control plants were at a late vegetative stage (due to the absence of flowering). The shoot fresh and dry weights (dried at 60 °C for 48 h) were recorded at each harvest and all plants were cut back to 5–8 cm above the sand surface.

Table 1. Chemical composition of the water used in the six salinity treatments in this study. EC, electrical conductivity of irrigation water that defines each salinity level (in deciSiemen per meter); mmolc·L^{-1}, millimole of charge of each cation or anion listed.

Treatment	1	2	3	4	5
EC (dS·m^{-1})	3.1	7.2	12.7	18.4	24.0
Ion Concentration in mmolc·L^{-1}					
Ca^{2+}	6.4	19.2	25.0	29.4	28.4
Mg^{2+}	4.0	14.3	24.1	40.7	58.5
Na^+	15.5	54.2	101	169	229
K^+	6.4	6.4	6.2	6.4	6.6
SO_4^{2-}	15.3	53.3	85.0	132	182
Cl^-	8.0	31.8	62.9	104	133
PO_4^{3-}	0.3	0.3	0.3	0.4	0.5
NO_3^-	5.5	5.6	5.5	6.0	6.0

All salinity levels had the following added nutrients, (in mmolc·L^{-1}): 0.3 KH_2PO_4, 5.0 KNO_3, 3.1 $MgSO_4.7H_2O$, 3.0 $CaCl_2$, and 1.0 KCl. Table modified from [19]. Highest salinity level (30 dS·m^{-1}) not shown as all plants died at this level.

The levels of the macronutrients N, P, K, Ca, Mg, and total S, and of the micronutrients Fe, Cu, Mn, Zn, and Mo were determined from nitric acid digestions of the dried and ground plant material using Inductively Coupled Plasma Optical Emission Spectrometry (ICP-OES, 3300DV, Perkin-Elmer Corp., Waltham, MA, USA). There was insufficient plant material to analyze samples from the EC = 24 dS·m^{-1} treatment at 84 DAS, and there are no data from the EC = 30 dS·m^{-1} treatment as all plants died at this salinity level.

2.3. Oxygen Radical Absorbance Capacity (ORAC) and Total Phenolics (TP) Analyses

Ground dried samples (0.5 g) of alfalfa tops were mixed with 5 g of sand. Each mixture was then extracted in a pressurized stainless steel cell (ASE 350, Thermo Scientific/Dionex, Sunnyvale, CA, USA) using hexane to extract the lipophilic fraction and acetone:water:acetic acid (70:29.5:0.5 by volume) for the hydrophilic fraction. The extraction time was 5 min, followed by a 100% flush, a 60-s purge with 2 cycles, at 80 °C and 1500 psi. The hexane extract was evaporated to dryness with nitrogen in an evaporator (N-EVAP, Organomation, Berlin, MA, USA) at 37 °C and then redissolved in 10 mL of pure acetone; a 50-μL aliquot was collected for dilution and lipophilic ORAC analysis. After extraction with aqueous acetone by the ASE 350, the samples were made up to a volume of 25 mL in the acetone-water-acetic acid solution. A 150-μL aliquot of the aqueous acetone extracts was diluted for hydrophilic ORAC analysis. The ORAC assay is based on the inhibition of the peroxyl-radical-induced oxidation initiated by thermal decomposition of azo-compounds such as [2,2′-azobis(2-amidino-propane) dihydrochloride (AAPH)] [22]. Samples were analyzed for their antioxidant capacity (ORAC) in triplicate. The same ASE 350 aqueous acetone extracts were used for quantification of TP according to the Folin-Ciocalteu method [23,24] using gallic acid (cat. No. 398225, Sigma-Aldrich, Saint Louis, MO, USA) as the standard. A 20-μL aliquot of the extracts or a gallic acid standard solution was pipetted into a cell of a 96-cell microplate, followed by the addition of 100 μL of 0.4 N Folin Ciocalteu phenol reagent (stock solution F9252, Sigma-Aldrich, Saint Louis, MO, USA) and the addition of 80 μL of 0.94 M Na_2CO_3. The plate was covered with a plastic plate cover and allowed to develop color for 5 min at 50 °C. The absorbance was read at 765 nm using a microplate spectrophotometer (xMark™, BIO-RAD, Hercules, CA, USA).

2.4. Forage Quality

Shoots were dried at 60 °C for 48 h. Samples were ground to a size of 1.0 mm and analyzed for acid detergent fiber (ADF), neutral detergent fiber (NDF), and moisture by an independent laboratory (Analytical Feed & Food Laboratory, Visalia, CA, USA), according to AOAC International Methodology [25]. The parameters and analytical methods used were AOAC 973.18 for ADF, AOAC 2002.04 for NDF, and AOAC 930.15 for moisture. The parameters calculated according to ADF, NDF, and/or moisture include the net energy for lactation (NEL), calculated as NEL = 0.8611 − (0.00835 × ADF); relative feed value (RFV), calculated as RFV = (DMD × DMI)/1.29; dry matter intake (DMI), calculated as DMI = 120/NDF; and dry matter digestibility (DMD), calculated as DMD = 88.9 − (0.779 × ADF), according to National Forage Testing Association [26]. Crude protein (CP) was estimated as N% × 6.25 [27]. Nitrogen was determined by sample combustion in pure oxygen and measured by thermal conductivity detection (AOAC, 2000; ID 990.03) using a Vario Pyro Cube® (Elementar Americas, Inc., Mt. Laurel, NJ, USA).

2.5. Statistical Analysis

The nutrient composition data for each harvest were analyzed using a split-plot procedure, with the following statistical model:

$$Y_{ijk} = \mu + S_j + R_i + C_k + (SC)_{jk} + \varepsilon_{ijk}$$

where R, S and C represent the replicates (i = 1,...4), salinity level (j = 1,...5), and cultivars (k = 1,...4) respectively. All effects were considered as fixed. Thus, Y_{ijk} is the response to replicate i in S_j and C_k, μ is the overall mean; and ε_{ijk} represents the random error. The significance in the split-plot design was calculated by deriving the mean squares in the analysis of variance using the InfoStat program [28] with a completely randomized design (CRD). The significance of the main plot (salinity, S) was tested by S > R (salinity inside replicate) as an experimental error of the main plot, and the mean square error was used to test significance of the subplot (C) and the interaction S × C (salinity per cultivar). The mean differences were determined using the Fisher LSD test at $p \leq 0.05$. Chemical analyses for forage parameters were performed on two samples per cultivar, which were combined to represent each salinity level ($n = 8$) per harvest. These data (Figure 1) were subjected to a one-way (salinity) ANOVA with means compared by the Fisher LSD test. For total phenolics (TP) and antioxidant capacity (ORAC) analyses, samples were analyzed in triplicate, where total phenolics were quantified from a gallic acid standard curve. The effects of salt as a main plot, cultivar as a subplot, and the interaction between salt and cultivar (salt × cultivar) for ORAC and TP concentrations were analyzed at $p \leq 0.05$ using the GLM procedure with a standard split-plot test format in SAS (version 9.3; SAS Institute, Cary, NC, USA). The differences in ORAC and TP between the two harvests were analyzed at $p \leq 0.05$ using the T-test procedure in SAS (version 9.3; SAS Institute, Cary, NC, USA).

3. Results

3.1. Forage Quality

The impact of salinity on forage quality, expressed as the mean of the four cultivars at each salinity level per harvest, is presented in Figure 1. The parameters used to evaluate forage quality include acid detergent fiber (ADF), neutral detergent fiber (NDF), net energy for lactation (NEL), crude protein (CP), and relative feed value (RFV).

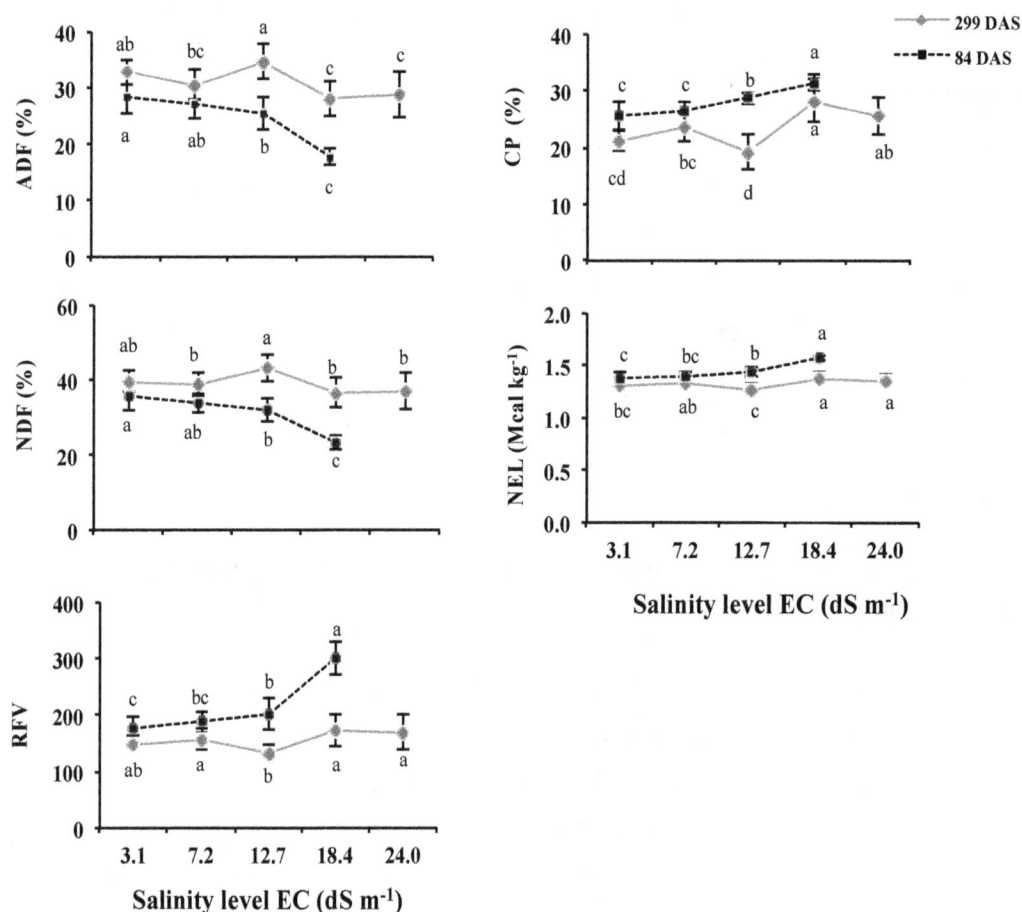

Figure 1. Impact of salinity increase on acid detergent fiber (ADF), neutral detergent fiber (NFD), net energy of lactation (NEL), crude protein (CP), and relative feed value (RFV) of salt-tolerant alfalfa. Data points represent the means (±SD) of the salinity-tolerant cultivars ($n = 8$). Means with the same letter are not significantly different according to a Fisher LSD test ($p \leq 0.05$). For the harvest at 84 DAS, the lack of data at 24 dS·m^{-1} was due to there being insufficient plant material for analysis because of growth limitations.

Salinity had a significant effect on the forage quality for both harvests ($p \leq 0.001$). At 84 DAS, there were no differences up to EC = 7.2 dS·m^{-1} for all parameters evaluated. Above that level, ADF and NDF decreased by approximately 8% and 9%, respectively, from 12.7 to 18.4 dS·m^{-1}. Consequently, the RFV (related to the ADF and NDF contents) increased sharply between those levels. CP increased by 5.2% from 7.2 to 18.4 dS·m^{-1} (Figure 1). In addition, the mean NEL increased as salinity increased. At 299 DAS, salinity also affected all forage parameters ($p \leq 0.05$). In contrast to 84

DAS, at 299 DAS significant differences between the control and salinity treatments generally were first observed at 12.7 dS·m^{-1} instead of at 7.2 dS·m^{-1} (Figure 1).

3.2. Nutrient Composition of Alfalfa

3.2.1. Macronutrients

The macronutrient (modified from [19]) data, including N and P, are expressed on a dry matter (DM) basis (Table 2). The main macronutrients found in alfalfa shoots (g·kg^{-1} DM) at both harvests were N, K, and Ca, while total S, Mg, and P were present at much lower levels (Table 2). Salinity had a significant effect on all macronutrients for both harvests, except for total S at 299 DAS. Nitrogen increased with salinity for both harvests, reaching levels that were significantly higher than those of the control at and above 12.7 dS·m^{-1} (84 DAS), and at and above 18.4 dS·m^{-1} (299 DAS). Shoot K decreased significantly ($P \leq 0.01$) for all cultivars and harvests as salinity increased. The calcium content remained constant up to 7.2 dS·m^{-1} (84 DAS) or up to 12 dS·m^{-1} (299 DAS), but decreased significantly for both harvests (more drastically at 299 DAS) as salinity increased. The Mg levels significantly increased for both harvests, with salinity, from the control to the highest level of salinity (84% and 48% increases for 84 DAS and 299 DAS, respectively). Sulfur concentrations increased with salinity, being significant ($p \leq 0.01$) at 84 DAS, but not at 299 DAS. Concentrations of P remained constant up to 12.7 dS·m^{-1}, but increased significantly ($p \leq 0.01$) above that salinity level for both harvests (Table 2). There was a significant ($p \leq 0.01$) cultivar effect for all macronutrients (except for N) at 84 DAS, while at 299 DAS, there was a significant cultivar effect only for Ca and Mg (both at $p \leq 0.05$). Both Na and Cl increased significantly ($p \leq 0.01$) in shoots with increasing salinity, but these and detailed data by cultivar and salinity are presented in a companion paper [19].

Table 2. Average macronutrients ($\pm$SE) in alfalfa shoot dry matter (DM) according to salinity levels. EC, electrical conductivity of irrigation water in deciSiemens per meter. ND, not determined (insufficient biomass). Modified from [19].

	N	P	K	Ca	Mg	Total S
			DM (g·kg^{-1})			
		EC dS·m^{-1} Second Harvest (84 DAS)				
3.1	40.8^{c} ± 1.43	2.6^{b} ± 0.09	46.4^{a} ± 1.05	14.1^{a} ± 0.4	2.6^{c} ± 0.14	3.5^{d} ± 0.08
7.2	42.1^{c} ± 1.04	2.7^{b} ± 0.09	41.4^{b} ± 0.94	13.5^{a} ± 0.5	2.7^{c} ± 0.16	3.9^{c} ± 0.10
12.7	46.0^{b} ± 0.56	2.9^{b} ± 0.08	38.6^{c} ± 0.62	13.0^{c} ± 0.69	3.4^{b} ± 0.22	4.8^{b} ± 0.20
18.4	50.5^{a} ± 0.80	3.8^{a} ± 0.13	34.3^{d} ± 0.88	12.1^{b} ± 0.24	4.8^{a} ± 0.07	7.4^{a} ± 0.17
24	ND	ND	ND	ND	ND	ND
		Seventh Harvest (299 DAS)				
3.1	34.1^{d} ± 1.07	3.4^{b} ± 0.17	40.3^{a} ± 1.12	18.0^{a} ± 0.51	2.5^{c} ± 0.08	3.8^{a} ± 0.12
7.2	37.6^{bc} ±1.37	3.1^{b} ± 0.06	30.4^{bc} ± 0.74	18.3^{a} ± 0.61	2.8^{bc} ± 0.12	4.6^{a} ± 0.20
12.7	30.8^{d} ± 1.77	2.8^{b} ± 0.14	31.0^{b} ± 0.68	16.7^{a} ± 0.51	3.2^{ab} ± 0.12	4.8^{a} ± 0.15
18.4	45.3^{a} ± 2.11	4.1^{a} ± 0.12	27.3^{cd} ± 0.56	12.1^{b} ± 0.45	3.0^{bc} ± 0.10	4.8^{a} ± 0.15
24	40.8^{a} ±1.92	4.3^{a} ±0.16	26.7^{d} ± 0.61	11.0^{b} ± 0.83	3.6^{a} ± 0.20	5.3^{a} ± 0.39

Different small letters within each column, and between EC levels, represent significantly different means according to Fisher's LSD test ($p \leq 0.05$), where $n = 16$ (except for N, $n = 8$) for EC levels.

3.2.2. Micronutrients

Shoot micronutrients analyzed for the four alfalfa cultivars were iron (Fe), copper (Cu), manganese (Mn), zinc (Zn), and molybdenum (Mo) (Table 3). At 84 DAS, there were no differences in mean Fe concentrations (ranging from 99.1 to 109.6 mg·kg^{-1} DM) or Cu (2.07–3.11 mg·kg^{-1} DM) as a function of increasing salinity (EC). Mean concentrations of Mn and Mo tended to increase with increasing salinity with significant ($p \leq 0.05$ and $p \leq 0.01$, respectively) differences between the control and the highest salinity level (18.4 dS·m^{-1}) at 84 DAS. There was a significant ($p \leq 0.01$) increase in Zn concentration at each level of salinity increase at 84 DAS. At 299 DAS, the Fe, Cu, Mn, and Zn levels remained mostly unchanged, but there was a small but significant ($p \leq 0.05$) decline (16%–28%) in the Fe levels between the 3.1 dS·m^{-1} control (116 mg·kg^{-1} DM) and the other saline treatments. Mn showed a transient increase of 42% (17.3 to 24.6 mg·kg^{-1} DM) as salinity increased from 3.1 to 7.2 dS·m^{-1}, and then declined to the salinity control levels. In general, the shoot Mo concentrations for all levels of salinity were significantly ($p \leq 0.05$) higher than those of the control (Table 3).

Table 3. Average micronutrient concentrations (±SE) in alfalfa shoot dry matter (DM), according to salinity levels. EC, electrical conductivity of irrigation water in deciSiemens per meter. ND, not determined (insufficient biomass).

	Fe	Cu	Mn	Zn	Mo
	DM (mg·kg^{-1})				
	EC dS·m^{-1} Second Harvest (84 DAS)				
3.1	104.0 [a] ± 6.29	2.1 [a] ± 0.27	25.5 [b] ± 3.38	40.9 [d] ± 1.32	2.0 [c] ± 0.09
7.2	99.1 [a] ± 4.90	2.3 [a] ± 0.10	31.7 [ab] ± 4.8	45.9 [c] ± 1.00	3.1 [b] ± 0.11
12.7	106.5 [a] ± 5.89	3.1 [a] ± 0.16	34.8 [a] ± 4.10	54.9 [b] ± 1.11	3.2 [b] ± 0.14
18.4	109.6 [a] ± 5.0	3.1 [a] ± 0.19	34.8 [a] ± 1.10	60.5 [a] ± 1.25	4.1 [a] ± 0.11
24	ND	ND	ND	ND	ND
	Seventh Harvest (299 DAS)				
3.1	116.1 [a] ± 6.35	5.8 [a] ±0.83	17.2 [b] ± 0.91	97.6 [a] ± 3.36	2.7 [c] ± 0.19
7.2	97.7 [b] ± 7.35	6.1 [a] ± 0.64	24.6 [a] ± 1.44	89.9 [a] ± 3.26	6.4 [a] ± 0.43
12.7	89.9 [b] ± 7.35	6.5 [a] ± 0.41	18.9 [b] ± 0.99	105.6 [a] ± 3.18	6.3 [a] ± 0.44
18.4	83.5 [b] ± 3.17	5.3 [a] ± 0.26	17.4 [b] ± 1.05	101.3 [a] ± 3.26	4.7 [c] ± 0.36
24	92.3 [b] ± 7.69	5.7 [a] ± 0.49	14.8 [b] ± 1.04	98.3 [a] ± 3.85	4.2 [c] ± 0.21

Different lower case letters within each column, and between EC levels, represent significantly different means according to Fisher's LSD test ($p \leq 0.05$), where $n = 16$.

3.3. Antioxidant Capacity of Alfalfa

Salinity had no effect ($p > 0.05$) on either the oxygen radical absorbance capacity (ORAC) or the total phenolic levels of the four alfalfa cultivars. The hydrophilic fractions of shoots had most (68%–99%) of the shoot total antioxidant capacity (Table 4). At early plant development (84 DAS), alfalfa shoots had hydrophilic ORAC (ORAC$_{Hydro}$) levels that ranged from 190–230 µmoles·TE·g^{-1} DM (Figure 2), while at 299 DAS, ORAC$_{Hydro}$ ranged from 229–274 µmoles·TE·g^{-1} DM, and the shoot total antioxidant capacity ranged from 244–287 µmoles·TE·g^{-1} DM (Figure 2, Table 4). Total phenolic (TP) concentrations ranged from 5.0–5.6 mg·GAE·g^{-1} DM for both harvests (Figure 2).

Table 4. Oxygen radical absorbance capacity of the lipophilic (ORAC$_{Lipo}$) and hydrophilic (ORAC$_{Hydro}$) fractions, and total antioxidant capacity (ORAC$_{Hydro}$ + ORAC$_{Lipo}$), in micromoles of trolox equivalents per gram of dry matter (μmoles·TE·g^{-1} DM) of alfalfa irrigated with water of different electrical conductivities (EC). Plants were sampled on 17 April 2012 (299 DAS). Data are means ± SE combined for the four cultivars with two replicated analyses per sample ($n = 8$).

EC	ORAC$_{Lipo}$	ORAC$_{Hydro}$	ORAC$_{Total}$
(dS·m^{-1})	(μmoles·TE·g^{-1} DM)		
3.1	15.0 ± 2.4	239.5 ± 12.8	254.5 ± 13.6
7.2	11.2 ± 1.5	252.1 ± 11.6	263.3 ± 11.0
12.7	13.4 ± 2.5	273.6 ± 14.3	286.9 ± 14.2
18.4	16.4 ± 1.7	268.4 ± 14.0	284.8 ± 15.5
24.0	15.3 ± 1.0	228.8 ± 18.3	244.1 ± 18.2

There was no effect of salinity (expressed as EC), cultivar, or the salt × cultivar interaction.

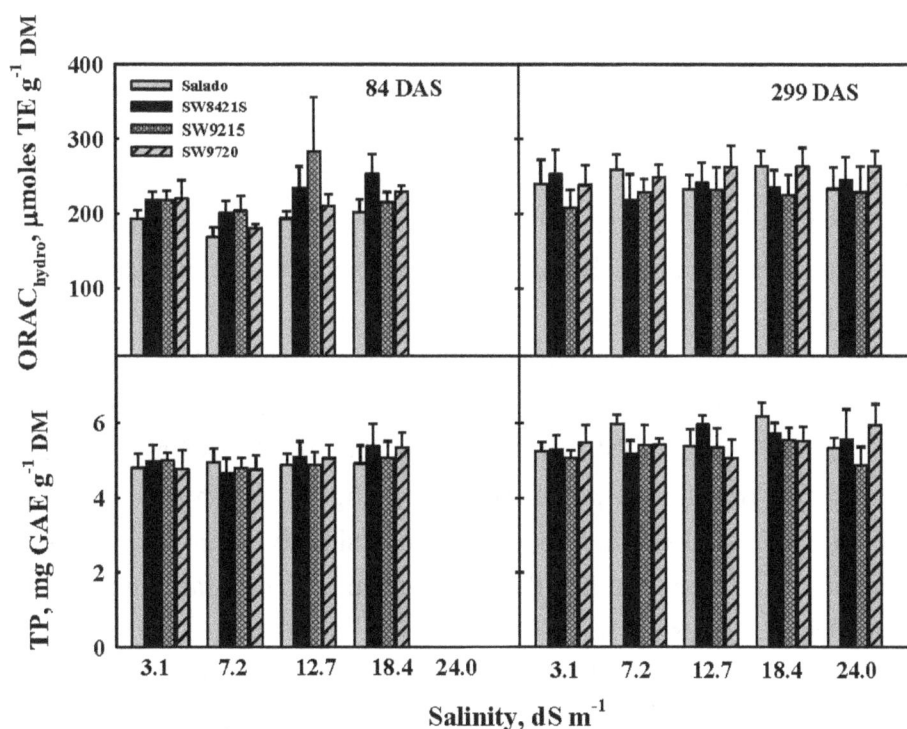

Figure 2. Total phenolics (TP) and hydrophilic shoot oxygen radical absorbance capacity (ORAC) of four salinity-tolerant alfalfa cultivars irrigated with saline water with different electrical conductivity levels. ORAC was measured in micromoles of trolox equivalents per gram of dry matter (μmoles·TE·g^{-1} DM). TP was measured as mg of gallic acid equivalents per gram of dry matter (mg·GAE·g^{-1} DM). Bars represent means (±SD), where $n = 4$. Plants were sampled at 84 and 299 days after sowing. For the harvest at 84 DAS, the lack of data at 24 dS·m^{-1} was due to growth limitations.

4. Discussion

4.1. Forage Quality

Forage quality was based on laboratory analyses of shoot biomass and evaluated in relation to recommended forage standards for livestock production output (e.g., milk, body weight gain) for animals consuming alfalfa of similar nutritional value and energy content [3,29]. Lower NDF translates into both increased DMI and milk yield within a forage family [3]. Regarding alfalfa protein, approximately 80% is degraded in the rumen of polygastric animals, but addition of tannins to alfalfa feed decreases rumen protein degradability and increases protein absorption [30].

Plant maturity is the main factor affecting forage quality [31], but the interaction between environmental and agronomic factors with maturity will influence the quality of alfalfa, even if harvested at the same stage of development [32]. Similarly, approaching harvest time, any stress that delays or accelerates alfalfa maturation affects the leaf-to-stem ratio and consequently, forage quality. The stems contain mostly structural components and are low in N, while the leaves contain mainly photosynthetic components and are richer in N than the stems. As a result, leaves have two to three times more CP than stems [33]. Increased leaf N leads to increased leaf area, thus increasing the leaf/stem ratio [34,35], but this could also be accounted for by the reduced stem height caused by salinity. The leaf-to-stem ratio increase leads to decreases in both ADF and NDF. Decreased ADF and NDF and increased shoot N lead to higher shoot CP levels in alfalfa irrigated with saline water. As reported in a previous study [19], plant height was significantly reduced by salinity only at 84 DAS, with the average difference in plant height between the control and EC = 18.4 dS·m^{-1} being 23 cm. Thus, we hypothesize that the decrease in height (shorter internodes) in salt-affected plants may have increased the leaf-to-stem ratio, shoot N, and CP by 61 g·kg^{-1} DM (6%). This decreased height of salt-affected plants also led to decreases in ADF and NDF of 107 and 122 g·kg^{-1} DM (10.7% and 12.2%) at 84 DAS and of 2.5% and 4% at 299 DAS, respectively, improving forage potential quality (Figure 1). This is in agreement with a previous report that salinity increased alfalfa leaf-to-stem ratio, slightly improving forage quality [36].

Al-Khatib and collaborators [7] reported that the leaf-to-stem ratio of alfalfa increased while forage mass decreased in response to increasing NaCl until 20 dS·m^{-1} (200 mM NaCl). At 299 DAS, there was also a significant increase in CP of 42.1 g·kg^{-1} DM (4.2%) between the control plants and those under 24 dS·m^{-1} (reflecting the increased accumulation of leaf N with increased salinity). This increase in CP was observed at both 84 and 299 DAS because N accumulation in shoots increased by 23% and 33%, respectively, in response to increased salinity (Table 2, Figure 1). Although plants had a fairly constant supply of N from NO_3^- in all irrigation treatments (Table 1), shoots significantly accumulated NO_3^--N, leading to higher CP. This could be due to morphological changes (e.g., increased leaf-to-stem ratio) under salinity stress or because the roots in the sand tanks were found to be associated with rhizobia. Despite the differences in developmental stages between the second and seventh harvest, there was a tendency for CP to increase with salinity levels up to 18.4 dS·m^{-1}. Although plants irrigated with salinity levels higher than the control had different stages of maturity, plant height has been used to predict forage parameters under field conditions [33,37].

Differences in forage parameters changed more sharply at 84 DAS (late summer) with salinity than at 299 DAS (early spring). These changes were likely caused by differences in climatic conditions combined with salinity [19]. Both climate and intervals between harvests (24 days before the second and 54 days before the seventh harvest) have a direct impact on maturity [33,38]. The RFV of alfalfa shoots in this experiment were similar to the values reported for alfalfa cultivars grown under field conditions with EC values ranging from 4–16 $dS \cdot m^{-1}$, although RFV did not change with salinity [39].

According to the classification of alfalfa hay [40], and judging from the parameters evaluated in this study, alfalfa herbage grown at the highest tolerated salinity fell within the "supreme" category. In comparison, forage grown at control salinity levels would be classified as "good" and "premium". Hence, our results indicated that forage quality improved with increasing salinity (despite some variation), independently of the changes between harvest seasons. Similar increases in CP and decreases in ADF in the salinity-tolerant cultivars Salado and SW9720 under salinity stress have been reported [9,11]. An increase in CP of alfalfa cultivars less tolerant to salinity was also reported when salinity increased from 2.1 to 7.8 $dS \cdot m^{-1}$ [41] or when salinity ranged from 0.3–4.5 $dS \cdot m^{-1}$ in one out of three years of cultivation [42]. Both drought and salinity restrict the growth of alfalfa, and mild drought also improves the forage quality of alfalfa [43]. These authors explained that the increase in quality with drought was due to a delay in plant maturation and an increase in the leaf-to-stem ratio; the latter is related to a reduction in stem length. However, the results of a 90-day pot experiment indicated that there were no differences in CP or N concentrations in alfalfa shoots when an EC of 15 $dS \cdot m^{-1}$ was applied using only NaCl [44].

The NEL values of alfalfa irrigated with increasing salinity, and ranging from 1.38–1.58 $Mcal \cdot kg^{-1}$ for the second harvest (84 DAS) and from 1.3 to 1.37 $Mcal \cdot kg^{-1}$ for the seventh harvest (299 DAS), were within the average (1.47 $Mcal \cdot kg^{-1}$) required for lactating cows [29], although some supplementation may be required to maintain the required energy levels.

4.2. Mineral Nutrient Composition

When irrigated with non-saline water, the predominant macronutrients in alfalfa are N, K, Ca, Mg, P, and S [45]. In our plants, which were fertilized to achieve the desired macro and micronutrients concentrations for ideal crop growth, and irrigated with saline water, the three main shoot macronutrients were also N, K, and Ca, followed by Cl and Na (data presented in [19]) and S, as these were added to the irrigation water to achieve high salinity, then followed by Mg and P at similar concentrations (Table 2). This suggests that alfalfa plants were provided adequate nutrients for growth, and our results express mostly the effects of salinity in a properly fertilized crop. The discussion on macro- and micronutrient requirements is based on the specifications for lactating dairy cattle provided by the Nutrient Requirements of Dairy Cattle [29]. The NRC requirement level for animals producing 35 kg $milk \cdot day^{-1}$ (Holstein or Jersey) was used, based on the average milk production for 2012 in California [46].

Macronutrients and sodium—Although adequate mineral nutrition alone will not prevent animal diseases, susceptibility to infectious diseases in response to malnourishment has been recognized for several centuries [47]. Thus, it is important to know if crop stress induced by salinity alters the nutrient composition of alfalfa.

The lowest Ca concentration in shoots in response to salinity (11 $g \cdot kg^{-1}$) was still above the daily dietary requirement (6.1 $g \cdot kg^{-1}$) for dairy cattle [29], while the highest Ca concentrations (18 $g \cdot kg^{-1}$)

were observed at ECs of 3.1 and 7.2 dS·m^{-1} at 299 DAS (Table 2). While dietary Ca concentrations above 10 g·kg^{-1} have been associated with reduced dry matter intake (Miller, 1983, in [29]), diets as high as 18 g·kg^{-1} have been fed to non-lactating dairy cows without problems (Beede *et al.*, 1991, in [29]). Feeding Ca in excess of daily dietary requirements is suggested to improve performance, mainly when cows are fed corn silage diets [29]. Potassium is the third most abundant element in mammals and is important for cellular osmotic balance. The cellular homeostasis of Na and K is maintained by Na^{+}/K^{+} pumps located inside the cell membrane. These two cations play an important role in electrical activity of nerve and muscle cells, in the acid-base balance, and in water retention. Potassium is a cofactor for the activation of enzymes, including those involved in protein synthesis and carbohydrate metabolism [48]. Because of increasing levels of Cl^{-} in irrigation water, shoot absorption of potassium decreased significantly ($p \leq 0.01$) for both harvests (by 26%–33%). Sodium significantly increased (by 60%), both with salinity and harvest date (presented elsewhere [19]), which was expected due to its elevated concentration in the saline treatment water. The levels of K across harvests and salinity (2.6%–4.6%) were well above the required levels (1.04%) for average lactating cows [29]. However, diets supplemented with potassium carbonate increased K from 1.6% to 4.6% (w/w) and decreased milk yield and feed intake [49]. Thus, K levels in alfalfa shoots irrigated with saline water containing 6 to 6.5 mmolc·L^{-1} could be of concern, depending on forage intake.

A continuous supply of Mg from feed is desirable because a high K level in forage decreases Mg absorption from the rumen and can lead to tetany [50]. The frequency of tetany in cows, triggered by low Mg and/or Ca, and high K in forage, increases when the ratio of K: (Ca + Mg) exceeds 2.2 [51]. In our results, the ratio of K: (Ca + Mg) was higher than 2.2 at 84 DAS, but lower than 2.2 at 299 DAS, suggesting that Mg levels should be monitored in alfalfa irrigated with saline water. Thus, although our results indicate that salinity can lead to a small, but significant accumulation of Mg by alfalfa shoots, Mg supplementation is still a must due to its poor absorption (13% to 16% from ration) by cows [52].

Sulfur (S) is an important component of cysteine and methionine, of many enzymes, and of antioxidants such as glutathione and thioredoxin, but elevated concentrations of S in alfalfa shoots can be detrimental to animal feed intake and function. Although we discuss the concentrations of S in shoots of different ages, the saline water used here was sulfate-dominant to mimic the drainage waters of California's Central Valley. Thus, levels of S might not be of concern where waters are Cl^{-} dominant. However, the S levels in our experiment remained similar at 299 DAS across salinity treatments. The lack of significant S uptake at 299 DAS may be explained by cooler temperatures and lower evapotranspiration before that harvest. The S concentration in shoots ranged from 0.38%, at the lowest EC, to 0.54% at the highest EC observed at 299 DAS. Regardless of season, a decrease in S in a later harvest (as seen here) was reported previously for alfalfa irrigated with sulfate-dominant water at both 15 and 25 dS·m^{-1} [53]. The authors reported an S range in alfalfa of 0.5%–0.9% at 25 dS·m^{-1}. In the S range recorded at 299 DAS for this study, and considering that the average consumption of alfalfa is 4.26 kg·cow^{-1} [3], the S consumption would be 16.2 to 23.0 g·day^{-1}, well below the 32 g S·day^{-1} upper limit recommended for a mature grazing beef cow [54], but 1.9 to 2.7 times above the 8.52 g S·day^{-1} (0.2% S/day) required for dairy cows [29]. Although no S toxicity has been reported [29], it is important to balance the diet in order to maintain S intake at a safe level (below 0.4% of DM daily), as levels of S of 0.4% in bailed alfalfa can lead to molybdenosis and reduced uptake of Cu and Se in beef cattle if alfalfa is the only source of feed [45].

The P requirement in the daily diet of average-producing dairy cows is 0.35% [29], but P levels regarded as adequate in alfalfa shoots are 0.08% to 0.15% [45]. P deficiency will lead to osteomalacia (softening of the bones) and fragile bones. The average levels of P in our alfalfa shoots at 299 DAS (0.28% to 0.44% DW) are considered to be high for shoot levels, relative to alfalfa grown in soils of the Mediterranean and desert zones [45]. In addition, according to nutrient tables presented by these authors, our Mg levels (0.25%–0.37%) were adequate, while shoot K and S were high.

Salinity significantly increased Na and Cl levels for both harvest dates by 40%–60%, as presented in a companion paper [19], resulting in shoot Na levels two to five times higher than the level required (0.23%) for average-producing lactating dairy cows [29]. Our data showed that alfalfa accumulates more Na and Cl$^-$ over time, even at the same irrigation salinity level. As previously reported [19], shoot Na ranged from 3.5–10 g·kg^{-1}, and Cl from 7–14 g·kg^{-1}, across salinity levels and harvest times. We found no reference reporting Na toxicity to livestock, but increasing Na in the diet from 5.5–8.8 g·kg^{-1} caused no reduction of feed intake, milk yield, or toxicity (Schneider *et al.* 1986, in [29]). NaCl, often added to feed mixes, can be tolerated up to 3% (lactating cows) or 4.5% (growing animals) of dietary dry matter. Thus, Na and Cl levels in alfalfa irrigated with saline water present no safety concern.

Micronutrients—Micronutrients and some vitamins are essential for animals to achieve optimal immune function, growth, and reproduction. Cattle can have sufficient amounts of these minerals for growth and reproduction, but not have enough for optimal immune function [47]. Examples are Cu and Zn, which are required for the activity of the antioxidant enzymes Cu-Zn superoxide dismutase (SOD) [55].

The average iron concentration was not affected by salinity and ranged from 83.5–116 mg·kg^{-1} across harvests, regardless of salinity treatment. Concentrations of 50 to 100 mg·kg^{-1} of Fe in a basal ration are within the requirements for the growth of grazing cattle [47,56] and concentrations of 15 mg·kg^{-1} in daily feed are recommended for average lactating cows [29]. Iron is essential for the formation of new red blood cells and only levels ≥4000 mg·kg^{-1} affect weight gain and cause diarrhea in young calves [47].

Copper (Cu) and zinc (Zn) are important micronutrients for immune function, and levels of 20 mg·kg^{-1} Cu and 40–60 mg·kg^{-1} Zn were suggested as optimal for feeding in the total diet of dairy cattle [57], while levels of 11 mg·kg^{-1} Cu and 48 mg·kg^{-1} Zn are recommended for average lactating dairy cows [29]. The Cu levels found in shoots for both harvests were below 7.0 mg·kg^{-1}, indicating the need for supplementation. In addition, the ratio of Cu to Mo in shoots was always approximately 1:1, well below the ratio of 10:1 that is considered a threshold for potential Cu toxicity [58].

Salinity significantly increased the Zn concentration in young plants (84 DAS) but not in established alfalfa plants (299 DAS), with concentrations ranging from 90–106 mg·kg^{-1}. Considering that a minimum Zn concentration of 48 mg·kg^{-1} is required for average lactating cows [29], our plants contained levels more than adequate to support a healthy immune function in livestock [57]. Manganese levels in alfalfa shoots were the third highest, after Fe and Zn. Manganese is important for its role in enzymatic systems but it is poorly absorbed (14%–18%) and if deficient, can reduce fertility and delay estrous [56]. This author mentions that Mn deficiency can lead to abortion and deformed calves at birth, but elevated Mn in the diet is generally not toxic. Levels of Mn in our alfalfa cultivars were at least 14 mg·kg^{-1}, as recommended for average lactating cows (NRC 2001). However, considering the poor absorption of Mn, mineral supplementation would be recommended.

4.3. Antioxidant Capacity of Alfalfa

Antioxidant flavonoids in the diet are believed to have health-promoting benefits to both humans and animals. In addition to protein, alfalfa is a rich source of flavonoid antioxidants and phytoestrogens including luteolin, coumestrol, and apigenin [59]. Phenolic compounds (including flavonoids) protect plants against the damaging effects of excessive reactive oxygen species (ROS) triggered by abiotic stresses, including salinity [60,61]. Although oxygen radical absorbance capacity (ORAC) has been widely accepted by industry to gauge the total antioxidant capacity of fruits, vegetables, spices, and other items consumed by humans, ORAC has only recently been used to estimate the antioxidant capacity of plants destined for livestock consumption [62–64]. The total antioxidant capacity is the sum of the lipophilic (ORAC$_{Lipo}$) and hydrophilic (ORAC$_{Hydro}$) fractions extracted from plants by hexane (lipophilic) and 70:30 acetone:aqueous buffer (hydrophilic). Our ORAC data (Table 4) confirmed those of others [63,64] who reported that the hydrophilic fractions of plant extracts contain most (68%–99%) of the total antioxidant capacity of shoots. Alfalfa shoots grown with saline water had 94%–96% of the total antioxidant capacity in the hydrophilic fraction with only 4%–6% in the lipophilic fraction, indicating that alfalfa shoots are low in lipophilic antioxidants such as tocopherols, carotenes, and fatty acids. The oven-dried alfalfa plants in our study had ORAC$_{Hydro}$ values that ranged from 229–274 μmoles·TE·g^{-1} DM (Table 4, Figure 2). Although these values may seem small compared with those of other leguminous forages, such as *Lespedeza cuneata* (ORAC$_{Hydro}$ = 530 μmoles·TE·g^{-1} DM), previously reported [63] alfalfa flavonoids and isoflavonoids present in hydrophilic (aqueous) extracts reduced oxidative stress and exerted hepatoprotective activity in rats treated with the liver-damaging compound carbon tetrachloride [65]. These results indicate that when animals consume alfalfa on a regular basis, it can provide benefits other than nutritional value.

The values for both ORAC and total phenolics (TP) remained unaltered by increased salinity, without differences for either ORAC or TP among cultivars (Figure 2). Our results agree with a previous report where there were no differences in antioxidant compounds among different cultivars of alfalfa in the absence of salt stress [15]. These authors also reported that the major antioxidants in alfalfa shoots, determined by HPLC, were tricin and apigenin glycosides (each approximately 40% of the total HPLC peaks), and luteolin and chrysoeriol glycosides (10% or less of the total HPLC peaks). Our results suggest that the salinity levels tested did not highly stress these salt-tolerant alfalfa cultivars. Previously, mostly the aglycons (flavonoids stripped of sugar moieties by acidic or enzymatic hydrolysis) have been determined, but the determination of full glycosidic forms (flavonoid plus sugar moieties) has also been conducted [59]. Flavonoids from alfalfa have the typical structure of several other flavonoids reported as beneficial to human diets and found in fruits and vegetables. Although sun drying (used to produce alfalfa hay) drastically decreased the antioxidant capacity of the antioxidant herb *Artemisia annua*, oven drying at 45 °C only slightly reduced the antioxidant capacity compared with freeze drying [66]. Thus, we consider that our oven-dried alfalfa shoots had an antioxidant capacity close to that of freeze-dried (or fresh) shoots. We could not find any published work on the antioxidant capacity of alfalfa shoots determined by ORAC or TP, except that the total ORAC (ORAC$_{Hydro+Lipo}$) of alfalfa hay was 171 μmoles·TE·g^{-1}, and the ORAC$_{Lipo}$ was only 3% of the total ORAC [63]. The antioxidant capacity of all cultivars used here was not affected by salinity, thus expanding the value of alfalfa beyond its contents of CP and minerals.

Although the value of antioxidants in animal and human nutrition is still debated by some, several benefits (e.g., anti-cancer, anti-inflammatory, *etc.*) of antioxidant-rich diets have been proposed. Dairy cows supplemented daily with 500 g of oregano (2082 μmoles·TE·g^{-1} DM) increased their milk fat concentration, feed and milk NEL efficiencies, and fat-corrected milk yield by 3.5% [67]. Although oregano has an ORAC value 8–9 fold higher than our oven-dried alfalfa shoots (225 to 256 μmoles·TE·g^{-1}), the average consumption of alfalfa shoots by cows is 5.4 kg·day^{-1}, which is 10-fold higher than the 500 g·day^{-1} oregano supplement from the above-mentioned study. Thus, daily alfalfa consumption can provide as much antioxidant flavonoid intake as oregano, thus adding to the forage value of alfalfa.

5. Conclusions

The effect of salinity in irrigation water on the suitability of alfalfa as a forage was based on shoot levels of macro- and micronutrients, and the forage quality estimated from ADF, NDF, and CP. Additional forage value was based on the antioxidant capacity and total phenolics in response to salinity. The nutrient composition of alfalfa can vary with salinity. Although our saline irrigation waters provided 27%–87% more SO$_4$ than Cl and 60%–94% more Na than Cl, alfalfa shoots contained 20%–190% more Cl than total S and 20%–120% more Cl than Na. Although Na and Cl in shoots increased with salinity, reducing the K concentration by 26%–32% and Ca by 15%–32% in shoots, shoot K and Ca were considered high and adequate [1,45], respectively, at all salinity levels. Increased salinity also increased shoot N (23%–33%), P (21%–46%), Mg (20%–84%), and total S (100%–110%) for both harvests. In general, the levels of macro- and micronutrients were adequate or high for alfalfa forage [1,29,45] regardless of salinity. However, when irrigation water was sulfate-dominant, the S concentrations in alfalfa were close to the upper limits recommended for safe animal consumption and require monitoring for water EC higher than 12.7 dS·m^{-1}. Regarding forage potential quality, shoots from plants irrigated with salinity levels higher than the control remained unaltered, or slightly improved compared with the salinity control levels, with NDF and CP at levels recommended for various classes of milking cows, but below the NDF values required for bulls and dry cows [39]. The antioxidant capacity was 15–23 fold higher for hydrophilic than for lipophilic fractions, but remained mostly unaltered by salinity, indicating that total antioxidant compounds, including phenolics and flavonoids (postulated to neutralize reactive oxygen species triggered by salinity stress), may remain fairly constant in alfalfa cultivars that are tolerant to salinity. These constant antioxidant levels, regardless of salinity stress, may play an extra beneficial role in helping to maintain animal health, as accepted for antioxidants in humans. Except for numeric values (such as reduced K and increased S), salinity levels up to 24 dS·m^{-1} did not alter the potential nutritional value and antioxidant capacity of alfalfa for livestock. The nutrient composition and antioxidant capacity of alfalfa are expected to play a dual role in the maintenance of health, body index, and milk production in dairy cows. This is the first report we are aware of, which has determined the total antioxidant capacity of alfalfa in response to salinity. Further studies involving animal performance are required to confirm the potential feed value of salt-stressed alfalfa under field conditions.

Acknowledgments

We acknowledge Nedda Saremi for help with the macro- and micronutrient analyses and Nahid Vishteh for determining the chemical composition of the saline water used in this study.

Author Contributions

Jorge Ferreira was responsible for the antioxidant method (ORAC), the data interpretation and discussion of forage nutritional value and antioxidant capacity, and the writing of the manuscript with Monica Cornacchione and Donald Suarez. Monica Cornacchione conducted the experiments, analyzed the data, and helped write the manuscript. Xuan Liu performed the tests for antioxidant activity (ORAC) and total phenolics (TP) and helped write the experimental section. Donald Suarez developed the experimental design, including the composition of the saline water, and assisted with the writing of the manuscript.

Conflicts of Interest

The authors declare no conflict of interest. The mention of proprietary brands and names is solely for the convenience of the reader and does not imply endorsement by the authors or the USDA versus similar products. The USDA is an equal-opportunity employer.

Abbreviations

ADF, acid detergent fiber; NDF, neutral detergent fiber; NEL, net energy for lactation; CP, crude protein; RFV, relative feed value; ORAC, oxygen radical absorbance capacity; TP, total phenolics.

References

1. Radović, J.; Sokolović, D.; Marković, J. Alfalfa—Most important perenial forage legume in animal husbandry. *Biotechnol. Anim. Husb.* **2009**, *25*, 465–475.
2. USDA-ARS. Roadmap for Alfalfa Research. Available online: http://ars.usda.gov/SP2UserFiles/Place/54281000/alfalfaroadmap2.pdf (accessed on 11 March 2014).
3. DePeters, E. Forage Quality: Important Attributes & Changes on the Horizon. In the Proceedings of California Alfalfa and Grains Symposium, Sacramento, CA, USA, 10–12 December 2012. UC Cooperative Extension, Plant Sciences Department, University of California, Davis: Davis, CA, USA, 2012.
4. Khorasani, G.R.; Janzen, R.A.; McGill, W.B.; Kennelly, J.J. Site and extent of mineral absorption in lactating cows fed whole-crop cereal grain silage of alfalfa silage. *J. Anim.Sci.* **1997**, *75*, 239–248.
5. Smedema, L.K.; Shiati, K. Irrigation and salinity: A perspective review of the salinity hazards of irrigation development in the arid zone. *Irrig. Drain. Syst.* **2002**, *16*, 161–174.
6. Ashraf, M.; Harris, P.J.C. Potential biochemical indicators of salinity tolerance in plants. *Plant Sci.* **2004**, *166*, 3–16.

7. Al-Khatib, M.; McNeilly, T.; Collins, J.C. The potential of selection and breeding for improved salt tolerance in lucerne (*Medicago sativa* L.). *Euphytica* **1992**, *65*, 43–51.

8. Mass, E.V.; Grattan, S.R. Crop yields as affected by salinity. In *Agricultural Drainage*; Agron. Monograph 38; Skaggs, R.W., van Schilfgaarde, J., Eds.; ASA, CSSA, SSA: Madison, WI, USA, 1999; pp. 55–108.

9. Robinson, P.H.; Grattan, S.R.; Getachew, G.; Grieve, C.M.; Poss, J.A.; Suarez, D.L.; Benes, S.E. Biomass accumulation and potential nutritive value of some forages irrigated with saline-sodic drainage water. *Anim. Feed Sci. Technol.* **2004**, *111*, 175–189.

10. Grattan, S.R.; Grieve, C.M.; Poss, J.A.; Robinson, P.H.; Suarez, D.L.; Benes, S.E. Evaluation of salt-tolerant forages for sequential water reuse systems: I. Biomass production. *Agric. Water Manag.* **2004**, *70*, 109–120.

11. Suyama, H.; Benes, S.E.; Robinson, P.H.; Grattan, S.R.; Grieve, C.M.; Getachew, G. Forage yield and quality under irrigation with saline-sodic drainage water: Greenhouse evaluation. *Agric. Water Manage.* **2007**, *88*, 159–172.

12. Steppuhn, H.; Acharya, S.N.; Iwaasa, A.D.; Gruber, M.; Miller, D.R. Inherent responses to root-zone salinity in nine alfalfa populations. *Can. J. Plant Sci.* **2012**, *92*, 235–248.

13. Rubio, M.C.; Bustos-Sanmamed, P.; Clemente, M.R.; Becana, M. Effects of salt stress on the expression of antioxidant genes and proteins in the model legume *Lotus japonicus*. *New Phytol.* **2009**, *181*, 851–859.

14. Mhadhbi, H.; Fotopoulos, V.; Mylona, P.V.; Jebara, M.; Elarbi Aouani, M.; Polidoros, A.N. Antioxidant gene–enzyme responses in *Medicago truncatula* genotypes with different degree of sensitivity to salinity. *Physiol. Plant.* **2011**, *141*, 201–214.

15. Stochmal, A.; Oleszek, W. Seasonal and structural changes in flavones in alfalfa (*Medicago sativa*) aerial parts. *Int. J. Food Agric. Environ.* **2007**, *5*, 170–174.

16. Choi, K.C.; Hwang, J.M.; Bang, S.J.; Kim, B.T.; Kim, D.H.; Chae, M.; Lee, S.A.; Choi, G.J.; Kim, D.H.; Lee, J.C. Chloroform extract of alfalfa (*Medicago sativa*) inhibits lipopolysaccharide-induced inflammation by downregulating ERK/NF-κB signaling and cytokine production. *J. Medic. Food* **2013**, *16*, 410–420.

17. Bora, K.S.; Sharma, A. Phytochemical and pharmacological potential of *Medicago sativa*: A review. *Pharm. Biol.* **2011**, *49*, 211–220.

18. Agati, G.; Biricolti, S.; Guidi, L.; Ferrini, F.; Fini, A.; Tattini, M. The biosynthesis of flavonoids is enhanced similarly by UV radiation and root zone salinity in *L. vulgare* leaves. *J. Plant Physiol.* **2011**, *168*, 204–212.

19. Cornacchione, M.V.; Suarez, D.L. Emergence, forage production, and ion relations of alfalfa in response to saline waters. *Crop Sci.* **2015**, *55*, 444–457.

20. Suarez, D.L.; Simunek, J. Unsatchem: Unsaturated water and solute transport model with equilibrium and kinetic chemistry. *Soil Sci. Soc. Am. J.* **1997**, *61*, 1633–1646.

21. Kalu, B.A.; Fick, G. Quantifying morphological development of alfalfa for studies of herbage quality. *Crop Sci.* **1981**, *21*, 267–271.

22. Prior, R.L.; Hoang, H.; Gu, L.; Wu, X.; Bacchiocca, M.; Howard, L.; Hampsch-Woodill, M.; Huang, D.; Ou, B.; Jacob, R. Assays for hydrophilic and lipophilic antioxidant capacity [oxygen radical absorbance capacity (ORAC)] of plasma and other biological and food samples. *J. Agric. Food Chem.* **2003**, *51*, 3273–3279.

23. Singleton, V.L.; Rossi, J.A. Colorimetry of total phenolics with phosphomolybdic-phosphotungstic acid reagents. *Am. J. Enol. Vitic.* **1965**, *16*, 144–158.

24. Slinkard, K.; Singleton, V.L. Total phenol analysis: Automation and comparison with manual methods. *Am. J. Enol. Vitic.* **1977**, *28*, 49–55.

25. AOAC. *Official Methods of Analysis of AOAC International*, 17th ed.; Association of Official Analytical Chemists: Gaithersburg, MD, USA, 2000; p. 2000.

26. National Forage Testing Association. Forage Analysis Procedures. Available online: http://www.foragetesting.org/files/LaboratoryProcedures.pdf (accessed on 27 May 2013).

27. Atwater, W.O.; Bryant, A.P. *The Chemical Composition of American Food Materials*; USDA Office of Experiment Stations, Ed.; US Government Printing Office: Washington, DC, USA, 1906; p 87.

28. Di Rienzo, J.A.; Casanoves, F.; Balzarini, M.G.; González, L.; Tablada, M.; Robledo, C.W. Infostat. Grupo Infostat; FCA Universidad Nacional de Córdoba, Argentina. Available online: Http://www.Infostat.Com.Ar (accessed on 30 August 2013).

29. NRC. *Nutrient Requirements of Dairy Cattle*, 7th ed.; National Academy Press: Washington, DC, USA, 2001.

30. Getachew, G.; Pittroff, W.; DePeters, E.J.; Putnam, D.H.; Dandekar, A.; Goyal, S. Influence of tannic acid application on alfalfa hay: *In vitro* rumen fermentation, serum metabolites and nitrogen balance in sheep. *Animal* **2008**, *2*, 381–390.

31. Minson, D.J. In *Forage in Ruminant Nutrition*. Academic Press: San Diego, CA, USA, 1990; p. 463.

32. Buxton, D.R. Quality-related characteristics of forages as influenced by plant environment and agronomic factors. *Anim. Feed Sci. Technol.* **1996**, *59*, 37–49.

33. Putnam, D.H.; Robinson, P.; DePeters, E. Forage quality and testing. In *Irrigated Alfalfa Management for Mediterranean and Desert Zones*, Publication 3512; Summers, C.G., Putnam, D.H., Eds.; University of California/Agricultural and Natural Resources: Davis, CA, USA, 2008; pp. 241–264.

34. Lemaire, G.; Avice, J.C.; Kim, T.H.; Ourry, A. Developmental changes in shoot N dynamics of lucerne (*Medicago sativa* L.) in relation to leaf growth dynamics as a function of plant density and hierarchical position within the canopy. *J. Exp. Bot.* **2005**, *56*, 935–943.

35. Lemaire, G.; Khaity, M.; Onillon, B.; Allirand, J.M.; Chartier, M.; Gosse, G. Dynamics of accumulation and partitioning of N in leaves, stems and roots of lucerne (*Medicago sativa* L.) in a dense canopy. *Ann. Bot.* **1992**, *70*, 429–435.

36. Hoffman, G.J.; Maas, E.V.; Rawlins, S.L. Salinity-ozone interactive effects on alfalfa yield and water relations. *J. Environ. Qual.* **1975**, *4*, 326–331.

37. Mueller, S.C.; Teuber, L.R. Alfalfa growth and development. In *Irrigated Alfalfa Management for Mediterranean and Desert Zones*, Publication 3512; Summers, C.G., Putnam, D.H., Eds.; University of California/Agricultural and Natural Resources: Davis, CA, USA, 2008; pp. 31–38.

38. Orloff, S.B.; Putnam, D.H. Harvest strategies for alfalfa. In *Irrigated Alfalfa Management for Mediterranean and Desert Zones*, Publication 3512; Summers, C.G., Putnam, D.H., Eds.; University of California/Agricultural and Natural Resources: Davis, CA, USA, 2008; pp. 197–207.

39. Yurtseven, S. The nutrient and energy contents of medicago varieties growth in salt-affected soils of the harran plain. *Hayvansal Üretim* **2011**, *52*, 39–45.

40. USDA-CO, D.O.A.M.N.S. California hay report. Available online: http://www.ams.usda.gov/mnreports/ml_gr311.txt (accessed on 9 May 2013).

41. Hussain, G.; Al-Jaloud, A.A.; Ai-Shammary, S.F.; Karimulla, S. Effect of saline irrigation on the biomass yield, and the protein, nitrogen, phosphorus, and potassium composition of alfalfa in a pot experiment. *J. Plant Nutr.* **1995**, *18*, 2389–2408.

42. Isla, R.; Aragüés, R. Response of alfalfa (*Medicago sativa* L.) to diurnal and nocturnal saline sprinkler irrigations. I: Total dry matter and hay quality. *Irrig. Sci.* **2009**, *27*, 497–505.

43. Halim, R.A.; Buxton, D.R.; Hattendorf, M.J.; Carlson, R.E. Water-stress effects on alfalfa forage quality after adjustment for maturity differences. *Agron. J.* **1989**, *81*, 189–194.

44. Pessarakli, M.; Huber, J.T. Biomass production and protein synthesis by alfalfa under salt stress. *J. Plant Nutr.* **1991**, *14*, 283–293.

45. Meyer, R.D.; Marcum, D.B.; Orloff, S.B.; Schmierer, J.L. Alfalfa fertilization strategies. In *Irrigated Alfalfa Management for Mediterranean and Desert Zones*, Publication 3512; Summers, C.G., Putnam, D.H., Eds.; University of California/Agricultural and Natural Resources: Davis, CA, USA, 2008; pp. 73–87.

46. te Velde, G. Milking Jersey's vs. Holstein's on a Commercial Dairy in California: Milk Production, Feed Efficiency, Intake, Costs, and Advantages; BS, California Politechnic State University: San Luis Obispo, CA, USA, 2013.

47. Koong, L.-J.; Wise, M.B.; Barrick, E.R. Effect of elevated dietary levels of iron on the performance and blood constituents of calves. *J. Anim. Sci.* **1970**, *31*, 422–427.

48. Ammerman, C.B.; Goodrich, R.D. Advances in mineral nutrition in ruminants. *J. Anim. Sci.* **1983**, *57*, 519–533.

49. Fisher, L.J.; Dinn, N.; Tait, R.M.; Shelford, J.A. Effect of level of dietary potassium on the absorption and excretion of calcium and magnesium by lactating cows. *Can. J. Anim. Sci.* **1994**, *74*, 503–509.

50. Grattan, S.R.; Grieve, C.M.; Poss, J.A.; Robinson, P.H.; Suarez, D.L.; Benes, S.E. Evaluation of salt-tolerant forages for sequential water reuse systems: III. Potential implications for ruminant mineral nutrition. *Agric. Water Manage.* **2004**, *70*, 137–150.

51. Grunes, D.L.; Stout, P.R.; Brownell, J.R. Grass tetany of ruminants. In *Advances in Agronomy*; Brady, N.C., Ed.; Academic Press: London, UK, 1970; Volume 22, pp. 331–374.

52. Jittakhot, S.; Schonewille, J.T.; Wouterse, H.; Focker, E.J.; Yuangklang, C.; Beynen, A.C. Effect of high magnesium intake on apparent magnesium absorption in lactating cows. *Anim. Feed Sci. Technol.* **2004**, *113*, 53–60.

53. Grieve, C.M.; Poss, J.A.; Grattan, S.R.; Suarez, D.L.; Benes, S.E.; Robinson, P.H. Evaluation of salt-tolerant forages for sequential water reuse systems: II. Plant–ion relations. *Agric. Water Manage.* **2004**, *70*, 121–135.

54. Arthington, J. Know the Sulfur Content of Your Forage—Test It. Available online: http://rcrec-ona.ifas.ufl.edu/pdf/publications/ona-reports/2013/5%202013/or5-13.html (accessed on 9 May 2013)

55. Spears, J.W.; Weiss, W.P. Role of antioxidants and trace elements in health and immunity of transition dairy cows. *Vet. J.* **2008**, *176*, 70–76.

56. Corah, L. Trace mineral requirements of grazing cattle. *Anim. Feed Sci. Technol.* **1996**, *59*, 61–70.

57. Scaletti, R.W.; Amaral-Phillips, D.M.; Harmon, R.J. *Using Nutrition to Improve Immunity Against Disease in Dairy Cattle: Copper, Zinc, Selenium, and Vitamin E*; University of Kentucky: Lexington, KY, USA, 1999; pp. 1–4.

58. Jones, M.; van der Merwe, D. *Copper Toxicity in Sheep is on the Rise in Kansas and Nebraska*; Kansas State University/Veterinary Medical Teaching Hospital: Manhattan, KS, USA, 2008; p. 5.

59. Stochmal, A.; Piacente, S.; Pizza, C.; De Riccardis, F.; Leitz, R.; Oleszek, W. Alfalfa (*Medicago sativa* L.) flavonoids. 1. Apigenin and luteolin glycosides from aerial parts. *J. Agric. Food Chem.* **2001**, *49*, 753–758.

60. Petridis, A.; Therios, I.; Samouris, G.; Tananaki, C. Salinity-induced changes in phenolic compounds in leaves and roots of four olive cultivars (*Olea europaea* L.) and their relationship to antioxidant activity. *Environ. Experim. Bot.* **2012**, *79*, 37–43.

61. Tattini, M.; Remorini, D.; Pinelli, P.; Agati, G.; Saracini, E.; Traversi, M.L.; Massai, R. Morpho-anatomical, physiological and biochemical adjustments in response to root zone salinity stress and high solar radiation in two mediterranean evergreen shrubs, *Myrtus communis* and *Pistacia lentiscus*. *New Phytol.* **2006**, *170*, 779–794.

62. Brisibe, E.A.; Umoren, U.E.; Brisibe, F.; Magalhäes, P.M.; Ferreira, J.F.S.; Luthria, D.; Wu, X.; Prior, R.L. Nutritional characterisation and antioxidant capacity of different tissues of *Artemisia annua* L. *Food Chem.* **2009**, *115*, 1240–1246.

63. Ferreira, J.F.S. Artemisia Species in Small Ruminant Production: Their Potential Antioxidant and Anthelmintic Effects. In *Appalachian Workshop and Research Update: Improving Small Ruminant Grazing Practices*; Morales, M., Ed.; Mountain State University/USDA: Beaver, WV, USA, 2009; pp. 53–70.

64. Katiki, L.M.; Ferreira, J.F.S.; Gonzalez, J.M.; Zajac, A.M.; Lindsay, D.S.; Chagas, A.C.S.; Amarante, A.F.T. Anthelmintic effect of plant extracts containing condensed and hydrolyzable tannins on *Caenorhabditis elegans*, and their antioxidant capacity. *Vet. Parasitol.* **2013**, *192*, 218–227.

65. Al-Dosari, M.S. *In vitro* and *in vivo* antioxidant activity of alfalfa (*Medicago sativa* L.) on carbon tetrachloride intoxicated rats. *Am. J. Chin. Med.* **2012**, *40*, 779, doi:10.1142/S0192415X12500589.

66. Ferreira, J.F.S.; Luthria, D.L. Drying affects artemisinin, dihydroartemisinic acid, artemisinic acid, and the antioxidant capacity of *Artemisia annua* L. Leaves. *J. Agric. Food Chem.* **2010**, *58*, 1691–1698.

67. Tekippe, J.A.; Hristov, A.N.; Heyler, K.S.; Cassidy, T.W.; Zheljazkov, V.D.; Ferreira, J.F.S.; Karnati, S.K.; Varga, G.A. Rumen fermentation and production effects of *Origanum vulgare* L. Leaves in lactating dairy cows. *J. Dairy Sci.* **2011**, *94*, 5065–5079.

12

Finding Ways to Improve Australia's Food Security Situation

Quentin Farmar-Bowers

17 The Grange, East Malvern, Victoria 3145, Australia;
E-Mail: Quentin.farmarbowers@deakin.edu.au

Academic Editor: Stephen J. Herbert

Abstract: Although Australia exports more than half of its agricultural production, there are food security problems as the current food supply systems in Australia fail to deliver healthy diets to all Australians and fail to protect the natural resources on which they depend. In addition, the food systems create "collateral damage" to the natural environment including biodiversity loss. In coming decades, Australia's food supply systems will be increasingly challenged by resource price inflation and falling yields due to climate change. Government and business are aiming to increase production and agricultural exports. This will increase pressure on agricultural resources and exacerbate "collateral" damage to the environment. The Australian public has an ongoing interest in issues associated with the food systems including the environment, education, health and sustainability. A health-giving diet is essential for a full life and over a life-time people need food security. Currently economy development and social planning is undertaken through the pragmatic application of a set of ideas, such as relying on markets and deregulation, collectively referred to as neoliberalism. This paper contends that the neoliberal approach is not solving the current and developing problems in food security and agriculture more generally and suggests that more emphasis should be given to alternatives approaches. Seven alternatives approaches are suggested that could be used to identify gaps and guide the creation of overarching goals in economic development and social planning to improve food security and secure the other material goods and social arrangements that all Australians require to live full lives. However, changing large systems such as those involved in food supply is difficult because vested interests in the existing arrangements make the current systems resilient to change. There are a range of leverage points that have differing abilities to change systems. The paper points out that goals and information flows are good leverage points and suggests establishing overarching goals for the systems relevant to food and restructuring the flow of information about these systems will help reform the food supply systems in Australia.

Keywords: agriculture; Australia; diet; food security; natural resources; system change

1. Introduction

Australia is a continent comparable in size to the contiguous states of the United States but it is principally dry land with only the eastern seaboard and south-west Western Australia supporting arable farming. Despite this, Australia exports more than half of its agricultural production. This might give the impression that Australians are food secure and that agricultural production is sustainable; this is not the case. Australia's commercial food supply and agricultural systems have a negative impact on some people through, for example, poor diet leading to ill health, (an intra-generational equity issue) and have a negative impact on ecosystems through, for example, land and water degradation and biodiversity loss (an inter-generational equity issue).

These negative impacts are likely to be exacerbated by population growth which is predicted to double in sixty years [1], by increasing economic inequity and by the growing impacts of climate change; including changes to rainfall, with more heatwaves and greater storm intensities [2–4].

Reducing the negative impacts of the domestic food supply systems and agriculture more generally is the collective responsibility of the Australian Commonwealth and State governments and the industries concerned. This collective responsibility would be effectively discharged if these organisations collaborate on the development and implementation of joint goals and work from the perspective of the whole system. It is unlikely that credible long-term solution can be found by dealing with issues independently. It is also unlikely that the current approach to economic development and social planning that encourages reliance on markets and deregulation referred to as neoliberalism will be able to solve the food security issues.

The next five sections of this paper outline the issues and failings involved in the food systems in Australia. Section 7 concerns how these issues are being addressed and provides a very brief outline of neoliberalism; the approach being taken to manage these issues over the last decades. Section 8 asks the rhetorical question "do the systems related to food need to change" and goes on to outline seven alternatives approaches to neoliberalism that could be used to identify how the systems related to food ought to change. These approaches are: (1) human rights; (2) national securities; (3) human needs; (4) authentic happiness; (5) capabilities; (6) sustainable development; and (7) environmental ethics. Section 9 outlines ways of changing systems and suggests developing overarching goals and changing the flow of information about the systems related to food security. Section 10 is the conclusions.

2. Current Issues for Consumers in the Commercial Food Supply System

Although about half of surveyed Australian households grow some of their own food [5,6] the vast majority of Australians depend on the commercial food supply systems. The ability of families to eat well is thus substantially related to family income. About 5% of Australian families suffer food insecurity [7,8] and many suffer diet-related health problems through malnourishment [9], including those related to obesity [10–13] and diabetes [14]. In 2011–2012, 63% of Australian adults and 25% of children were classified as overweight or obese [15]. "Unfortunately our diet has changed pretty dramatically over 30 years. The composition [of] what we're eating has changed, the really high levels of sugar and fat in our food." [16].

There is evidence that some sectors, such as older Australians [17,18], Aboriginal and Torres Strait Islander people, rural and remote Australians [19] and Australians with low household incomes are more at risk than others. For instance, Ward *et al.* [20] indicated that low-income families in Adelaide would have to spend about 30% of household income on eating healthily whereas high-income households needed to spend about 10%. Barosh *et al.* [21] studied the inequity in western Sydney in purchasing a health and sustainable diet (H&S diet) and noted that "The most disadvantaged groups in the region, both at the neighbourhood and household level, experience the greatest inequality in affordability of the H & S diet."

There are long-term consequences for not eating well, notably for young children [22]. Poor diet is currently implicated in over half of all deaths in Australia [23]. Mathers *et al.* [24] in 1999 noted that "the most disadvantaged 20% of Australians have a mortality burden that is 35% higher than that for the least disadvantaged 20%". The total direct cost for overweight and obesity in 2005 was $21 billion with estimated indirect costs of $35.6 billion per year, resulting in an overall total annual cost of $56.6 billion [25].

This health situation is mirrored in other developed countries [26]. Although obesity occurs at every socio-economic level in society, Drewnowski and Darmon [27] found that "Highest rates of obesity and diabetes in the United States are found among the lower-income groups" and Levine [28] noted that obesity tracks poverty in the USA. The situation in Australia may be similar as Burns [29] noted a relationship between obesity and poverty. Siahpush *et al.* [30] found that prolonged financial stress was a predictor of obesity in Australia. The Australian health survey data show that more adult women living in areas of most dis-advantage were overweight or obese (63.8%) compared with women living in areas of least disadvantage (47.7%). This difference was not seen in data for men [11]. While the causes of the relationship between poverty and unhealthy diet are not clear it would seem that the relative high cost of a healthy diet to income (as suggested by the work of Ward *et al.* [20]) is likely to be a major factor.

3. Issues of Resource Degradation in Australian Agriculture

Agricultural development in Australia has long been associated with land and water degradation [31]. About two thirds of agricultural land in Australia is degraded [32]. The problems range from soil loss and salinization to wetland loss, degradation of river systems and biodiversity decline [32,33–38]. Australia has the worse extinction rate for mammals in the world (29 species since European settlement) and more than 1700 species of plants and animals are listed by the Australian Government as being of risk of extinction [39]. Land and water degradation damage agricultural productivity and also result in "collateral" damage to ecosystems services through reduced functionality of ecosystems and biodiversity loss, including in associated marine areas [40,41].

While some natural resource management issues are within the capacity of individual landholders to deal with, many, perhaps most issues are related to the operation of larger systems both social systems and ecological systems, very often in tandem. For example, social systems have promoted the development of irrigation schemes in Australia which have led to the degradation of river systems such as the

Murray-Darling system in eastern Australia [42]. This case demonstrates the social and environmental complexity of repairing this major system as a consequence of agricultural development and restoring productivity that will take decades and is likely to require perpetual large scale adjustments as well as individual landholder adaptation [43].

4. Future Issues

In addition to the continuation of current issues, increasing demand from the growing Australian population for agricultural products and competition for resources used in agriculture (such as land and water) and climate change will exacerbate these existing problems and introduce new problems.

The income poverty rate (below 60% of median income) in Australia has been consistently around 21% in the 2001–2010 decade. A capability approach based on Sen's capabilities approach, [44] indicated that marginal and deep social exclusion rates were about 25% in 2010 up a little from the decade low point in 2008 [45]. To some degree, the rate of poverty is related to the level of government support payments available [46,47]. There seems to be no progress on poverty reduction in the first decade of this century. This is a strong indication that the food issues related to poverty will continue.

The impact of climate change on agriculture is complex and likely to vary from year to year. Overall the physical production is likely to be reduced because of increasing severity of droughts, heatwaves, storms and changes in the pattern of rainfall (rainfall is likely to be less in much of Australia) [48–51]. Agriculture in other parts of the world is also impacted in similar ways [52].

Climate change will not only reduce the harvest from agriculture but policies to mitigate climate change must, in the long-term, be aimed at reducing the greenhouse gas pollution from agricultural production. This is because about 15% of greenhouse gas production in Australia is estimated to come from agricultural activities (enteric fermentation 65.1%, agricultural soils 17.8%, prescribed burning of savannahs 12.3% and manure management 3.9%) [53,54]. Mitigation policies must also reduce the greenhouse gases involved in the resources used in agriculture especially nitrogen fertilizers and fossil fuels and in the production of machinery and equipment used in agriculture and in the manufacturing and retail aspects of the food supply chains.

The current response to climate change in Australian agriculture is principally adaptation; incremental but also transformative change is being discussed [55]. There are moves by the current Australian government from 2013 onwards, to encourage carbon farming [56] to sequester carbon in soils and vegetation as a principal mitigation measure for climate change as part of their emissions reduction fund [57]. A review in 2013 [58] indicated that the potential of improved agricultural management practices to store carbon in soils was limited to the surface 10 cm of soil and diminished with time. They found that none of the carbon storage practices were financially attractive under Australia's Carbon Farming Initiative. The carbon farming initiative is now part of the emissions reduction fund.

Australian agriculture depends on oil and therefore production is vulnerable to future oil shocks and oil price rises [59–61]. Agricultural production depends on resource inputs such as irrigation water which is already overcommitted [62], phosphate and nitrogen fertilizers [63] and land. It is likely that prices for phosphates will rises because of increasing demand and supply limitation [64,65]. The price of nitrogen fertilizers is dependent on the cost of natural gas which has seen significant price fluctuations in recent times. It is possible that nitrogen fertilizer prices may stabilise with increasing gas availability and proposed increased manufacturing capacity [66,67].

Land is being lost to agricultural production through land degradation, such as salinity and erosion, but also through city expansion into peri-urban areas [68] and coal seam gas and mining [69]. The outcome of price rises in agricultural inputs, coupled with increasing domestic demand from Australia's rapidly growing population [1] and exports is likely to lead to food price inflation which will have a negative impact on the affordability of a healthy diet for Australians.

Australia's production of agricultural commodities is much greater than the demand from its population of 23 million, allowing Australia to export over half of the physical production of agriculture [70,71]. There are claims (which seem exaggerated) that Australian agriculture feeds about 60 million people world-wide [72]. Business and government want to expand exports further by increasing agricultural production to take advantage of the international demand for food from a growing middle class especially in Asia. The previous Australian Government; The Labor Rudd/Gillard Government 2007–2013, supported an increase in agricultural production and exports setting out their ideas in a National Food Plan [73,74]. The National Food Plan is no longer government policy but the idea of agricultural expansion is still supported. A controversial aspect of agricultural expansion is the proposed development of a "food bowl" in northern Australia [75–77] that has long been held as infeasible [78,79]. Expanding agricultural production (a proposed doubling) will greatly exacerbate the problems of "collateral" environmental damage [80], especially when farming moves into more marginal production areas and increases its use of resources, such as water, fertilisers and fuels.

5. Public Interests

The topics that catch and sustain the interest of the Australian public are the ones likely to receive political attention in the longer term. The Australian public's growing interest in food security and sovereignty [81,82].is not spawned by a single issue.

Current issues of public concern around food and diet can be grouped roughly under four headings: consumption, production, environmental and control issues.

Consumption issues include: diet-related health problems, social inequities, poverty, poor diets (especially for children), urbanisation and food deserts, population growth, food insecurity, food imports, making ends meet, ageing population, food waste, food contamination and food safety, premature deaths of Indigenous Australians, GM foods, food quality, food labelling, and food literacy.

Production issues include: vulnerability of production to climate change (droughts, floods, storms and heat waves), oil price rises (e.g., peak oil), fertiliser price rises (e.g., peak phosphate), land and water degradation and pollution, declining growth in agricultural productivity, over-fishing, loss of agricultural resources (land and water) to urbanisation and mining, biofuel production, aquifer pollution from coal seam gas production [83], over allocation of irrigation resources [84], static agricultural research budget, animal welfare, live animal exports, loss of the family farm, the exodus of your people from agriculture coupled with a low turnout of agricultural graduates.

Environmental issues include: river health, biodiversity loss, loss of forests and natural areas, loss of wetlands, by-catch in wild fisheries [85], damage to coral reefs and marine areas [86] and greenhouse gas production.

Control issues include: foreign ownership of agricultural land and water resources, foreign control of agribusinesses, super-trawlers, globalisation (with the increasing influence of international markets and the increasing power of multinational agribusinesses), supermarket control of food chains, and food providence information. Farmar-Bowers *et al.* [87] provides an overview of food security in Australia.

The public's concern and mistrust around these issues exists for a range of reasons. There is confusion in the general public about what really constitutes a health providing diet [88,89]. The increasing (physical and emotional) distance between farmers and urban consumers leads to a poor general appreciation of where food comes from and what foods might contain [90]. The nature of property rights and industrial agricultural production methods with a focus on profits suggest that producers have low levels of obligations to neighbours including consumers and the natural environment [91]. Industrial agricultural production methods used in Australia may not guarantee food security [92,93]. The sustainability of modern agricultural production is in question in Australia and internationally especially as intensification is likely to lead to a tradeoff between production and more environmental damage [94]. The level of relative poverty in Australia [46], especially among children, contrasts with traditional Australian values of mateship and empathy [95,96]. There is also concern in regard to the political influence of the super-rich and large corporations (including the market dominance of the super-market chains) in promoting their own affairs. Seemingly, governments have an economic focus on facilitating increased agricultural production for export with less concern about social and environmental issues at home [97]. There is also concern that the Australian government is not taking climate change seriously as it has dropped nearly all the existing arrangements for climate change mitigation in its first term [98].

6. How These Failing in the Food System Relate to Preferred Outcomes

The Australia food system seems to be failing on two counts: (1) in its ability to provide a healthy diet for all Australians; and (2) in its ability to sustain natural resources and protect biodiversity. In addition, it does not seem to be gearing up to deal with the changes that the future is known to bring, especially climate change, population growth and the increasing range of public concerns. Other things being equal, the Australian public would probably prefer to have heathy diets, produced in a way that sustains natural resources, maintains biodiversity and have agriculture and food systems able to copy with coming change.

Healthy diets: Ensuring Australia has a well-nourished population should be seen as an internationally accepted moral obligation and human right within the International Covenant on Economic, Social and Cultural Rights (ICESCR) [99] and also an economic necessity. Roetter and Van Keulen [100] noted that: "The ultimate aim….[of]….food security is to arrive at a healthy and well-nourished population that can take on, to the maximum of its capacities, the development of its own community, area or country". The Australian health ministers accepted that "Improving Australians' diets therefore has (sic) the potential to cut health care costs and improve quality of life" [101]. Although the costs aspects are important, they should be secondary or an additional reason for ensuring people have access to a health giving diet. Issues about health giving diets are reviewed by Katz and Meller [88].

Sustain natural resources: Maintaining current levels of production and moving towards sustainability depend inter alia on: (1) re-developing agriculture (and fisheries) in ways that conserves the natural resources on which production depends (e.g., soil and water quality and fish habitat and stocks);

(2) reducing agriculture's dependence on fossil fuels, Haber-Bosch nitrogen fertilizer and resources such as mined rock phosphate [102]; (3) reducing the carbon foot-print of the food systems; and (4) improving the control of pests and diseases and biosecurity. By protecting productivity, natural resource conservation has significant prudential value for the future nutrition of Australians and the agricultural economy [103]. Food and agricultural production needs to be carried out within planetary boundaries. This implies the need for considerable change in production methods and resource use [104].

Sustain biodiversity: Natural areas and native biodiversity in Australia has long been valued domestically and internationally. For example, in 2000, Australia was identified as one of 17 mega-diverse countries by the World Conservation Monitoring Centre. In 1993 Australia ratified the 1992 Convention on Biological Diversity and developed a national reserve system project. Australia is periodically adding reserves to the system. An important aspect of conservation is working to reduce agriculture's impact on the natural environment [105] and increase private conservation efforts [106]. In addition to Australia's international legal obligations, the natural environment is accepted by Australian society as having economic, social, cultural and national prestige values. The "collateral damage" to the natural environment from existing agricultural development continues and the pressure for expansion of commercial agriculture with increased resource use is substantial.

Overcoming resource and climate change problems: Australian agriculture is adaptable, notably in regard to weather variability, and there is scope for further adaptation [107,108]. However, there is a limit to adaptation and climate change is likely to lead to declining agricultural yields. Increasing resource costs will lead to higher product prices. These costs may be passed on easily to high value overseas markets but higher food prices may reduce the ability of Australians to purchase healthy diets. The option of selling overseas will tend to move Australian domestic food prices upwards to export parity. The proposed expansion of production is very likely to lead to an increased rate of damage to natural resources and biodiversity.

7. Will These Failings Lead to New Government Policies and Programs?

How important these failings are viewed by government now and in future and whether they lead to goal setting and policy development will depend on the political ideas in vogue at the time. Historically, recognition of their political importance has led to government action. For example, the social security and welfare benefits systems [109], help low income Australians buy food; the public health system [110], is paying for ameliorating some of the illnesses due to poor diet; government programs such as "Caring for our Country" [111], is making a contribution towards the restoration of land and water degraded by agricultural production and the national reserve system and environmental protection programs are funding initiatives to reduce agriculture's collateral damage to the environment [112]. Currently the Australian Government proposes a winding back of aspects of these programs, such as reducing or delaying welfare payments [113], introducing medical co-payments [114] and reducing action on climate change and environmental issues [115].

Resolution of the remaining issues is left to private citizens such as landholders for biodiversity conservation on private land, and non-government organisations, such as the Salvation Army, Food Bank Australia, Conservation Volunteers and commercial businesses through their corporate social responsibility programs.

In the last 30 years, neoliberal ideas [116,117] have become more dominant in all Australian governments' approach to public policy. Neoliberal ideas put faith in markets and the maintenance of the value of money (via interest rate control) to provide for the advancement of human welfare. Application of these ideas has led to deregulation, privatization, establishment of markets in many new areas, smaller government, the withdrawn of the state from the provision of many social programs and the acceptance of personal responsibility for events often even in the face of clear system failures. The practical application of neoliberal ideas has varied from the original theory across time and between countries making a clear definition of neoliberalism difficult. Despite this, neoliberal ideas have become the "norm" for thinking about problems and solutions in Australian society. Consequently intra-generational inequity (e.g., income inequity) and inter-generational inequity (e.g., low level of climate change mitigation) are considered to be "natural" with less urgency for ameliorative action from government as it is assumed that solutions will come from the "market".

The move towards embracing neoliberal ideas has led to the reduced relative spending power of lower wage earners and welfare recipients [118] and has consequences in health outcomes [119] Although increasing equity appears to be beneficial for social outcomes [120], Whiteford [121] suggested that "Rising inequality and rising prosperity are therefore characteristic of Australia's experience over recent decades". And "while all income groups have benefited from increases in real incomes, the benefits have gone mainly to the better off". Like Australia, inequity has increased in most OECD countries in the last 30 years [122,123]. Harvey [124] maintains that in a sense neoliberalism is about the concentration of wealth and power in the hands of a concentrated part of society and this power facilitates the perpetuation of policies that maintain this situation. Certainly, the neoliberal ideas in play in Australia have facilitated expansion of big business and wealth at the top of the social scale. They also result in food security, natural resources protection and biodiversity maintenance, being low on the Australian Government's agenda although there is lobbying by civil society and minor political parties [125] and a number of local government councils have food security plans and programs [126]. The Australian Government issued a green paper in 2014 on agricultural competitiveness and will release a white paper in late 2015. The green paper notes that "The Australian Government's agricultural policy is driven by one key objective: to achieve a better return at the farm gate to ensure a sustainable and competitive Australian agriculture sector" [127]. The Australian Government has nine policy principles related to Agriculture which are set out in their green paper on Agricultural Competitiveness; the ninth principle is "Maintains access for all Australians to high-quality and affordable fresh food" [127]. The green paper relies on neoliberal ideas by suggesting the current issues of food insecurity in Australia are problems the individual should deal with. "Australia has a high level of food security due to our income level and trade surplus in food. Australia is ranked the 15th most food secure country on the assessment of 109 nations by the Economist Intelligence Unit [128]. Pockets of food insecurity for some individuals and communities in Australia remain due to low income or remoteness." [127].

Issues relating to food/health, natural resources and biodiversity seem important for the future. If we rely on neoliberal ideas then it would seem that we are passing responsibilities for "fixing" the failings of the food system listed in Section 6 above to the "market". There is nothing wrong with this if this approach is effective but evidence suggests otherwise. As a democracy, ideas about these vital issues and how to proceed should be on the public agenda for "good sense" debate. The following three questions are suggested as "debate starters":

- Do the systems related to food need to change? And if so,
- What levers might be useful in making the changes?'
- How can the people and organisations use the levers to make changes?'

8. Finding out "What Ought to be": Is System Change Needed?

This section addresses the first of the three questions above: Do the systems related to food need to change? The following section (Section 9) addresses the other two questions.

From a neoliberal perspective "what ought to be" is set out in the Federal government's green paper mentioned in above; basically better returns to agriculture (higher profits). The green paper is a step in a public discussion process. High on the requirements for a "good sense" public discussion is reliable and full information about the systems involved and about the criteria needed to evaluate the systems. Although there is information about the food systems and closely related systems such as employment, education, health, welfare, natural resources, information is needed on how the systems interact. However, just knowing about the systems involved and their impacts on people and the environment is not adequate; these has to be a moral objective or ethical systems to evaluate this information to identify gaps between current reality and "what ought to be". This implies that the most pressing issue now is deciding what moral or ethical systems should we use to decide "what ought to be" so we can identify the gaps that exist between "what ought to be" and the current situation. Should we rely on neoliberal ideas or are there other approaches such as moral objectives or ethical systems we can use. Once we have decided on an approach we can identify gaps and hence what overarching goals we should be develop to close these gaps. Once we have overarching goals, it may be possible to determine a suite of secondary goals for individual issues and develop programs to implement the goals. Developing and using appropriate goals is vital as goals help to direct and energise effort, encourage persistence, and also help people to use and develop skills and knowledge [129].

Many alternative ideas to neoliberalism exist that could be used to identified the gaps between "what is" and "what ought to be". Seven alternative sets of ideas from which we could identify gaps and hence develop overarching goals are considered in this paper. The alternative ideas considered are: (1) human rights; (2) national securities; (3) human needs; (4) authentic happiness; (5) capabilities; (6) sustainable development; and (7) environmental ethics. The distinction between these ideas is not always sharp and there are ways of using ideas together to determine gaps and overarching goals.

Human rights: The Australian Government ratified the 1966 International Covenant on Economic, Social and Cultural Rights in 1975. Article 11 Sections 1 and 2a and 2b, refers to food "right of everyone to an adequate standard of living for himself and his family, including adequate food, clothing and housing, and to the continuous improvement of living conditions" and Article 12 refers to health "the right of everyone to the enjoyment of the highest attainable standard of physical and mental health" [130]. Adequate food is thus a human right. As a consequence of the commitments made in 2011 during Australia's Universal Periodic Review at the United Nations and Australia's Human Rights Framework 2010, the government produced its third National Human Rights Action Plan in 2012 [131]. The plan addresses poverty as one of the 19 topics discussed which included mention of the "Food for All Tasmanians: A Food Security Strategy 2012" [132] developed as advice to the Tasmanian Government. The Action Plan also included references to health for Indigenous Australians in the

"Closing the Gap" program, and national food security was mentioned in terms of climate change programs. While the Tasmanian advisory strategy and the work with Indigenous Australians represent an important appreciation of the need for change, it seems that this approach to "human rights" is not proving to be the catalyst for substantial and rapid change in Australia.

National securities: Most countries have some form of national security agenda. The Australian National Security Statement [133] sets out five national security objectives. They concern; border integrity, political sovereignty, protecting Australia's national interests and promoting global stability in Australia's interests. The third objective of the five national security objectives is: "preserving a cohesive and resilient society and strong economy". As a flow on from this third national security objective, the Australian Government has developed a "Critical Infrastructure Resilience Strategy" which identifies seven critical sectors within the Australian economy that must not fail if this third national security objective is to be met [134]. The food chain is one of the seven critical infrastructures that must be maintained. "The aim of this Strategy is the continued operation of critical infrastructure in the face of all hazards, as this critical infrastructure supports Australia's national defense and national security, and underpins our economic prosperity and social wellbeing. More resilient critical infrastructure will also help to achieve the continued provision of essential services to the community" [134]. The other sectors that must be maintained are; energy, water, communications, transport, health and banking.

Although the food chain is only one of the seven critical infrastructures identified in the Critical Infrastructure Resilience Strategy, it is easy to see that the other sectors; energy, water, communication and transports, as well has having their own values, are essential for maintaining food supply chains, and food, in the sense of good nutrition, is essential for health (the remaining critical infrastructure objective). The critical infrastructure approach is relevant for government's planning for crises such as cyclones and acute financial problems that might occur in groups of people (such as farmers). Citizens, as well as government, would want these "infrastructures" to be maintained in a crisis but citizens probably want these "infrastructures" to be satisfied throughout our lives. So while the securities approach may be useful for crisis evaluation, in its current form, it may not provide useful criteria for evaluating ongoing day to day issues such as healthy diets and protection of natural resources for food production. In contrast to the situation with food, the "strong economy" aspect in the third objective of national security agenda is a central and daily focus of government, quite apart from its security aspect.

Human needs: Human needs were given a central position in the Brundtland report on sustainable development [135]. Max-Neef *et al.* [136] listed nine fundamental human needs as: subsistence, protection, affection, understanding, participation, leisure, creation, identity and freedom. Max-Neef's notion is that people or societies are "poor" when one or more of these needs are not met. "Security" is about ensuring each of these needs will always be fulfilled in the future. Most of these needs are psychological, but to achieve them we have to use materials and establish social arrangements and processes. For example, "leisure" requires places for recreation as well as the social acceptance of "free time".

Instead of trade-offs between needs, the "needs approach" asks us to seek synergies so that one action can secure as many human needs as possible. A balance is required in policy and within the functioning of society to avoid the situation in which the provision of a specific need hinders the provision of people's other needs. For example, too much "protection" could mean a police state (loss of freedom); too much "subsistence" in the form of high calorie food could mean obesity and ill health (loss of protection).

"Freedom" is an especially difficult need to balance; e.g., do we sanction freedom to burn fossil fuels or freedom to enjoy clean air and a stable climate.

In terms of psychological needs, "Self-determination Theory" postulates that social conditions can either facilitate or forestall the satisfaction of three innate psychological needs; competence, autonomy and relatedness. Self-determination theory holds that satisfaction of these needs enhances self-motivation and mental health, conversely, not meeting these needs leads to lower motivation and reduced well-being [137,138]. In other words, people need to have the feeling of competence (based on real rather than imagined competence), to be able to make and implement their own decisions and have friendships to be mentally healthy and actively motivated. Motivation and mental health impact people's activities in all areas of their lives. In terms of psychological well-being self-determination theory may provide useful criteria along with Max-Neef's fundamental human needs to evaluate the current situation to identify gaps and devise overarching goals.

Authentic Happiness: Perhaps "authentic happiness" would also provide a useful criterion for evaluating food systems as authentic happiness has a central role in people's well-being. But it has to be "authentic", that is real and not the consequence of mis-information; "happiness has no prudential payoff unless it is fully informed…." [139]. One is not going to be authentically happy if suffering a terminal illness as a consequence of an unhealthy diet. Haybron [140] supports this notion and proposes that welfare centres on self-fulfilment, part of which is authentic happiness. Haybron notes that: "A full-blooded account of welfare would likely incorporate goods other than happiness, depending on how we view our natures. Obviously important in this regard…..is a person's identity. And it is highly plausible that self-fulfilment will involve, not just being happy, but success as well in relation to those commitments that define who we are and lend meaning to our lives…..welfare consists partly …..on how things go with respect to the commitments that shape one's identify……we should consider incorporating identity-related fulfilments into an account of well-being as part of fulfilment" [140]. Haybron used the example of the loss of a spouse as a commitment but probably there are a wide range of commitments that would be "identity-related" such as having well-nourished children or conserving native biodiversity on your farm. Haybron [140] points out that well-being (and the authentic happiness theory) is an evaluative concept by asking if people are happy with their life (or with their well-being). Consequently we could use it to evaluate the current systems around food to determine their ability to deliver authentic happiness or not. The United Nations' World Happiness Reports provide a practical example of applying happiness as a evaluative criterion [141].

Capabilities: the capabilities approach asks the question: What are people actually able to do and to be? The answer is the set of capabilities, or real opportunities, that a person has and is able to count on in future (capability security). The "capabilities approach" is the framework used in the Human Development Reports published by the United Nations Development Programme (UNDP) and is used in many state and regional development reports. "Capabilities are important human entitlements, inherent in the idea of basic social justice, and can be viewed as one species of a human rights approach" [142]. Nussbaum [143] has proposed a tentative set of ten central human capabilities that provide a basic set for social justice. Their fulfilment requires affirmative action and government support for their creation and maintenance. They are: (1) Life, living a normal length; (2) Bodily health; (3) Bodily integrity; free from violence; (4) Senses, imagination, and thought, ranging from education, religion to freedom of expression; (5) Emotions, being able to express emotions; (6) Practical reason, being able to plan one's life;

(7) Affiliation, (A) social interaction, (B) to have social respect; (8) Other species, have concern for the world of nature; (9) Play, enjoy recreational activities; (10) Control over one's Environment, (A) political, (B) material. Nussbaum [142] has the view that governments have work to do in regards to capabilities: "If a capability really belongs on the list (or, if a given human right really belongs in a list of human rights), then governments have the obligation to protect and secure it, using law and public policy to achieve this end".

In discussing the capability approach and sustainability Sen [144] suggested that "A fuller concept of sustainability has to aim at sustaining human freedoms, rather than only at our ability to fulfil our felt needs." Sen suggested that determining needs (perhaps done by experts) would reduce freedom to choose; "There is a big issue of individual choice here. There is also a related "social choice" problem, in determining the priorities between different kinds of freedoms, or, for that matter, even in the identification of different types of needs and the priorities between them" [144]. He holds that prioritised (crucial) freedoms are those that people have reason to value. Crucial freedoms may include: "such liberties as freedom from hunger, from illiteracy, from avoidable ill-health, from escapable mortality, as well as the freedom to achieve dignity and respect, among other critical emancipations" [144]. These freedoms require broad based social support programs to become operational. The notion is to give certain freedoms priorities and that such freedoms are not likely to be adapted downwards by deprivation as needs might be (e.g., being destitute but thankful/happy for small mercies). Debate is suggested as the process for identifying what "crucial freedoms" are, and presumably "having reason to value" is the key. It seems that Sen's approach could be used to provide priorities in Nussbaum's list of ten central capabilities in particular cases through debate among the people involved. Debate would allow the people freedom to choose.

Sustainable Development: The definition of sustainable development from the Brundtland report is: "Sustainable development is development that meets the needs of the present without compromising the ability of future generations to meet their own needs. It contains within it two key concepts:

(1) the concept of "needs", in particular the essential needs of the world's poor, to which overriding priority should be given; and development [152]
(2) the idea of limitations imposed by the state of technology and social organization on the environment's ability to meet present and future needs." [135]

Sustainable development takes a system thinking approach to integrate environmental, social and economic pillars and as such provides a good approach to identifying gaps and developing overarching goals for the food systems. Applying sustainable development ideas has had mixed success in Australia perhaps because the economic pillar, using neoliberalism, has been favoured over the social and environmental pillars [145].

Environmental ethics: Environmental ethics addresses the notion that the future of humanity is linked with the future of the functioning of ecosystems. Maintaining the productive capacity of ecosystems is an important aspect of ensuring the ongoing production of health-giving diets for all people (*i.e.*, natural resource and biodiversity maintenance aspects of agriculture and the food supply systems). In addition to this instrumental value of nature Pelenc *et al.* [146] points out that many people recognize intrinsic values of nature. Therefore only giving nature an instrumental value for humanity is an underestimation of what many people feel and value in nature.

Instrumental and intrinsic valuations of nature can provide assessment criteria. For example, Leopold's land ethic, "that something is right when it maintains the integrity, stability and beauty of the biotic community" is relevant [147]; or the eight point platform for deep ecology that sets out the intrinsic value and connectedness of human and nonhuman life and indicates that the flourishing of human life and cultures is compatible with a substantial decrease in human population [148]. The platform raises a basic problem as it asks for a balance between human vital needs and the interests of the rest of nature [149,150] Meyer [151] also highlights the balance between people and nature and suggests that the problem of how much of the Earth's resources should be for humans and how much for other forms of life, may be one of the most important ethical questions of our times. Liu *et al.* [95] suggest that this issues needs to be on the international agenda for debate.

Which approach is best for evaluating food security? It is likely that any or all of these seven approaches would help governments identify gaps between "what is" and "what ought to be" in terms of food security. Perhaps Max-Neef's nine human needs or Nussbaum ten central human capabilities being more clearly defined might expedite gap analysis and the development of overarching goals but all approaches discussed above could provide assessment criteria to identify gaps and then develop overarching goals for the food systems.

However, in studying these alternative approaches it is obvious that there are more needs, more capabilities, more freedoms, more ecological values or more human rights than those that relate only to the food supply systems. While we want food security we, as a society, also want these other values to be available to us over a life time (and for future generations); they too must be secure implying that we need a number of securities. Food systems can and do modify these other values such as ecological systems, biodiversity and health. So food security is not independent of these other securities but rather they are all integral parts of larger systems. The seven alternative approaches outlines provide people with the opportunity to see food security in the context of the other securities we need in the long term in order to have a flourishing human society. It seems unlikely that neoliberalism with its reliance on markets and concentration of power and wealth provides such opportunity.

We would want to know, for example, if implementing a program or policy would protect biodiversity and protect the integrity of ecosystems. Each of the overarching goal could represent a "security" in that they concern material goods (such as a health-giving diet) or a process (such as access to justice) that people would want to know are available and a "secure feature" of society into the future.

It is clear that the food supply systems ought to change because currently they are not delivering all the needs or freedoms or rights or securities people require and are not protecting nature on which humans depend. That is, the food systems are not sustainable in social and environmental terms. Nor are food systems sustainable in national economic terms because poor diets lead to ill health and a reduced working life.

Substantial change is unlikely to happen any time soon in Australia because vested interests in the current situation have sufficient economic and political power to suppress these ideas. Perhaps another way of viewing this is to note that ideas about needs and securities (such as food security, or ecological security/conservation) have been around for decades yet have not been sufficiently viable politically to lead to substantial change. It may be that the diversity of approaches dissipates intellectual energy away from effective change. Undoubtedly the ideas have made a difference but they have to overcome the very strong vested interests and a persuasive neoliberal turn in politics that has created powerful organisations and people wishing to continue on the current policy trajectories. The deck is stacked against reforming the systems related to food in any substantial way as the current commercial approach to food production is part and parcel of the apparently successful current approach to all industrial development [152].

Perhaps climate change and population growth will force change. Hopefully, the necessary changes can be developed through public discourse democratically over the coming years and not come out of crisis as emergency measures.

9. What Levers Might be Useful in Making the Changes?

Should we manage to agree on what the changes to the systems ought to be, we must then find levers that will work. While there are no levers that might act as a "silver bullet" to change the systems overnight, some levers may be more effective than others. So this section addresses the two questions: "what levers might be useful in making the changes" and "how can people and organisations use the levers to make changes".

A three steps iterative process is suggested to work out which levers might be the most effective. The first step is to obtain an understanding of what to address the gaps exist between "what is" and "what ought to be".

The second step is to obtain a deeper appreciation of the systems involved and how they behave and interact. The systems involved include more than just the food supply systems. The commercial food supply systems sit within a wider framework of human and ecological systems such that the food systems are part of larger social–ecological systems in which people interact with the world's biophysical landscapes and processes [153]. Causes and solutions for problems in the food systems, such as poor health from poor diets, may exist outside the immediate food supply systems. Poor diet may have its origins in a wide range of factors that are influenced by many other systems. These may include poverty (employment system), lack of nutritional education (education system), and/or lack of access (transport or retail systems). Gunasekera *et al.* [154] put forward an economic perspective for research on food security within the "human-earth" system. However, an even wider frame may be needed for many long-term issues. People do not want the food system to "fail", so resilience may be a useful idea in food security research [155,156]. The notion in resilience is that system failure (losing the things we value that the system produces) might be avoided if we can maintain innovation to transform the system, or parts of it, to keep the benefits coming and the failure point somewhere in the future. Unfortunately there are many systems involved in Australia's domestic food supply problems in addition to the actual food supply chains and this complexity is further enhanced by globalisation as many of the systems are influenced or controlled by systems that operate globally. The global influence may be physical, such as oil or fertilizer supply, or social, such as ideas about economics or conventions on biodiversity and human rights. This makes understanding the operation, interaction and impacts of these systems on the Australian food issues very difficult. To help address the difficulties in understanding globally complex systems Anderies *et al.* [157] reviewed the key concepts of sustainability, resilience, and robustness. Their aim was to clarify the meaning of these terms and suggest how they can be used in tandem to address the complexities involved in national and global social–ecological systems. A better understanding of how the systems operate will provide the basis for the next step of finding an effective mechanism to resolve Australia's food issues. Information gathered in this second step will facilitate the development of overarching goals that are capable of closing the gaps identified in step one; the gaps between "what is" and "what ought to be".

The third step is about finding effective change mechanisms. Meadows [158] listed twelve ways in which to change systems in a declining order of effectiveness (Table 1). She notes that the order in the list is "slippery" because there are exceptions which can move the leverage point up or down the list. The higher the leverage point (the lower the number in this list) the more effective the lever but the more the system will resist change.

Table 1. How to change systems [158].

Parts of Systems	Effectiveness
12. Constraints, parameters and numbers (such as subsidies, taxes and standards)	least effective
11. Size of buffers and other stabilizing stocks relative to their flows	
10. Structure of material stock and flows (such as transport networks and population age structures)	
9. Length of delays relative to rate of system change	
8. Strength of negative feedback loops relative to the impacts they are trying to control	
7. The gain around driving positive feedback loops	
6. The structure of information flows (who does and does not have access to what kinds of information)	
5. The rules of the system (such as incentives, punishments and constraints)	
4. The power to add, change, evolve or self-organise system structure	
3. The goals of the system	
2. The mindset or paradigm out of which the system—its goals, structure, rules, delays and parameters—arises	
1. The power to transcend paradigms	
	most effective

To change the systems in Australia to deal with poor diets, natural resource and biodiversity losses. We should be looking for a lever at a low number, perhaps at six or below in Meadows's list. The first step, mentioned above, was to work out the gaps between "what is" and "what ought to be" and step 2 involved developing overarching goals to address these gaps. Goals come in at No 3 in Meadows' list so implementing these overarching goals aimed at filling the gaps would be very effective. However, developing goals that motivate people to take action may involve considerable iteration and wide consultation with people involved directly in the food systems and in related systems such as health, employment and education (see for guidance: [129]).

Development of renewable energy (e.g., wind and solar) in Australia is a demonstration of the powers of these alternative levers. The fossil fuel electricity generators are able to change the rules of the system (such as changing the electricity tariffs and costs for installing and running renewable grid connected energy). This is lever No. 5. The renewable electricity industry has the goal of climate mitigation, reducing electricity costs and a changed mindset in terms of becoming independent of the multinational fossil fuel electricity companies and providing home grown power. These include a range of levers such as Nos. 6 and 7 but also include No. 2; "people's mindsets". It would seem that renewable electricity industry has the more effective lever at No, 2 if they can apply it by changing enough "mind sets", but the fossil fuel companies have political and financial power [159–162].

Having the ability to use a lever (such as No. 2 *changing mindsets*) and actually applying it, so that a significant proportion of the population develops this mindset, are two different things. In a democratic country like Australia, Meadow's No. 6; *The structure of information flows*, could be an important lever for changing the systems to ensure the problems of the food supply systems (food security, health, and the maintenance of natural resources and biodiversity) are resolved in a reasonable time period. Arrangements need to be made to ensure that reliable and comprehensive information about the issues flows through authoritative organisations to critical people in leadership positions in ways that will get their attention and influence their decisions. In addition, paying attention to these matters in education would improve the structure of information flows for the longer term [163].

Changing the structure of information flow is a tactic used by government. For example, the current Australian Government tried to reduce the flow of information on climate change in a number of ways; for example, by abolishing the Climate Change Commission that was established by the former government to provide public information on climate change [164] and by trying to keep climate change off the agenda of the G20 heads of government summit help in Brisbane in November 2014 [165]. In the former case the employees established a crowd funded information provider, the Climate Council, and in the second case other world leaders decided to discuss climate change at the summit.

10. Conclusions

The food supply systems in Australia fail to deliver healthy diets to all Australians and protect the natural resources on which agricultural production depends. The operation of the systems also creates "collateral damage" to the natural ecosystems including biodiversity loss. The food supply systems and food prices will be challenged by resource price increases and production decline due to climate change. Food prices will be further challenged by increasing domestic demand for food and increasing demand for the resources used in agriculture and by increasing exports to the expanding middle and upper classes mainly in Asia. The expansion of Australian agriculture to increase exports is likely to increase the collateral damage to the environment and have a negative impact on the sustainability of resources used in agriculture. The food supply systems are also nationally economically unsustainable because the poor diets they deliver lead to ill health and a reduced capacity for work.

There is a range of evaluative approaches that could be used to address these issues. The paper briefly discusses seven of these, which are: human rights, national securities, human needs, authentic happiness, capabilities, sustainable development and environmental ethics. These approaches could help to clearly identify the gaps between what currently exists and what ought to exist in the food systems and develop overarching goals to address these gaps. These seven approaches contain powerful ideas and already have had an impact on what happens, but so far they have not been applied in ways that resolve the food issues that exist in Australia's food supply systems. Implementing the overarching goals to close the gaps depends on being able to successfully apply policy levers to change how the systems operate and create more acceptable outcomes. Obtaining and employing effective leverage points for changing the food systems is difficult because change has to overcome resilience in the existing systems coming from economically and politically powerful vested interests. Perhaps changing the structure of information flows would be a feasible leverage point facilitated by Australia's democratic political system.

The food supply systems ought to facilitate the consumption of a health giving diet for all Australians throughout their lives. However, the seven approached described in this paper identify other "needs", "capabilities" and "rights" as well, such as ecological stability, health and leisure that are important for humanity and ought to be secured for the long-term to enable individuals and societies to flourish. In improving food security in Australia, we ought to opt for policies and programs that deliver a full range of securities synergistically rather than competitively.

The existence and continuation of the problems in the food supply systems in Australia are a black mark against society as options available in Australia, as a wealthy and well-endowed country, are such that these issues could be substantially resolved within a generation.

Acknowledgments

I would like to thank John Martin for encouraging my interest in food security and Andrew Parratt for his support and my visitor status at Deakin University.

Conflicts of Interest

The author declares no conflict of interest.

References

1. ABS. *Catalogue 3101.0—Australian Demographic Statistics, June 2014*; Australian Bureau of Statistics: Canberra, ACT, Australia, 2014.

2. BoM. State of the Climate, Bureau of Meteorology and CSIRO, 2014. Available online: http://www.bom.gov.au/state-of-the-climate/documents/state-of-the-climate-2014_low-res.pdf? ref=button (accessed on 26 February 2015).

3. Steffen, W. *Quantifying the Impact of Climate Change on Extreme Heat in Australia*; Climate Council of Australia Limited: Canberra, ACT, Australia, 2015.

4. Rosenzweig, C.; Elliottb, J.; Deryng, D.; Ruanea, A.C.; Müllere, C.; Arneth, A.; Booteg, K.J.; Folberthh, C.; Glotteri, M.; Khabarovj, N.; *et al.* Assessing agricultural risks of climate change in the 21st century in a global gridded crop model inter-comparison. *Proc. Natl. Acad. Sci. USA* **2014**, *111*, 3268–3273.

5. Wise, P. *Grow Your Own: The Potential Value and Impacts of Residential and Community Food Gardening*; The Australia Institute: Canberra, ACT, Australia, 2014.

6. O'Kane, G. What is the real cost of our food? Implications for the environment, society and public health nutrition. *Public Health Nutr.* **2011**, *15*, 268–276.

7. Temple, J.B. Severe and moderate forms of food insecurity in Australia: Are they distinguishable? *Aust. J. Soc. Issues* **2008**, *43*, 649–668.

8. Coles-Rutishauser, I.; Penm, R. Monitoring food habits and food security: Australia. In *1995–1996. Food and Nutrition Monitoring Unit Working Paper No. 96.3*; Australian Institute of Health and Welfare: Canberra, ACT, Australia, 1996.

9. AIHW. Australia's Health 2010. Australia's Health No. 12. Cat. No. AUS 122. Available online: http://www.aihw.gov.au/WorkArea/DownloadAsset.aspx?id=6442452962 (accessed on 24 March 2012).

10. ABS. *Catalogue 4719.0—Overweight and Obesity in Adults, Australia, 2004–2005*; Australian Bureau of Statistics: Canberra, ACT, Australia, 2008.

11. ABS. *Catalogue 4338.0—Profiles of Health, Australia, 2011–2013*; Australian Bureau of Statistics: Canberra, ACT, Australia, 2013.

12. Haby, M.; Markwick, A. Future Prevalence of Overweight and Obesity in Australian Children and Adolescents, 2005–2025. Available online: http://docs.health.vic.gov.au/docs/doc/768FD9A 0683F9259CA2578EC0081AD6A/$FILE/future_overweight_prevalence_report.pdf (accessed on 27 August 2013).

13. Zimmet, P.Z.; Phillip, W.; James, T. The unstoppable Australian obesity and diabetes juggernaut. What should politicians do? *Med. J. Aust.* **2006**, *185*, 187–188.

14. AIHW. National Indicators for Monitoring Diabetes: Report of the Diabetes Indicators Review Subcommittee of the National Diabetes Data Working Group. Diabetes Series no. 6. Cat. no. CVD 38. Available online: http://www.aihw.gov.au/WorkArea/DownloadAsset.aspx?id=6442455061 (accessed on 20 February 2015).

15. ABS. *Catalogue 4364.0.55.001—Australian Health Survey: First Results, 2011–2012*; Australian Bureau of Statistics: Canberra, ACT, Australia, 2012.

16. ABC. Australian Obesity Rates Climbing Faster than Anywhere Else in the World, Study Shows. Interviews on News Breakfast Including Professor Rob Moodie. Available online: http://www.abc. net.au/news/2014-05-29/australian-obesity-rates-climbing-fastest-in-the-world/5485724 (accessed on 25 February 2015).

17. Radermacher, H.; Feldman, S.; Bird, S. Food security in older Australians from different cultural backgrounds. *J. Nutr. Educ. Behav.* **2010**, *42*, 328–336.

18. Rosier, K. Food Insecurity in Australia: What is It, Who Experiences It and how can Child and Family Services Support Families Experiencing It? Australian Institute of Family Studies, Child and Family Community Group, CAFCA Practice Sheet—August 2011. Available online: https://aifs.gov.au/cfca/sites/default/files/publication-documents/ps9.pdf (accessed on 19 February 2015).

19. AIHW. *Australia's Food & Nutrition 2012. Cat. No. PHE 163*; Australian Institute of Health and Welfare: Canberra, ACT, Australia, 2012.

20. Ward, P.R.; Verity, F.; Carter, P.; Tsourtos, G.; Coveney, J.; Wong, K.C. Food Stress in Adelaide: The relationship between low income and the affordability of healthy food. *J. Environ. Public Health* **2013**, *2013*, 1–10.

21. Barosh, L.; Friel, S.; Engelhardt, K.; Chan, L. The cost of a healthy and sustainable diet—Who can afford it? *Aust. N. Zeal. J. Public Health* **2014**, *38*, 7–12.

22. The First Thousand Days. Available online: http://www.thousanddays.org (accessed on 21 March 2015).

23. NHMRC. Australian Dietary Guidelines Incorporating the Australian Guide to Healthy Eating, Providing the Scientific Evidence for Healthier Australian Diets, Draft for Public Consultation, December 2011. National Health and Medical Research Council. Available online: https://www.eatforhealth.gov.au/sites/default/files/files/public_consultation/n55_draft_australian_dietary_guidelines_consultation_111212.pdf (accessed on 10 March 2012).

24. Mathers, C.; Vos, T.; Stevenson, C. *The Burden of Disease and Injury in Australia—Summary Report*; Australian Institute of Health and Welfare: Canberra, ACT, Australia, 1999.

25. Colagiuri, S.; Lee, C.M.Y.; Colagiuri, R.; Magliano, D.; Shaw, J.E.; Zimmet P.Z.; Caterson, I.D. The cost of overweight and obesity in Australia. *Med. J. Aust.* **2010**, *192*, 260–264.

26. Marmot, M.; Wilkinson, R. *Social Determinants of Health*; Marmot, M., Wilkinson, R., Eds.; Oxford University Press: Oxford, UK, 2005.

27. Drewnowski, A.; Darmon, N. The economics of obesity: Dietary energy density and energy cost. *Am. J. Clin. Nutr.* **2005**, *82*, 265S–273S.

28. Levine, J.A. Poverty and obesity in the U.S. *Diabetes* **2011**, *60*, 2667–2668.

29. Burns, C. A Review of the Literature Describing the Link between Poverty, Food Insecurity and Obesity with Specific Reference to Australia. Available online: http://secondbite.org/sites/default/files/A_review_of_the_literature_describing_the_link_between_poverty_food_insecurity_and_obesity_w.pdf (accessed on 2 October 2013).

30. Siahpush, M.; Huang, T.T.K.; Sikora, A.; Tibbits, M.; Shaikh, R.A.; Singh, G.K. Prolonged financial stress predicts subsequent obesity: Results from a prospective study of an Australian national sample. *Obesity* **2014**, *22*, 616–621.

31. Gretton, P.; Salma, U. Land Degradation and the Australian Agricultural Industry, Staff Information Paper. Available online: http://www.pc.gov.au/research/completed/land-degradation/landdegr.pdf (accessed on 20 February 2015).

32. Lumb, M. Land Degradation Facts Sheet, the Australian Collaboration. Available online: http://www.australiancollaboration.com.au/pdf/FactSheets/Land-degradation-FactSheet.pdf (accessed on 23 March 2015).

33. ABS. Australia's Biodiversity. In *1301.0—Year Book Australia, 2009–2010*; Australian Bureau of Statistics: Canberra, ACT, Australia, 2010.

34. Bradshaw, C.J.A. Little left to lose: Deforestation and forest degradation in Australia since European colonization. *J. Plant Ecol.* **2012**, *5*, 109–120.

35. Chappell, M.J.; LaValle, L.A. Food security and biodiversity: Can we have both? An agroecological analysis. *Agric. Hum. Values* **2011**, *28*, 3–26.

36. EA. *Salinity and Water Quality*; Department of Sustainability, Environment, Water, Population and Communities: Canberra, ACT, Australia, 2012.

37. SoE. *Australia State of the Environment 2011—In Brief*; State of the Environment 2011 Committee: Canberra, ACT, Australia, 2011.

38. Whittington, J.; Liston, P. *Australia's Rivers. 1301.0—Year Book Australia*; Bureau of Statistics: Canberra, ACT, Australia, 2003.

39. Australian Wildlife Conservancy. Available online: http://www.australianwildlife.org/wildlife.aspx (accessed on 21 March 2015).

40. Cardinale, B.J.; Duffy, E.J.; Gonzalez, A.; Hooper, D.U.; Perrings, C.; Venail, P.; Narwani, A.; Mace, G.M.; Tilman, D.; Wardle, D.A.; *et al.* Biodiversity loss and its impact on humanity. *Nature* **2012**, *486*, 59–67.

41. Kroon, F.J.; Schaffelke, B.; Bartley, R. Informing policy to protect coastal coral reefs: Insight from a global review of reducing agricultural pollution to coastal ecosystems. *Mar. Pollut. Bull.* **2014**, *85*, 33–41.

42. Holland, J.E.; Luck, G.W.; Finlayson, C.M. Threats to food production and water quality in the Murray-Darling Basin of Australia. *Ecosyst. Serv.* **2015**, *12*, 55–70.

43. CoA. *Basin Plan 2012*; Commonwealth of Australia: Canberra, ACT, Australia, 2012.

44. Robeyns, I. The Capability Approach, the Stanford Encyclopedia of Philosophy (Summer 2011 Edition) 2011. Available online: http://plato.stanford.edu/entries/capability-approach/ (accessed on 9 October 2013).

45. Azpitarte, F. Social Exclusion Monitor Bulletin December 2012, Research Bulletin. Brotherhood of St Laurence and the Melbourne Institute of Applied Economic and Social Research 2012. Available online: http://www.bsl.org.au/pdfs/Azpitarte_Social_exclusion_monitor_bulletin_ Dec2012.pdf (accessed on 2 October 2013).

46. Australian Council of Social Service. *Poverty in Australia: ACOSS Paper 194 (Updated March 2013)*; Australian Council of Social Service: Strawberry Hills, NSW, Australia, 2012. Available online: http://www.acoss.org.au/uploads/ACOSS%20Poverty%20Report%202012_Final.pdf (accessed on 6 October 2013).

47. Goldie, C. *For a Real New Start, Stop Miring People in Poverty*; Australian Council of Social Service (ACOSS): Redfern, NSW, Australia, 2013.

48. Cleugh, H.; Smith, M.S.; Battaglia, M.; Graham, P. *Climate Change: Science and Solutions for Australia*; Cleugh, H., Stafford Smith, M., Battaglia, M., Graham, P., Eds.; CSIRO Publishing: Collingwood, VIC, Australia, 2011.

49. Garnaut, R. *Garnaut Climate Change Review Update 2011, Update Paper Four, Transforming Rural Land Use*; Commonwealth of Australia: Canberra, ACT, Australia, 2011.

50. Kingwell, R. Climate change in Australia: Agricultural impacts and adaptation. *Austr. Agribus. Rev.* **2006**, *14*, Paper 1.

51. Qureshia, M.E.; Hanjrac, M.A.; Ward, J. Impact of water scarcity in Australia on global food security in an era of climate change. *Food Policy* **2013**, *38*, 136–145.

52. Beddington, J.; Asaduzzaman, M.; Clark, M.; Fernández, A.; Guillou, M.; Jahn, M.; Erda, L.; Mamo, T.; van Bo, N.; Nobre, C.A.; *et al. Achieving Food Security in the Face of Climate Change: Final Report from the Commission on Sustainable Agriculture and Climate Change*; CGIAR Research Program on Climate Change, Agriculture and Food Security (CCAFS): Copenhagen, Denmark, 2012.

53. AG. *Australian National Greenhouse Accounts, National Inventory Report 2011 Volume 1, The Australian Government Submission to the United Nations Framework Convention on Climate Change April 2013*; Department of Industry, Innovation, Climate Change, Science, Research and Tertiary Education: Canberra, ACT, Australia, 2013.

54. CoA. *National Inventory by Economic Sector 2011–2012: Australia's National Greenhouse Accounts*; Commonwealth of Australia: Canberra, ACT, Australia, 2014.

55. Rickards, L.; Howden, S.M. Transformational adaptation: Agriculture and climate change. *Crop Pasture Sci.* **2012**, *63*, 240–250.

56. Carbon Farming. Available online: http://www.cleanenergyregulator.gov.au/Carbon-Farming-Initiative/Pages/default.aspx (accessed on 21 March 2015).

57. Emissions Reduction Fund. Available online: http://www.environment.gov.au/climate-change/emissions-reduction-fund (accessed on 20 March 2015).

58. Lam, S.K.; Chen, D.; Mosier, A.R.; Roush, R. The potential for carbon sequestration in Australian agricultural soils is technically and economically limited. *Sci. Rep.* **2013**, *3*, 2179.

59. Alexander, S. Peak Oil is Alive and Well, and Costing the Earth. The Conversation, 9 September 2013. Available online: http://theconversation.com/peak-oil-is-alive-and-well-and-costing-the-earth-17542 (accessed on 5 October 2013).

60. Murray, J.W.; Hansen, J. Peak oil and energy independence: Myth and reality. *Eos* **2013**, *94*, 245–252.

61. Pfeiffer, D.A. *Eating Fossil Fuels: Oil, Food and the Coming Crisis in Agriculture*; New Society Publishers: Gabriola Island, BC, Canada, 2006.

62. Australian Government. *Water for the Future; Restoring the Balance in the Murray-Darling Basin*; Australian Government: Canberra, ACT, Australia, 2010.

63. Stewart, W.M.; Dibb, D.W.; Johnston, A.E.; Smyth, T.J. The contribution of commercial fertilizer nutrients to food production. *Agron. J.* **2005**, *97*, 1–6.

64. Cordell, D.; Drangert, J.; White, S. The story of phosphorus: Global food security and food for thought. *Glob. Environ. Chang.* **2009**, *19*, 292–305.

65. Cordell, D.; Jackson, M.; White, S. Phosphorus flows through the Australian food system: Identifying intervention points as a roadmap to phosphorus security. *Environ. Sci. Policy* **2013**, *29*, 87–102.

66. Pratt, S. Future Fertilizer Supplies Expected to Send Prices Down. Available online: http://www.producer.com/2013/04/future-fertilizer-supplies-expected-to-send-prices-down/ (accessed on 3 October 2013).

67. Lawrence, D. Why Cheap Natural Gas May Be a Boost to Farmers. Available online: http://www.theglobeandmail.com/report-on-business/breakthrough/why-cheap-natural-gas-is-a-boost-to-farmers/article14227180/ (accessed on 3 October 2013).

68. Buxton, M.; Alvarez, A.; Butt, A.; Farrell, S.; O'Neill, D. *Planning Sustainable Futures for Melbourne's Peri-Urban Region*; RMIT University: Melbourne, Australia, 2008.

69. Lloyd, D.J.; Luke, H.; Boyd, W.E. Community perspectives of natural resource extraction: Coal seam gas mining and social identity in South Eastern Australia. *Coolabah* **2013**, *10*, 144–164.

70. Department of Agriculture, Fisheries and Forests. *Australian Food Statistics 2010–2011*; Department of Agriculture Fisheries and Forests: Canberra, ACT, Australia, 2012.

71. National Farmers' Federation. *NFF Farm Facts: 2012*; National Farmers' Federation: Canberra, ACT, Australia, 2013.

72. Department of Agriculture, Fisheries and Forests. *National Food Plan: Our Food Future*; Department of Agriculture, Fisheries and Forests: Canberra, ACT, Australia, 2013.

73. Australian Government. *Australia in the Asian Century, White Paper*; Australian Government: Canberra, ACT, Australia, 2012.

74. Department of Agriculture, Fisheries and Forests. *National Food Plan Green Paper 2012*; Department of Agriculture, Fisheries and Forestry: Canberra, ACT, Australia, 2012.

75. Abbott, T. The Coalition's 2030 Vision for Developing Northern Australia June 2013. Available online: http://www.tonyabbott.com.au/LinkClick.aspx?fileticket=ymP4ynYQKOA%3D&tabid= 86 (accessed on 29 August 2013).

76. Napthine, D. Growing Food Exports into Asia, Victoria Well Positioned to be Part of "Food Boom", Speech at the Global Food Forum. Available online: http://www.theaustralian.com.au/ business/in-depth/growing-food-exports-into-asia/story-fni2wt8c-1226623320373 (accessed on 26 August 2013).

77. Pratt, A. Australia, the "Clean Green Food Bowl of Asia", Keynote Speech at the Global Food Forum. Available online: http://www.theaustralian.com.au/business/in-depth/australia-the-clean-green-food-bowl-of-asia/story-fni2wt8c-1226623265405 (accessed on 26 August 2013).

78. NALWTF. *Northern Australia Land and Water Science Review 2009 Chapter Summaries*; Northern Australian Land and Water Task Force; Department of Infrastructure; Transport; Regional Development and Local Government: Canberra, ACT, Australia, 2009.

79. Davidson, B. *The Northern Myth: Limits to Agriculture and Pastoral Development in Tropical Australia*, 3rd ed.; Melbourne University Press: Carlton, VIC, Australia, 1972.

80. Tilman, D.; Cassman, K.G.; Matson, P.A.; Naylor, R.; Polasky, S. Agricultural sustainability and intensive production practices. *Nature* **2002**, *418*, 671–677.

81. Patel, R. What does food sovereignty look like? *J. Peasant Stud.* **2009**, *36*, 663–706.

82. Menezes, F. Food sovereignty: A vital requirement for food security in the context of globalization. *Development* **2001**, *44*, 29–33.

83. Hamawand, I.; Yusaf, T.; Hamawand, S.G. Coal seam gas and associated water: A review paper. *Renew. Sustain. Energy Rev.* **2013**, *22*, 550–560.

84. Bark, R.; Kirby, M.; Connor, J.D.; Crossman, N.D. Water allocation reform to meet environmental uses while sustaining irrigation: A case study of the Murray-Darling Basin, Australia. *Water Policy* **2014**, *16*, 739–754.

85. Penney, A.; Kirby, D.; Cheshire, K.; Wilson, M.; Bray, S. *Technical Review for the Commonwealth Policy on Fisheries Bycatch: Risk Based Approaches, Reference Points and Decision Rules for Bycatch and By-Product Species*; Australian Bureau of Agricultural and Resource Economic; Department of Agriculture, Fisheries and Forests: Canberra, ACT, Australia, 2013.

86. Grech, A.; Bos, M.; Brodie, J.; Coles, R.; Dale, A.; Gilbert, R.; Hamann, M.; Marsh, H.; Neil, K.; Pressey, R.L.; *et al.* Guiding principles for the improved governance of port and shipping impacts in the Great Barrier Reef. *Mar. Pollut. Bull.* **2013**, *75*, 8–20.

87. Farmar-Bowers, Q.; Higgins, V.; Millar, J. *Food Security in Australia: Challenges and Prospects for the Future*; Farmar-Bowers, Q., Higgins, V., Millar, J., Eds.; Springer: New York, NY, USA, 2013.

88. Katz, D.L.; Meller, S. Can we say what diet is best for health? *Annu. Rev. Public Health* **2014**, *35*, 83–103.

89. NHMRC. Australian Dietary Guidelines. Available online: http://www.eatforhealth.gov.au/ sites/default/files/files/the_guidelines/n55_australian_dietary_guidelines.pdf (accessed on 20 March 2013).

90. Meyer, S.B.; Coveney, J.; Henderson, J.; Ward, P.R.; Taylor, A.W. Reconnecting Australian consumers and producers: Identifying problems of distrust. *Food Policy* **2012**, *37*, 634–640.

91. Burdon, P. What is good land use? From rights to relationships. *Melb. Univ. Law Rev.* **2010**, *34*, 708–735.

92. Lawrence, G.; Richards, C.; Lyons, K. Food security in Australia in an era of neoliberalism, productivism and climate change. *J. Rural Stud.* **2013**, *29*, 30–39.

93. Selvey, L.A.; Carey, M.G. Australia's dietary guidelines and the environmental impact of food "from paddock to plate". *Med. J. Aust.* **2013**, *198*, 18–19.

94. Liu, Y.; Pan, X.; Li, J. Current agricultural practices threaten future global food production. *J. Agric. Environ. Ethics* **2015**, *28*, 203–216.

95. Biddle, N.; Bursian, O. FactCheck: Is Poverty on the Rise in Australia? Available online: http://theconversation.com/factcheck-is-poverty-on-the-rise-in-australia-17512 (accessed on 5 October 2013).

96. Bursian, O. We should be Shamed by Our Record on Child Poverty. Available online: http://theconversation.com/we-should-be-shamed-by-our-record-on-child-poverty-15698 (accessed on 5 October 2013).

97. Department of Agriculture, Fisheries and Forestry. *Queensland's Agriculture Strategy: A 2040 Vision to Double Agricultural Production. Report CS2320, 06/13*; Department of Agriculture, Fisheries and Forestry: Queensland, Australia, 2013.

98. Hannam, P. Tony Abbott's Government is "Recklessly Endangering" the Future on Climate, Says UK Chief. The Sydney Morning Herald 2014. Available online: http://www.smh.com.au/ federal-politics/political-news/tony-abbotts-government-is-recklessly-endangering-the-future-on-climate-says-uk-chief-20140708-zszx4.html (accessed on 20 February 2015).

99. International Covenant on Economic, Social and Cultural Rights. Available online: http://www.ohchr.org/EN/ProfessionalInterest/Pages/CESCR.aspx (accessed on 21 March 2015).

100. Roetter, R.P.; van Keulen, H. Food Security. In *Science for Agriculture and Rural Development in Low-Income Countries*; Roetter, R.P., van Keulen, H., Verhagan, J., Kuiper, M., Verhagen, J., van Laar, H.H., Eds.; Springer Science + Business Media B.V.: Dordrecht, The Netherlands, 2008; Chapter 3, pp. 27–56.

101. SIGNAL. Eat Well Australia, A Strategic Framework for public Health Nutrition, 2000–2010. Strategic Inter-Governmental Nutrition Alliance of the National Public Health Partnership 2001. Available online: http://www.health.vic.gov.au/archive/archive2014/nphp/ (accessed on 26 August 2013).

102. Neset, T.S.; Cordell, D. Global phosphorus scarcity: Identifying synergies for a sustainable future. *J. Sci. Food Agric.* **2012**, *92*, 2–6.

103. Lindenmayer, D.; Cunningham, S.; Young, A. *Land Use Intensification: Effects on Agriculture, Biodiversity and Ecological Processes*; Lindenmayer, D., Cunningham, S., Young, A., Eds.; CSIRO Publishing: Collingwood, VIC, Australia, 2012.

104. Steffen, W.; Richardson, K.; Rockström, J.; Cornell, S.E.; Fetzer, I.; Bennett, E.M.; Biggs, R.; Carpenter, S.R.; de Vries, W.; de Wit, C.A.; *et al.* Planetary boundaries: Guiding human development on a changing planet 2015. *Science* **2015**, *347*, doi:10.1126/science.1259855.

105. Norton, D.; Reid, N. *Nature and Farming: Sustaining Native Biodiversity in Agricultural Landscapes*; CSIRO Publishing: Collingwood, VIC, Australia, 2013.

106. Gunn, I. Private Land is an Important Piece of the Conservation Jigsaw. Available online: http://theconversation.com/private-land-is-an-important-piece-of-the-conservation-jigsaw-11572 (accessed on 10 October 2013).

107. Stokes, C.J.; Howden, S.M. *An Overview of Climate Change Adaptation in Australian Primary Industries—Impacts, Options and Priorities*; CSIRO: Canberra, ACT, Australian, 2008.

108. Barlow, K.M.; Christy, B.P.; O'Leary, G.J.; Riffkin, P.A.; Nuttal, J.G. Simulating the impact of extreme heat and frost events on wheat crop production: A review. *Field Crops Res.* **2015**, *171*, 109–119.

109. An Outline of Social Security. Available online: http://www.humanservices.gov.au/customer/dhs/centrelink (accessed on 20 March 2015)

110. An Outline of Medicare Services. Available online: http://www.humanservices.gov.au/customer/subjects/medicare-services (accessed on 21 March 2015).

111. An outline of Commonwealth Conservation Arrangements on Private Land. Available online: http://www.environment.gov.au/biodiversity/conservation/index.html#private (accessed on 20 March 2015).

112. NRMMC. *Australia's Biodiversity Conservation Strategy 2010–2030*; Commonwealth of Australia: Canberra, ACT, Australia, 2010.

113. Ireland, J. Canberra Moves to Make People Wait a Week For Welfare. Available online: http://www.smh.com.au/federal-politics/political-news/canberra-moves-to-make-people-wait-a-week-for-welfare-20140520-38mnr.html (accessed 20 February 2015).

114. Australian Medical Association. Comments on Co-Payments. Available online: http://www.abc.net.au/news/2014-12-17/gp-co-payment-opposed-by-australian-medical-association/5974342 (accessed on 20 March 2015).

115. A political View of the Abbot's Government's Actions on Climate Change. Available online: http://larissa-waters.greensmps.org.au/abbotts-attacks-environment (accessed on 21 March 2015).

116. George, S. A Short History of Neo-liberalism: Twenty Years of Elite Economics and Emerging Opportunities for Structural Change. In Proceedings of the Conference on Economic Sovereignty in a Globalising World, Bangkok, Thailand, 24–26 March 1999.

117. Harvey, D. *A Brief History of Neoliberalism*; Oxford University Press: Oxford, UK, 2005.

118. Lansley, S. Inequality, the crash and the ongoing crisis. *Political Quart.* **2012**, *83*, 754–761.

119. APS. *Submission to the Senate Community Affairs References Committee Inquiry into the Extent of Income Inequality in Australia*; Australian Psychological Society: Melbourne, VIC, Australia, 2014.

120. Wilkinson, R.G.; Pickett, K. *The Spirit Level: Why More Equal Societies Almost Always Do Better*; Allen Lane: London, UK, 2009.

121. Whiteford, P. Australia: Inequality and Prosperity and Their Impacts in a Radical Welfare State. Social Policy Action Research Centre, Crawford School of Public Policy, the Australian National University 2013. Available online: https://crawford.anu.edu.au/public_policy_community/content/doc/Australia_Inequality-and-Prosperity_final-15-March-13.pdf (accessed on 3 October 2013).

122. NSC. Sustainable Australia Report 2013, Conversation with the Future—In Brief. National Sustainability Council, Australian Government 2013. Available online: http://www.environment. gov.au/system/files/resources/e55f5f00-b5ed-4a77-b977-da3764da72e3/files/sustainable-australia-report-2013-summary.pdf (accessed on 23 March 2015).

123. OECD. *Divided We Stand: Why Inequality Keeps Rising*; OECD Publishing: Paris, France, 2011.

124. Primrose, D. Contesting capitalism in the light of the crisis: A conversation with David Harvey. *J. Aust. Political Econ.* **2013**, *71*, 5–25.

125. The Australian Greens Policy on Sustainable Agriculture. Available online: http://greens.org.au/ policies/sustainable-agriculture (accessed on 20 March 2015).

126. CoD. Food Security Policy, City of Darebin, Victoria, Australia 2010. Available online: http://www.darebin.vic.gov.au/~/media/cityofdarebin/Files/Darebin-Living/CaringfortheEnviron ment/SustainableLiving/Food-Security-Policy-Final.ashx?la=en (accessed on 25 February 2015).

127. Commonwealth of Australia. *Agricultural Competitiveness Green Paper*; Commonwealth of Australia: Canberra, ACT, Australia, 2014.

128. Economist Intelligence Unit. The Global Food Security Index. Available online: http://foodsecurityindex.eiu.com (accessed on 22 March 2015).

129. Locke, E.A.; Latham, G.P. Building a practically useful theory of goal setting and task motivation: A 35-year odyssey. *Am. Psychol.* **2002**, *57*, 705–717.

130. AHRC Webpage. Australian Human Rights Commission. Available online: https://www.humanrights.gov.au/ (accessed on 21 May 2015).

131. AG. Australia's National Human Rights Action Plan 2012. Available online: http://www.ag.gov.au/Consultations/Documents/NationalHumanRightsActionPlan/National%20 Human%20Rights%20Plan.pdf (accessed on 10 October 2013).

132. Tasmanian Food Security Council. Food for All Tasmanians: A Food Security Strategy (2012). Available online: http://www.dpac.tas.gov.au/__data/assets/pdf_file/0005/159476/Food_for_all_ Tasmanians_-_A_food_Security_Strategy.PDF (accessed on 8 October 2012).

133. Rudd, K. House of Representatives National Security Speech. Available online: http://parlinfo. aph.gov.au/parlInfo/genpdf/chamber/hansardr/2008-12-04/0045/hansard_frag.pdf;fileType= application%2Fpdf (accessed on 26 March 2012).

134. AG. Critical Infrastructure Resilience Strategy. Available online: http://www.tisn.gov.au/ Documents/Australian+Government+s+Critical+Infrastructure+Resilience+Strategy.pdf (accessed on 1 October 2013).

135. WCED. *Our Common Future (Brundtland Report)*; United Nations World Commission on Environment and Development; Oxford University Press: Oxford, UK, 1987.

136. Max-Neef, M. *With Contributions from Antonio Elizalde and Martin Hopenhayn. Human Scale Development Conception, Application and Further Reflections*; The Apex Press: New York, NY, USA; London, UK, 2012.

137. Deci, E.L.; Ryan, R.M. The Darker and Brighter Side of Human Existence: Basic Psychological Needs as a Unifying Concept. *Psychol. Inq.* **2000**, *11*, 319–338.

138. Ryan, R.M.; Deci, E.L. Self-determination theory and the facilitation of intrinsic motivation, social development, and well-being. *Ame. Psychol.* **2000**, *55*, 68–78.

139. Sumner, L.W. *Welfare, Happiness and Ethics*; Clarendon Press: Oxford, UK, 1996.

140. Haybron, D.M. *The Pursuit of Unhappiness, the Elusive Psychology of Well-Being*; Oxford University Press: Oxford, UK; New York, NY, USA, 2008.

141. Helliwell, J.; Layard, R.; Sachs, J. World Happiness Report 2013. Available online: http://unsdsn.org/wp-content/uploads/2014/02/WorldHappinessReport2013_online.pdf (accessed on 20 May 2015)

142. Nussbaum, M.C. Capabilities, entitlements, rights: Supplementation and critique. *J. Hum. Dev. Capab.* **2011**, *12*, 23–37.

143. Nussbaum, M.C. *Women and Human Development: The Capabilities Approach*; Cambridge University Press: Cambridge, UK, 2000.

144. Sen, A. The ends and means of sustainability. *J. Hum. Dev. Capab.* **2013**, *14*, 6–20.

145. Farmar-Bowers, Q. *Making Sustainable Development Ideas Operational: A General Technique for Policy Development*; VDM Verlag Dr. Müller: Saarbrücken, Germany, 2008.

146. Pelenc, J.; Lompo, M.K.; Ballet, J.; Dubois, J. Sustainable human development and the capability approach: Integrating environment, responsibility and collective agency. *J. Hum. Dev. Capab.* **2013**, *14*, 77–94.

147. Leopold, A. *The Land Ethic. pp. 87–100 in Environmental Philosophy, From Animal Rights to Radical Ecology*, 2nd ed.; Zimmerman, M.E., Ed.; Prentice Hall: Upper Saddle River, NJ, USA, 1998.

148. McLaughlin, A. *Regarding Nature*; State University of New York Press: New York, NY, USA, 1993.

149. The Foundation for Deep Ecology. Available online: http://www.deepecology.org/platform.htm (accessed on 21 March 2015).

150. Devall, W.; Sessions, G. *Deep Ecology: Living as if Nature Mattered*; Peregrine Smith: Salt Lake City, UT, USA, 1985.

151. Meyer, J.L. *The State of the Global Environment. pp. 7–15 in Ethics and the Global Marketplace*; Dallmeyer, D.G., Ike, A.F., Eds.; The University of Georgia Press: London, UK, 1998.

152. Fieldman, G. Neoliberalism, the production of vulnerability and the hobbled state: Systematic barriers to climate adaptation. *Clim. Dev.* **2011**, *3*, 159–174.

153. Ericksen, P.J. What is the vulnerability of a food system to global environmental change? *Ecol. Soc.* **2008**, *13*, 14.

154. Gunasekera, D.; Newth, D.; Finnigan, J. Reconciling the competing demands in the Human-Earth system: Ensuring food security. *Econ. Papers* **2011**, *30*, 296–306.

155. Resilience Alliance. Assessing Resilience in Social-Ecological Systems: Workbook for Practitioners, Version 2.0, 2010. Available online: http://www.resalliance.org/3871.php (accessed on 26 March 2012).

156. Walker, B.; Salt, D. *Resilience Thinking, Sustaining Ecosystems and People in a Changing World*; Island Press: Washington, DC, USA, 2006

157. Anderies, J.M.; Folke, C.; Walker, B.; Ostrom, E. Aligning key concepts for global change policy: Robustness, resilience, and sustainability. *Ecol. Soc.* **2013**, *18*, 8.

158. Meadows, D. *Leverage Points Places to Intervene in a System*; The Sustainability Institute: Hartland, VT, USA, 1999.

159. Marsden, J. Fossil Fuel Industry's Tired Battle against Clean Energy Is Also a Losing One. Available online: http://www.forbes.com/sites/edfenergyexchange/2014/04/12/fossil-fuel-industrys-tired-battle-against-clean-energy-is-also-a-losing-one/ (accessed on 21 March 2015).

160. IRENA. Renewable Power Generation Costs in 2014. Available online: http://www.irena.org/DocumentDownloads/Publications/IRENA_RE_Power_Costs_2014_report.pdf (accessed on 20 February 2015).

161. Parkinson, G. Fossil Fuels Win Battle over RET, but Will They Win the War? Available online: http://reneweconomy.com.au/2014/fossil-fuels-win-battle-over-ret-but-will-they-win-the-war-60416 (accessed on 20 February 2015).

162. Readfearn, G. Gods and Faith *versus* Coal in Name of Climate Change. Available online: http://www.theguardian.com/environment/planet-oz/2014/dec/05/gods-and-faith-versus-coal-in-name-of-climate-change (accessed on 20 February 2015).

163. Weaver-Hightower, M.B. Why education researchers should take school food seriously. *Educ. Res.* **2011**, *40*, 15–21.

164. Arup, T. Abbott Shuts Down Climate Commission, 19 September 2013, Sydney Morning Herald. Available online: http://www.smh.com.au/federal-politics/political-news/abbott-shuts-down-climate-commission-20130919-2u185.html (accessed on 26 February 2015).

165. White, A. Climate change "off the G20 agenda" as Australia Prepares to Abolish the Carbon Price. Available online: http://www.theguardian.com/environment/southern-crossroads/2014/jun/05/g20-climate-change-agenda-obama-abbott (accessed on 26 February 2015).

13

Fermented Apple Pomace as a Feed Additive to Enhance Growth Performance of Growing Pigs and Its Effects on Emissions

Chandran M. Ajila [1], Saurabh J. Sarma [1], Satinder K. Brar [1,*], Stephane Godbout [2], Michel Cote [2], Frederic Guay [3], Mausam Verma [2,4] and Jose R. Valéro [1]

[1] INRS-ETE, Université du Québec, 490, Rue de la Couronne, QC G1K 9A9, Canada;
E-Mails: ajilaa@yahoo.com (C.M.A.); Saurabh_Jyoti.Sarma@ete.inrs.ca (S.J.S.);
josevalero@videotron.ca (J.R.V.)

[2] Institut de recherche et de développement en agroenvironnement inc (IRDA), 2700 rue Einstein,
QC G1P 3W8, Canada; E-Mails: stephane.godbout@irda.qc.ca (S.G.);
michel.cote@irda.qc.ca (M.C.); mausamverma@yahoo.com (M.V.)

[3] Department of Animal Science and Center de Recherche en Biologie de la Reproduction,
Laval University, Sainte-Foy, QC G1K 7P4, Canada; E-Mail: frederic.Guay@fsaa.ulaval.ca

[4] CO2 Solutions Inc., 2300, rue Jean-Perrin, QC G2C 1T9, Canada

* Author to whom correspondence should be addressed; E-Mail: Satinder.Brar@ete.inrs.ca or
satinderbrar2003@yahoo.ca

Academic Editor: Stephen R. Smith

Abstract: Apple pomace is a by-product from the apple processing industry and can be used for the production of many value-added compounds such as enzymes, proteins, and nutraceuticals, among others. An investigation was carried out to study the improvement in the protein content in apple pomace by solid-state fermentation using the fungus *Phanerochaete chrysosporium* by tray fermentation method. The effect of this protein in terms of how it enriched apple pomace as animal feed for pigs has also been studied. There was a 36% increase in protein content in the experimental diet with 5% w/w fermented apple pomace. The efficiency of conversion of ingested food was increased from 43.5 ± 2.5 to 83.1 ± 4.4 in the control group and the efficiency of conversion of feed increased from 55.4 ± 4.5 to 92.1 ± 3.6 in the experimental group during the animal feed experiment. Similarly, the effect of a protein enriched diet on odor emission and greenhouse gas emission has also been studied. The results demonstrated that the protein enrichment of apple pomace by solid state cultivation of the fungus *P. chrysosporium* makes it possible to use it as a dietary supplement for pigs.

Keywords: apple pomace; *P. chrysosporium*; fermentation; protein enrichment; animal feed; greenhouse gas emission

1. Introduction

The production and processing of fruits into different products, such as fruit juices, flavors and concentrates, result in the production of a large quantity of fruit processing by-products, such as pomace. Generally, the apple processing industry generates 25%–30% apple pomace and 5%–10% sludge. Apple pomace residues are normally rich in carbohydrates, and other functionally important bioactive compounds, such as polyphenols and other natural antioxidants. Fruit and vegetable residues have been successfully used for protein enrichment and for bioconversion into value-added products, such as enzymes and other metabolites [1]. Apple pomace is presently used to feed animal or simply added to soil as a fertilizer. Several factors adversely affect the value of apple pomace as an animal feed. Firstly, the residue has low digestibility due to high lignin/cellulose ratio. Secondly, the protein, vitamin and mineral contents of apple pomace are low, which contribute to the low nutritional level and consequently lower commercial value of the residue. From an animal nutrition point of view, apple pomace is not a suitable feed as it is deficient in digestible protein [2]. Recently, *Saccharomyces cerevisiae* was used to increase protein levels of pineapple waste by Solid State fermentation (SSF) with and without nitrogen supplementation [3].

The apple pomace can be used for the production of the high protein enriched residue that can serve as animal feed [4]. It was reported that the growth of yeast on apple pomace increases protein and vitamin content. However, the lower level of fermentable sugars limits protein enrichment of the pomace by yeasts; a major portion of the pomace comprises lignocelluloses. The co-culture of *Candida utilis* and *Aspergillus niger* increased the protein content of dried and pectin-extracted apple pomace to 20% and 17%, respectively, under SSF conditions [5]. The combination of fermented apple pomace with standard feed in the ratio of 1:1 was found to be acceptable during animal experiment with rat model [6]. The apple pomace based feed had also been evaluated successfully in poultry, when mixed with standard poultry feed in 1:1. There was an increase in body weight of broiler from 6th week to 8th week in all the groups where the mean body weight increased from 270 to 537 g [7].

The agricultural sector is one of the main sources of greenhouse gas (GHG) emissions worldwide and the magnitude of its contribution changes from country to country. It was reported that globally, livestock are responsible for 18% of greenhouse gas emissions [8]. Reducing or mitigating greenhouse gases (GHG) from livestock systems can play a vital role in providing solutions to climate change obligations. The major greenhouse gases generated by cattle production include methane (CH_4) and nitrous oxide (N_2O), which has global warming potential of 21 and 310 times that of CO_2 respectively, making them very potent GHGs. Many attempts have been made to mitigate greenhouse gas emissions without altering animal performance by dietary manipulation [8]. It was reported that emissions per kilogram of livestock product seem to be lower for monogastric than for ruminant animals, partly because pigs and poultry have better feed-conversion efficiency than ruminants and also they do not emit enteric methane while digesting their feed [9]. Emissions of GHGs from pork are lower than for beef, and produce very small amount of methane in their feed digestion and production is dominated by nitrous oxide [10].

The alteration in protein concentration in animal feed has been proposed as a means of mitigating GHG, especially methane. It was reported that increasing dietary crude protein concentration from 120 to 150 g/kg DM (digestible matter) in dairy cows significantly reduced methane emission as a proportion of dry matter intake and milk yield and further increasing dietary crude protein concentration to 180 g/kg DM had no effect [11]. Increasing dietary concentrate level reduced methane emission as a proportion of feed intake and milk yield.

The objectives of the present investigation were to study the protein enrichment of apple pomace by the process of solid-state fermentation using *P. chrysosporium* and its addition as a protein supplement to the diet of pigs and study the growth performance of animals. The present investigation also studied the effect of these protein enriched diets on the reduction of GHG emissions.

2. Results and Discussion

2.1. Proximate Composition in the Control and Experimental Diet with Fermented Apple Pomace

The proximate composition of the control and experimental diet which was formulated with protein enriched apple pomace by *Phanerochaete chrysosporium* under solid state fermentation (SSF) is given in Table 1. There was 36% increase in protein content (15.85% to 24.81%) by the addition of 5% w/w protein enriched apple pomace by solid state fermentation. The analysis of most of other nutrients and minerals showed no significant difference ($p < 0.05$) between the control diet and experimental diet with protein enriched apple pomace.

Table 1. Proximate composition of control and experimental diet.

Component	On Dry Weight Basis	
	Control Diet	Experimental Diet
Dry matter (%)	89.50 ± 0.10	89.55 ± 0.15
Crude Protein (%)	15.85 ± 0.05	24.81 ± 0.2
Lipid (%)	5.62 ± 0.08	4.57 ± 0.01
Ash (%)	3.55 ± 0.29	4.06 ± 0.03
NDF (%)	9.82 ± 0.14	10.55 ± 0.45
ADF (%)	3.95 ± 0.05	3.88 ± 0.24
Phosphorus (mg/kg)	6032 ± 130	5241 ± 95
Potassium (mg/kg)	6757 ± 27	7053 ± 13
Calcium (mg/kg)	5946 ± 407	4712 ± 73
Sodium (mg/kg)	1551 ± 40	1847 ± 197
Boron (mg/kg)	09.37 ± 0.15	12.95 ± 0.95
Aluminum (mg/kg)	146 ± 3.5	146 ± 2
Copper (mg/kg)	27.35 ± 1.75	30.00 ± 1.2
Iron (mg/kg)	258 ± 3	231 ± 5
Magnesium (mg/kg)	1484 ± 19	1462 ± 1
Manganese (mg/kg)	62.45 ± 0.25	62.95 ± 0.05
Zinc (mg/kg)	321 ± 17	195 ± 12

All data are the mean ± SD of three replicates.

Earlier, it was reported that co-culture of *Candida utilis* and *Aspergillus niger* increased the protein content of dried and pectin-extracted apple pomace to 20% and 17%, respectively, under SSF conditions [5]. In solid state fermentation, the microorganisms hydrolyze the cellulose or hemicellulose component of the pomace by secreting extracellular enzymes (cellulases and xylanases) and then use the released sugar as carbon source for growth. This results in better utilization of the substrate than the microorganism could achieve independently. The higher yield of protein would have possibly resulted from the enzymatic hydrolysis of the lignocellulosic component of the pomace which was later used by the microorganisms. *Phanerochaete chrysosporium* has been already reported to produce ligninolytic and carbohydrate metabolizing enzymes during solid state fermentation of apple pomace [12].

Carbohydrate content of apple pomace could be as high as 85% [12] and Zheng and Shetty (1998) used it to produce a food rich in proteins by employing the fungus, *Rhizopus oligosporus* [13]. Bisaria *et al.* (1997) used the fungus *Pleurotus sajor-caju* in the bioconversion of rice straw and wheat straw and they found that supplementation of the solid residue with urea and ammonium nitrate increased the protein level from 2.87% to 6.3% (w/w) with rice straw and from 3.1% to 7.5% (w/w) with wheat straw [14]. Solid state fermentation of apple pomace blended with 10% molasses and 1.8% ammonium sulfate as a nitrogen source resulted in increased crude protein content [4]. Compared to these reports, as already mentioned, the protein content can be increase from 15.85% to 24.81% with the addition of 5% w/w protein enriched apple pomace from the present study.

2.2. Effect of Protein Enriched Diet on Growth Performance

With the objective of testing the apple pomace treated biologically with *P. chrysosporium* as a food supplement in animal feed, two diets were provided to the animals. The diets used over a period of four weeks were conventional feed and conventional feed amended with the addition of 5% (w/w) treated apple pomace (Table 2). Table 3 presents the effects of protein enriched diet on average daily gain on growing pigs and also on growth performance of the growing pigs during the experiment. The average daily gain for control diet group ranged from 0.72 ± 0.13 kg/day to 2.4 ± 0.11 kg/day and for the experimental diet from 0.64 ± 0.05 kg/day to 2.60 ± 0.08 kg/day during the experimental period. The results of comparisons of feed conversion ratio are presented in Figure 1. The feed conversion is calculated by dividing the amount of feed fed by amount of pig weight gain. The feed conversion ratio was found to be lower in the experimental diet group than the control group.

Table 2. Composition of the basic diet.

Component	On Dry Weight Basis (% w/w)
Maize	82.09
Soya	15.79
Limestone	0.95
Phosphate	0.42
Lysine-HCl	0.21
Vitamin Premix PorcCroiss	0.21
Phytase 5000 (Phyzyme)	0.01
NaCl	0.32

Control diet—95% w/w Basic diet + 5% w/w wheat; Experimental diet—95% w/w Basic diet + 5% w/w fermented apple pomace.

Table 3. Effect of diet on average daily gain and average daily feed intake.

Week	Average Daily Gain (kg/day)		Average Daily Feed Intake (kg/day)	
	Control Diet	Experimental Diet	Control Diet	Experimental Diet
1	0.721 ± 0.132 [a]	0.637 ± 0.046 [a]	2.601 ± 0.028 [x]	2.928 ± 0.024 [xy]
2	1.136 ± 0.423 [b]	1.267 ± 0.073 [b]	2.978 ± 0.032 [xy]	3.114 ± 0.042 [y]
3	1.815 ± 0.071 [c]	1.934 ± 0.075 [c]	3.011 ± 0.046 [y]	3.325 ± 0.046 [yz]
4	2.446 ± 0.113 [d]	2.601 ± 0.080 [d]	3.290 ± 0.054 [yz]	3.504 ± 0.062 [z]

All data are the mean ± SD of three replicates. Mean followed by different letters in the same column differ significantly ($p < 0.05$) between the control and experimental diet.

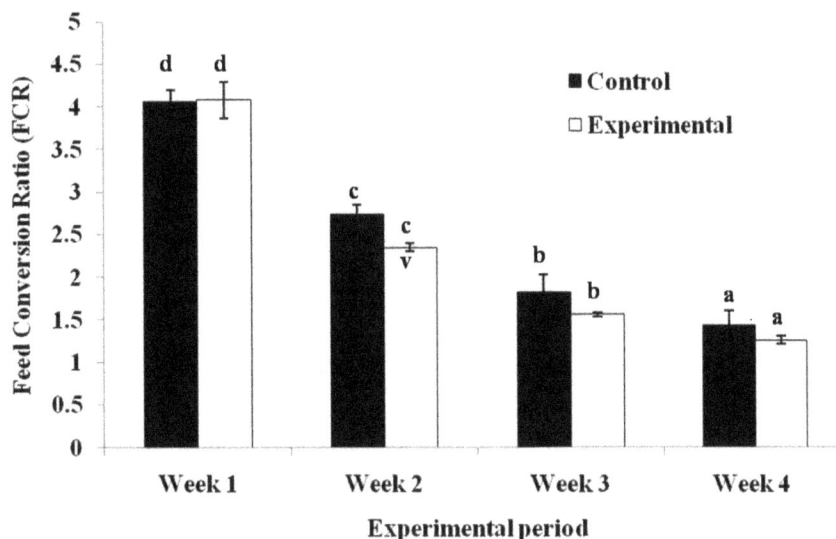

Figure 1. Effect of protein enriched diet on feed conversion ratio. All data are the mean ± SD of three replicates. Mean followed by different letters in the same column differ significantly ($p < 0.05$) between the control and experimental diet.

In the case of the experimental diet, the average daily gain was in the range of 0.637 ± 0.05 kg/day to 2.60 ± 0.08 kg/day (Table 3). There was slight increase of average daily gain in the experimental diet animal group compared to control diet group. This increase in average daily gain may be due to the higher content of protein in experimental diet.

The average daily feed intake during the animal experiment in control group was in the range of 2.60 ± 0.03 to 3.29 ± 0.05 kg/day and it was in the range of 2.93 ± 0.02 to 3.50 ± 0.06 kg/day for the treated group (supplied with 5% fermented pomace supplement). In general, it was noted that there was no aversion of the experimental diet formulated with fermented apple pomace. This is clear from the similar trend in the average daily feed intake by the animals in both the control diet and experimental diet groups.

The feed conversion ratio during the experiment was in the range of 4.06 to 1.46 in control group and 4.08 to 1.13 in the experimental diet group (Figure 1). Improving feed conversion is a matter of either decreasing the amount of feed fed or increasing the amount of weight gained or both the factors. It was also important that the feed conversion ratio was higher in the first two weeks of the animal experiment and gradually decreased in both control and experiment diet groups. This may be due to the change in growth rate during the first weeks of the growth period.

The term feed conversion ratio is utilized to indicate the quantity of feed required to lay down a unit of body tissue. The feed conversion ratio is used as indicators of the performance standard of a production system. The Feed Conversion Ratio (FCR) is a reflection of performance in animals. Feed intake and weight gain are the integral components which influence FCR. FCR performance within a group of animals can vary widely. In fact, FCR is not an index based on the variables of growth rate and feed intake but based on the factors affecting them. It is not just a calculated value but an independent biological factor that can be influenced by genetics, feeding practice, environmental control or health status. In general, the faster the growth, the better the conversion efficiency. The FCR is only a measure of the ability of the animal to convert diets into growth at that particular condition. Factors that affect growth rate and feed consumption and utilization will also affect efficiency of conversion.

These results demonstrated the fact that the protein produced by the fungus *P. chrysosporium* during protein enrichment of the apple pomace enabled its use as a feed supplement in diets for pigs, representing a useful value-addition for the apple pomace and other agro-industrial residues. Besides avoiding the dumping of this residue directly into the environment and thereby reducing the levels of environmental pollution, this application also reduces the costs of feed, which represent around 60% of the production cost [15].

2.3. Water Consumption and Feed Consumption during the Experiment

The water and feed consumption during the four weeks of experimental period for both the control and experimental diet group has been recorded daily. The water consumption by the control and experimental diet group animal are given in the Figure 2. The water consumption in the control group and the experimental group increased depending on the experimental period from 5.18 to 5.26 L/day and from 4.95 to 5.05 L/day respectively. The consumption of feed was determined as shown in the Figure 3 and the feed consumption index (C.I.) with respect to fresh weight was calculated from fresh weight of feed intake and fresh weight of animal.

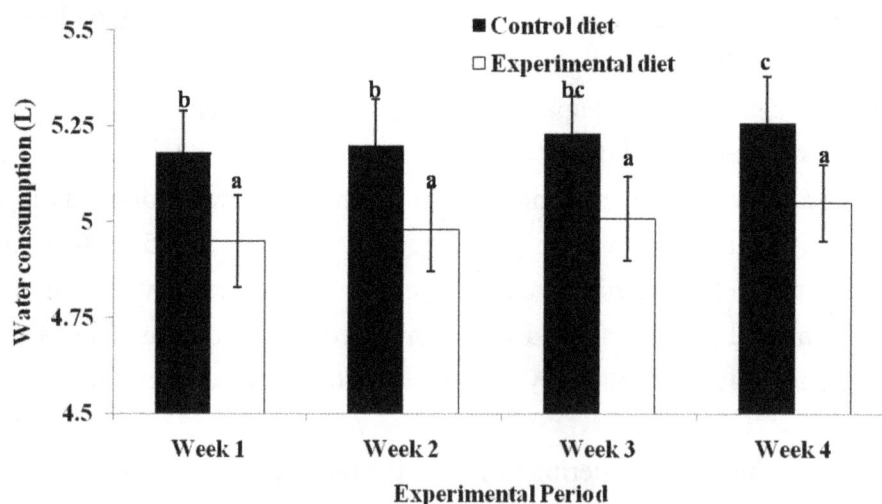

Figure 2. Water consumption during the experimental period. All data are the mean ± SD of three replicates. Mean followed by different letters in the same column differ significantly ($p < 0.05$) between the control and experimental diet.

The water consumption was higher in the control diet group compared to the experimental group (Figure 2). In the experimental group, 5% w/w of the normal diet, mainly wheat was replaced by fermented apple pomace. The fermented apple pomace is easily digestible than the normal grains, which probably in turn leads to lower consumption of water in the experimental group. This shows the favorable effect of the fermented pomace as a supplement in the normal feed.

The consumption of feed was determined as shown in Figure 3 and the feed consumption index (C.I.) is an indicator of relative intake of nutrients. The index as calculated underestimates absolute amount of food consumed due to the water content of the medium. No differences in consumption index were observed among the control and experimental diet group during the first two weeks. Later, there was a slight increase in the consumption index among the control and experimental diet group during the third and fourth week. This increase in consumption index during the third and fourth week may be due to the increase in growth of the animals and the easy availability of nutrients to the animal.

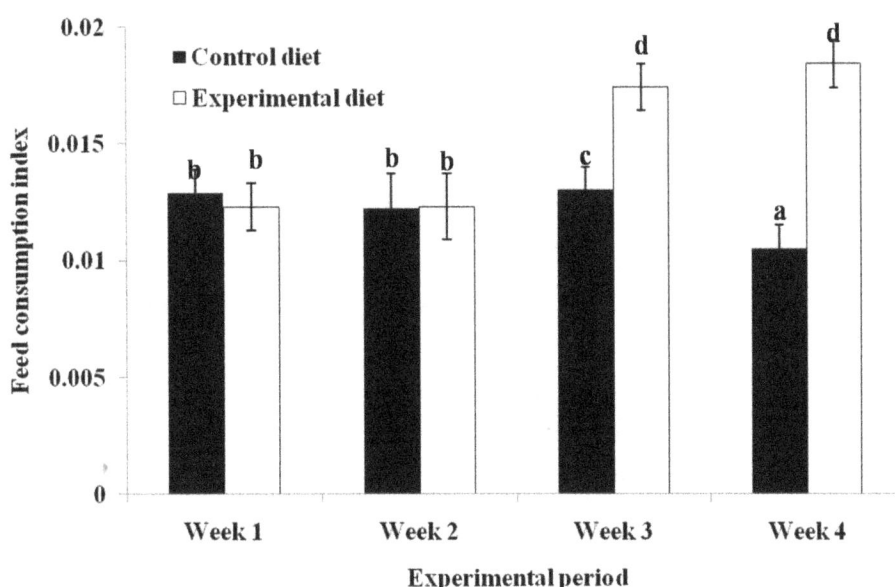

Figure 3. Feed Consumption index during the experiment. All data are the mean ± SD of three replicates. Mean followed by different letters in the same column differ significantly ($p < 0.05$) between the control and experimental diet.

Matoo et al., (2001) have reported better performance of broilers chickens fed on apple pomace diets supplemented with enzymes [16]. It was reported that apple by-products replaced the maize in normal diet of broiler chicken improved the growth performance and were also reported to reduce the cost of poultry feed [17].

2.4. Effect of Feed on Efficiency of Conversion of Ingested Feed (ECI)

The efficiency of conversion (ECI) of ingested feed increased with the experimental period in both the control and the experimental diet group and the ECI of both control and experimental diet group is given in the Table 4. The efficiency of conversion of ingested food increased from 43.5 ± 2.5 to 83.3 ± 4.4 during the experimental period in the control diet group, whereas in the case of experimental diet, the efficiency of conversion of feed increased from 55.4 ± 4.5 to 92.1 ± 3.6.

Table 4. Effect of feed on efficiency of conversion of ingested feed (ECI).

Week	Control diet	Experimental diet
1	43.46 ± 2.50 [a]	55.36 ± 4.50 [b]
2	72.98 ± 4.80 [c]	85.14 ± 6.20 [d]
3	79.14 ± 6.80 [e]	88.48 ± 8.40 [f]
4	83.14 ± 4.40 [f]	92.08 ± 3.60 [g]

All data are the mean ± SD of three replicates. Mean followed by different letters in the same column differ significantly ($p < 0.05$) between the control and experimental diet.

The term feed conversion efficiency (ECI) is utilized to indicate the quantity of feed required to lay down a unit of body tissue. ECI used as indicators of the performance standard of a production system. The higher the quality of the feed, in general, the better the animal will convert it to body tissue. The more palatable the feed, the higher the intake, which in turn gives better conversion rate. The faster the growth, the conversion efficiency will be better. Factors that affect growth rate and feed consumption and utilization will also affect efficiency of conversion.

It is noteworthy that the improving digestibility of nutrients leads to improvement in nutritive value of diet which in turn improves the growth performance of the animal. It has been reported that exogenous enzymes (cellulase, xylanase, alpha-amylase and polygalacturonase) treatment were good methods for enhancing the growth performance and digestibility of ration containing pearl millet without any hazard on buffalo health [18]. Rao *et al.* (2004) found that the performance of broiler chicks fed diets containing pearl millet (*Pennisetum glaucum*) with and without enzymes comprising amylase, hemicellulase, cellulase, proteinase and beta-glucanase improved the immunological traits and reduced total cholesterol in the tissue of broilers [19]. The effects of beta-glucosidase on feed usage, growth performance, and activities of digestive enzyme and a set of physiological parameters in broilers have been studied and showed positive impact on the growth performance of the broilers [20]. A study conducted by Omogbenigun *et al.* (2004) reported that an improvement in average daily gain had been observed in piglet fed diets based on corn and wheat supplemented with an enzyme cocktail containing cellulase, galactanase, mannanase, and pectinase [21]. Significant benefits to xylanase supplementation were demonstrated in growing pigs fed corn-based diets in terms of the growth rate [22]. Besides, Kim *et al.* (2001) observed that average daily growth was increased by 3% and 7%, respectively, with 0.1% enzyme complex supplementation to corn and soybean meal based diets for weaned piglets [23].

2.5. Nutrient Assimilation Efficiency in Control and Experimental Diets

The nutrient assimilation efficiency for different nutrients, such as protein, lipid and minerals has been determined and given in Table 5. The nutrient assimilation efficiency for protein increased from 91.15% ± 0.2% (control) to 93.46% ± 0.8% (experimental). This increase in protein assimilation in the experimental group may be due to the higher protein content, which was easily assimilated in the case of fermented pomace. The concentration of protein in fermented substrate (apple pomace) increased significantly ($p < 0.05$) during fermentation. There was not much difference in mineral assimilation efficiency between the control and experimental diet group.

Table 5. Nutrient assimilation efficiency in control and experimental diets fed to pigs.

Component	% of Nutrient Assimilation	
	Control Diet	**Experimental Diet**
Crude Protein	91.2 ± 2	93.5 ± 8
Lipid	93.1 ± 1	91.3 ± 2
Phosphorus	86.4 ± 1	84.6 ± 8
Potassium	77.5 ± 2	78.9 ± 4
Calcium	88.7 ± 4	84.9 ± 2
Magnesium	79.1 ± 8	78.7 ± 6
Boron	79.2 ± 6	81.2 ± 2
Aluminum	85.5 ± 5	84.0 ± 4
Copper	78.4 ± 8	72.8 ± 6
Iron	80.9 ± 3	74.0 ± 8
Manganese	76.6 ± 4	74.3 ± 6
Zinc	75.7 ± 1	75.8 ± 4
Sodium	79.4 ± 4	81.2 ± 6

All data are the mean $\pm$ SD of three replicates.

Diet choice should reflect selective feeding to maximize the rate of net energy gain [24]. Energy value of the diet, assimilation efficiency and nutrient content (in particular N or protein) are the factors generally considered when assessing food value relative to diet choice. Assimilation efficiencies were calculated by comparing organic and ash contents of the food with the corresponding fecal material, using ash as a non-absorbed reference marker. This increase in protein assimilation in the experimental diet group may be due to the higher protein content, which was easily assimilated in the case of fermented pomace. There was not much difference in mineral assimilation efficiency between the control and experimental diet group.

Earlier, it has been reported that the fermentation improved the nutritional levels of the apple pomace [25]. The concentration of protein fermented substrate increased significantly ($p < 0.05$) during fermentation. It was reported that total amino acids as well as the essential amino acids were increased remarkably reaching 46.8% and 71.7%, respectively [25].

2.6. Effect of Feed on Air Quality

The effect of feed on air quality has been determined in terms of odor and greenhouse gas emission. The odor emissions were measured for last two weeks and ranged from 4.8 to 7.6 ouE/s/head (Table 6) in control and experimental diet group. The results showed that the treatment had no significant impact on gas emissions (Table 7). The average CH_4 and CO_2 emissions were 0.0138 g/day/kg and 52 g/day/kg for the control diet and 0.0299 g/day/kg and 69 g/day/kg for the treatment diet. The emission of NH_3 during the experiment was 0.11 to 0.13 g/day/kg in the control and 0.11 to 0.14 g/day/kg in the experimental diet group.

Table 6. Effect of control and experimental diet on environmental parameter values.

	Temp (°C)		Ventilation L/s/pig		Water L/day/pig		Odor Emission ouE/s/pig		Odor Concentration ouE/m^3		Hedonic Tone	
	Cont	Expt	Cont	Expt	Cont	Expt	Cont	Expt	Cont	Expt	Cont	Expt
Week 1	19.5	19.5	19.6	18.1	4.1	4.6						
Week 2	18.9	18.9	19.7	18.4	4.4	4.5						
Week 3	18.5	18.5	20.4	19.2	5.1	4.8	6.5	7.6	341	410	0.9	−0.7
Week 4	18.1	18.0	21.2	19.7	5.4	5.4	6.4	4.8	310	261	1.0	−1.1

Table 7. GHG emission values for each diet treatment and growth period.

Weeks	Weight (kg)	CH$_4$		CO$_2$		N$_2$O		NH$_3$	
		Cont	Expt	Cont	Expt	Control	Expt	Control	Expt
1.0	52.5	0.0126	0.0324	70	95	2.02	2.20	0.13	0.14
2.0	62.2	0.0160	0.0373	66	85	2.13	2.28	0.13	0.14
3.0	69.1	0.0188	0.0346	59	80	2.24	2.12	0.12	0.13
4.0	77.0	0.0138	0.0299	52	69	2.07	2.33	0.11	0.11

CH$_4$, g/day/kg; CO$_2$, g/day/kg; N$_2$O, mg/day/kg; NH$_3$, g/day/kg.

2.6.1. Odor Emission

The treatment had no significant impact on odor emissions. The odor emissions were determined for 3–4 weeks of the experimental period and ranged from 4.8 to 7.6 ouE/s/head (Table 6) and were found to be in the same range as the literature values [26]. Due to the odor produced by apples as such, the hedonic tone was lower using the experimental diet than the control diet. However, due to the insufficient data, no statistical analysis has been performed.

2.6.2. Greenhouse Gas (GHG) Emission

The results showed that the treatment had no significant impact on gas emissions (Table 7). The ammonia emissions measured were 7.9 and 8.4 g/day/head and it was very close to values reported by Hamelin *et al.* (2009) [26]. On average, the emissions from growing finishing pig under Quebec condition are NH$_3$ = 5.48 g/day/head; CH$_4$ = 2.26, CO$_2$ = 1.940 kg/day/head and no N$_2$O emissions have been measured [26]. There may some other reason which played a role in GHG emission other than the nutrient management during the feeding trial, such as manure and farm house management. Further experiments are in progress which focuses on feed management to reduce crude protein intake keeping higher growth performance as well as lowering GHG emissions.

Many environmental organizations have established threshold values for NH$_3$ in ambient air. On average, the NH$_3$ chronic exposure limit varies from 100 to 300 ppb, whereas the acute exposure limit ranges from 1700 to 4500 ppb. The emission of NH$_3$ during the experiment was 0.11 to 0.13 g/day/kg in the control and 0.11 to 0.14 g/day/kg in the experimental diet group. This concentration of NH$_3$ does not reflect a health risk when compared to chronic and acute exposure limits.

Nutritional management can substantially reduce nitrogen excretion and ammonia emission by pigs. It is generally agreed that feeding low crude protein (CP) diets supplemented with synthetic amino acids lowers urea excretion by pigs, resulting in lower ammonia emission, and the nitrogen excretion is shifted from urea in the urine to bacterial protein in the feces when fibrous feedstuffs are included in the diet. It was found that nitrogen excretion in the urine of pigs fed a standard CP diet supplemented with dried apple pomace at the rate of 23.1% was 36% lower than with the standard CP diet. It was found that the protein assimilation in the experimental diet was higher (93.5%) than the control diet (91.2%) as shown in Table 5. The higher protein assimilation may indirectly reduce the nitrogen excretion and ammonia emission. A detailed study should be conducted to determine protein assimilation and its role in GHG emission in the future. Earlier, it was reported that when the productivity of animal improved, CH_4 emission per unit of product reduced as feed energy associated with maintenance of the animal is reduced [27]. It was reported that 50% of total nitrogen excreted by pigs could be reduced by modifying the composition of the diet without altering the animal performance by balancing amino acid composition of diets, and phase feeding [28].

3. Experimental Section

3.1. Materials

Apple pomace samples from the apple processing industry (Lassonde Inc., Rougemont, Montreal, QC, Canada) was collected and used as the solid substrate for the solid-state fermentation. All chemicals required for the experiments have been purchased from Fisher Scientific (Fisher Scientific Company, Ottawa, ON, Canada), VWR chemicals (VWR International, Québec, QC, Canada) and Sigma Chemicals (Sigma-Aldrich Canada Ltd., Ottawa, ON, Canada) and were of analytical grade.

3.2. Solid State Fermentation

Medium for fermentation: Apple pomace was used as natural substrate for solid-state fermentation [12]. The apple pomace solids were stored at −20 °C for its conservation prior to use. For the fermentation, apple pomace was treated with inducers, such as copper sulfate (2 mM), veratryl alcohol (2 mM) and Tween-80 (0.1% v/v) and the pH was adjusted to 4.5 and it was sterilized in an autoclave for 30 min at 121 ± 1 °C. The moisture content in the apple pomace was 72% w/v.

Microorganisms: *P. chrysosporium* (ATCC 24275) was selected as a suitable microorganism for bio-processing of solid state fermentation for its potential for higher enzyme production. *P. chrysosporium* was maintained on potato–dextrose–agar (PDA) medium at 4 ± 1 °C. The culture of *P. chrysosporium* was grown on PDA petri plate and incubated at 37 ± 1 °C (96 h). The spores were harvested from the sporulation medium plates and inoculated into sterile distilled water contained in test tubes and stored in the freezer until use. The fermented broth was initially filtered through glass wool to remove mycelial contamination and recover only the spores. The concentration of spore suspension used in the experiments was 2.5×10^6 spores/g of solid.

Solid-state fermentation in tray: The fermentation was carried out in 5L capacity plastic tray bioreactors (0.03 m × 0.018 m × 0.012 m). The sterilized medium containing 0.5 kg apple pomace was transferred into the sterilized tray in an aseptic condition. The inoculation was carried out using spore suspension. The fermentation was carried out in an environmental chamber (Percival Scientific, Perry, IA, USA) at 37 ± 1 °C and relative humidity ranged from 72%–74%. The fermentation was carried out for 10 days with occasional mixing for proper aeration and the samples were periodically drawn. The fermented apple pomace was dried at 60 °C for 48 h and the dried fermented apple pomace was milled and sieved using 3 mm size sieve. The powdered fermented apple pomace was ultimately used for the preparation of the experimental diet.

3.3. Diet

The main dietary ingredients in the basic diets were maize, soybean, mineral, and vitamin mix and phytase enzymes as shown in Table 2. Premixes were added to meet or exceed mineral and vitamin requirements [29]. For the control diet used for the experiment, 5% w/w of the basic diet was replaced by 5% w/w of wheat and for the experimental diet, 5% w/w of the basic diet was replaced by powdered fermented apple pomace. The animals showed aversion to the larger chunks of fermented apple pomace during the trial diet experiments with fermented apple pomace. Hence, for the final studies, powdered apple pomace was used.

3.4. Animal Experimental Design

The animal protocol for the study was followed as per the principles established by the Canadian Council on Animal Care (CCAC, 2009) [30]. The experiments were conducted at the experimental farm at research center of the Institut de recherche et de développement en agroenvironnement (IRDA, Deschambault, QC, Canada). In total, 12 pigs were weaned at 17 ± 1 day of age and healthy animals were selected for the experiment.

The experimental laboratory farm consisted of 12 identical and independent chambers laid out side-by-side. Six of them have been used for the project. Each room is equipped with a fully concrete slatted floor and has its own manure handling system. Two growing-finishing pigs were housed per chamber during seven weeks (2 weeks of accommodation and 4 weeks of testing).

The ventilation system consisted of an inlet and exhaust fan mounted in the ceiling of each chamber. The exhaust fan was able to vary the capacity from 14 to 75 L/s. The exhaust air was directed through a 204-mm iris orifice damper (Model 200; Continental fan manufacturer Inc., Buffalo, NY, USA). Its accuracy was rated at ±5%. A differential pressure transducer measured the pressure across the orifice plate. The temperature (17–18 °C) and the humidity (39%–52%) was maintained for a particular range during the entire experiment. The chambers were equipped with self-feeder and tabulated drinker. Pigs had free access to feed and water during the entire 4-week study. Three chambers were fed with control diet and three chambers were fed with experimental diet. All animal-related data (feed and water consumption, weight gain, volume of manure) as well as manure samples were taken weekly. Weekly, individual pigs were weighed to monitor weight gain and the amount of feed consumed was determined. Freshly voided feces were collected from each chamber. Feces were pooled by chambers and frozen at −20 °C. Upon completion of the growth trial, feces were thawed, homogenized, sub-sampled and used for the chemical analyses.

Feed consumed, water intake, weight gained, feed conversion ratio and mortality was regularly recorded. The efficiency of the feeds given to the pigs were analyzed on the basis of consumption index, growth rate, efficiency of conversion of ingested feed and digestibility of feed, as described by Waldbauer (1968) [31]. Assimilation efficiencies were calculated following a commonly used procedure in which the nutrient content of the food is compared with that of the feces using ash as an assumed non-absorbed marker [32]. The formulas used for the analysis has been given in Equations (1) to (5).

$$\text{Consumption index} = \text{Weight of feed eaten/Mean weight of animal} \times \text{Duration of experiment during feeding trial} \tag{1}$$

$$\text{Growth rate (GR)} = \text{Weight gained by the animals/Mean weight of animal} \times \text{Duration of feeding trial during feeding trial} \tag{2}$$

$$\text{Efficiency of conversion of ingested food to body substance (E.C.I.)} = 100 \times \text{wt gained/wt food intake} \tag{3}$$

$$\text{Efficiency of conversion of digested food to body substance (E.C.D.)} = 100 \times \text{wt gained/(wt food intake} - \text{wt of feces)} \tag{4}$$

$$\text{Nutrient assimilation efficiency (\%)} = 1 - (\text{\% of Ash in food/\% of Ash in Feces)} \times (\text{\% of nutrient in feces/\% of nutrient in food)} \times 100 \tag{5}$$

3.5. Chemical Analyses of Diet and Feces

Control and experimental diets and feces were analyzed for crude protein (CP), crude fat, ADF (Acid detergent fiber), NDF (Neutral Detergent Fiber) and ash (Association of Official Analytical Chemists, AOAC, 2006) [33]. Metal concentrations in the diets and feces were analyzed by ICP-AES (Varian Vista AX, Palo Alto, CA, USA) after a complete acid digestion of sample using HNO_3, $HClO_4$, HF and HCl. The quality control of elemental analysis by ICP-AES was performed with two certified liquid samples (multi-element standard, catalog number 900-Q30-100 (lot SC-8305871) and 990-Q30-101 (lot SC7256497), SCP Science, Lasalle, QC, Canada). The standards used for ICP-AES allowed the quantitative measurement of 22 elements in the sample sediments (Al, As, Ba, Ca, Cd, Co, Cr, Cu, Fe, K, Mg, Mn, Mo, Na, Ni, P, Pb, S, Si, Sn, Tn, and Zn).

3.6. Air Quality Measurements

The chambers were provided with uniform heating and ventilation rates, and with instrumentation to continuously measure temperature, relative humidity using a type T thermocouple and electronic humidity sensor (Model CHG-UGS, TDK Corporation of America, Mount Prospect, IL, USA). Each sensor was scanned every 10 sec and the average was recorded every 15 min. The concentration of methane (CH_4), carbon dioxide (CO_2) and nitrous oxide (N_2O) was measured with a gas chromatograph (Varian 3600, Palo Alto, CA, USA) equipped with a flame ionization detector (FID) for detection and quantification of CH_4 and an electron capture detector (ECD) for detection and quantification of CO_2 and N_2O. Ammonia was measured with a non-dispersive infra-red (NDIR) analyzer (Ultramat 6E, Siemens, Berlin, Germany) and the semi-quantitative evaluation of hydrogen sulphide (H_2S) was carried out using a UV fluorescence analyzer (M101E, Teledyne API, San Diego, CA, USA). Every two days, the analyzers would monitor ambient air and certified calibration gas.

Odor samples were collected one time during the two last weeks. Odor concentration and hedonic tone were evaluated by dynamic olfactometry. Air samples at the air room intake and exhaust were collected in 60-L flushed Nalophane bags for odor evaluation. The olfactometry (Nasal Ranger™,

St. Croix Sensory Inc., Stillwater, MN, USA) tests were carried out in a 24 h period from the sampling time.

3.7. Statistical Analyses

All the experiments were conducted in replicate and data presented are an average of replicates along with the standard deviation. The database was subjected to an analysis of variance (ANOVA) and multiple range tests among data were carried out using the Statistical Analysis System Software (STATGRAPHICS Centurion, XV trial version 15.1.02 year 2006, StatPoint, Inc., Warrenton, VA, USA) and the results which have $p < 0.05$ were considered as significant. The method used to discriminate among the means is Fisher's least significant difference (LSD) procedure. Using this method, there is a 5.0% risk of calling each pair of means significantly different when the actual difference equals 0.

4. Conclusions

Apple pomace was found to be a good substrate for the production of a protein feed rich in higher content of soluble proteins and the rapid colonization of the substrate by the fungus *P.chrysosporium* suggesting a potential use of this agro-industrial residue. There was a 36% increase in protein content in the experimental diet by using 5% fermented apple pomace. The fermented apple pomace incorporated into the diet of the pigs was a nutritive product and improved the performance of the animals. These results demonstrate that the protein produced by the fungus *P. chrysosporium* in the protein enrichment by solid-state cultivation of the apple pomace enhance its use as a dietary supplement in swine diet. From the improved performance of animals, it can be concluded that the microbial protein supplied the energy demand of the animals.

Modification of diets in terms of its nutrient content was found to have a dual impact on both enteric and manure GHG emissions and was cost effective in terms of feed and livestock management. Furthermore, there is a great deal of uncertainty associated with emissions from the manure management systems, including loss of N associated with different handling systems, storage and application and therefore reduction strategies cannot be effectively assessed.

Acknowledgments

The authors are sincerely thankful to the Natural Sciences and Engineering Research Council of Canada (Discovery Grant 355254, Canada Research Chair), FQRNT (ENC 125216), MAPAQ (No. 809051) and Projet Initiative Inde 2010 for financial support. The views or opinions expressed in this article are those of the authors.

Author Contributions

Experiments were designed, performed and samples were analyzed by C.M. Ajila. Data analysis and manuscript preparation were jointly completed by C.M. Ajila and S.J. Sarma. All other authors have equally participated in supervising the work and finalization the manuscript.

Conflicts of Interest

The authors declare no conflict of interest.

List of Abbreviations

Acid detergent fiber (ADF)

Association of Official Analytical Chemists (AOAC)

Canadian Council on Animal Care (CCAC)

Consumption index (C.I.)

Control (Cont)

Efficiency of conversion (ECI)

European Odor Units (ouE)

Experimental (Expt)

Feed Conversion Ratio (FCR)

Flame ionization detector (FID)

Greenhouse gas (GHG)

Growth rate (GR)

Least significant difference (LSD)

Neutral Detergent Fiber (NDF)

Non-dispersive infra-red (NDIR)

Solid State Bio-processing (SSF)

Temperature (Temp)

References

1. Shah, A.R.; Madamwar, D. Xylanase production under solid-state fermentation and its characterization by an isolated strain of *Aspergillus foetidus* in India. *World J. Microbiol. Biotechnol.* **2005**, *21*, 233–243.

2. Rumsey, T. Ruminal fermentation products and plasma ammonia of fistulated steers fed apple pomace-urea diets. *J. Anim. Sci.* **1978** *47*, 967–976.

3. Correia, R.; Magalhaes, M.; Macedo, G. Protein enrichment of pineapple waste with *Saccharomyces cerevisiae* by solid state bioprocessing. *J. Sci. Indus. Res.* **2007**, *66*, 259–262.

4. Joshi, V.; Sandhu, D. Preparation and evaluation of an animal feed byproduct produced by solid-state fermentation of apple pomace. *Bioresour. Technol.* **1996**, *56*, 251–255.

5. Bhalla, T.; Joshi, M. Protein enrichment of apple pomace by co-culture of cellulolytic moulds and yeasts. *W. J. Microbiol. Biotechnol.* **1994**, *10*, 116–117.

6. Devrajan, A.; Joshi, V.K.; Gupta, K.; Sheikher, C.; Lal, B.B. Evaluation of apple pomace based reconstituted feed in rats after solid state fermentation and ethanol recovery. *Braz. Archi. Biol. Technol.* **2004**, *47*, 93–106.

7. Joshi, V.; Gupta, K.; Devrajan, A.; Lal, B.; Arya, S. Production and evaluation of fermented apple pomace in the feed of broilers. *J. Food Sci.Technol.* **2000**, *37*, 609–612.

8.   Steinfeld, H.; Gerber, P.; Wassenaar, T.; Castel, V.; Rosales, M.; Haan, C.D. *Livestock's Long Shadow: Environmental Issues and Options*; Food and Agriculture Organization of the United Nations (FAO): Rome, Italy, 2006.

9.   Determining the Environmental Burdens and Resource Use in the Production of Agricultural and Horticultural Commodities. IS0205. Available online: http://randd.defra.gov.uk/Default.aspx? Menu=Menu&Module=More&Location=None&Completed=0&ProjectID=11442 (accessed on 27 January 2015).

10. Cederberg, C.; Sonesson, U.; Henriksson, M.; Sund, V.; Davis, J. *Greenhouse Gas Emissions from Swedish Production of Meat, Milk and Eggs 1990 and 2005*; SIK-Institutet för livsmedel och bioteknik: Gothenburg, Sweden, 2009.

11. Greenhouse Gas Reduction via Nutrition. Available online: http://www.allaboutfeed.net/ Home/General/2010/11/Greenhouse-gas-reduction-via-nutrition-AAF011538W/ (accessed on 27 January 2015).

12. Ajila, C.; Gassara, F.; Brar, S.K.; Verma, M.; Tyagi, R.; Valéro, J. Polyphenolic antioxidant mobilization in apple pomace by different methods of solid-state fermentation and evaluation of its antioxidant activity. *Food Bioproc. Technol.* **2012**, *5*, 2697–2707.

13. Zheng, Z.; Shetty, K. Cranberry processing waste for solid state fungal inoculant production. *Process Biochem.* **1998**, *33*, 323–329.

14. Bisaria, R.; Madan, M.; Vasudevan, P. Utilisation of agro-residues as animal feed through bioconversion. *Bioresour. Technol.* **1997**, *59*, 5–8.

15. Small Acreage. Available online: http://www.colostate.edu /Dept/CoopExt/Adams/sa/livestock.htm (accessed on 27 January 2015).

16. Matoo, F.; Beat, G.; Banday, M.; Ganaie, T. Performance of broilers fed on apple pomace diets supplemented with enzyme(s). *Ind. J. Anim. Nutr.* **2001**, *18*, 349–352.

17. Zafar, F.; Idrees, M.; Ahmed, Z. Use of apple by-products in poultry rations of broiler chicks in karachi. *Pak. J. Physiol.* **2005**, *1*, 1–2.

18. El-Kady, R.; Awadalla, I.; Mohamed, M.; Fadel, M.; El-Rahman, H.A. Effect of exogenous enzymes on the growth performance and digestibility of growing buffalo calves. *Int. J. Agric. Biol. (Pakistan)* **2006**, *8*, 355–359.

19. Effect of Supplemental Enzymes in Diets Containing Yellow Maize or Pearl Millet as a Tropical Source of Energy in Broiler Chicken. Available online: http://agris.fao.org/agris-search/search.do? recordID=IN2006001195 (accessed on 27 January 2015).

20. Qian, L.; Sun, J. Effect of β-glucosidase as a feed supplementary on the growth performance, digestive enzymes and physiology of broilers. *Asian-Austr. J. Anim. Sci.* **2009**, *22*, 260–266.

21. Omogbenigun, F.; Nyachoti, C.; Slominski, B. Dietary supplementation with multienzyme preparations improves nutrient utilization and growth performance in weaned pigs. *J. Anim. Sci.* **2004**, *82*, 1053–1061.

22. Fang, Z.; Peng, J.; Liu, Z.; Liu, Y. Responses of non-starch polysaccharide-degrading enzymes on digestibility and performance of growing pigs fed a diet based on corn, soya bean meal and Chinese double-low rapeseed meal. *J. Anim. Physiol. Anim. Nutr.* **2007**, *91*, 361–368.

23. Kim, S.; Mavromichalis, I.; Easter, R. Supplementation of alpha-1galactosidase and beta-1, 4-mannanase to improve soybean meal utilization by nursery pig. *J. Anim. Sci* **2001**, *79*, 106.

24. Horn, M. Optimal diets in complex environments: Feeding strategies of two herbivorous fishes from a temperate rocky intertidal zone. *Oecologia* **1983**, *58*, 345–350.

25. Zhong-Tao, S.; Lin-Mao, T.; Cheng, L.; Jin-Hua, D. Bioconversion of apple pomace into a multienzyme bio-feed by two mixed strains of *Aspergillus niger* in solid state fermentation. *Electron. J. Biotechnol.* **2009**, *12*, 2–3.

26. Hamelin, L.; Godbout, S.; Lemay, S.P. Baseline scenario for gas, odor, dust and particulate matter emissions from swine buildings in Québec-part I: Emissions inventory. In Proceedings of the CSBE/SCGAB 2009 Annual Conference Rodd's Brudenell River Resort, Prince Edward Island, Charlottetown, Canada, 12–15 July 2009.

27. Wittenberg, K.; Boadi, D. Reducing Greenhouse Gas Emissions from Livestock Agriculture in Manitoba. Manitoba Climate Change Task Force, Public Consultation Sessions 2001. Available online: https://www.iisd.org/taskforce/pdf/dept_animal_sci.pdf (accessed on 15 April 2015).

28. Pomar, C. *Potential for Reducing Ghg Emissions from Domestic Monogastric Animals*; Agriculture and Agri-Food Table Climate Change Workshop: Montreal, QC, Canada, 1998.

29. Nutrient Requirements of Swine. Available online: http://www.nap.edu/catalog/6016/nutrient-requirements-of-swine-10th-revised-edition (accessed on 28 January 2015).

30. CCAC Guidelines on: The Care and Use of Farm Animals in Research, Teaching and Testing. Available online: http://www.ccac.ca/Documents/Standards/Guidelines/Farm_Animals.pdf (accessed on 28 January 2015).

31. The Consumption and Utilization of Food by Insects. Available online: http://garfield.library.upenn.edu/classics1982/A1982MZ92300001.pdf (accessed on 28 January 2015).

32. Fris, M.B.; Horn, M.H. Effects of diets of different protein content on food consumption, gut retention, protein conversion, and growth of *Cebidichthys violaceus* (girard), an herbivorous fish of temperate zone marine waters. *J. Exp. Mar. Biol. Ecol.* **1993**, *166*, 185–202.

33. Association of Official Analytical Chemists, AOAC, 2006. Available online: http://www.eoma.aoac.org/ (accessed on 28 January 2015).

14

Linking Management, Environment and Morphogenetic and Structural Components of a Sward for Simulating Tiller Density Dynamics in Bahiagrass (*Paspalum notatum*)

Masahiko Hirata

Department of Animal and Grassland Sciences, Faculty of Agriculture, University of Miyazaki, Miyazaki 889-2192, Japan; E-Mail: m.hirata@cc.miyazaki-u.ac.jp

Academic Editor: Cory Matthew

Abstract: A model which describes tiller density dynamics in bahiagrass (*Paspalum notatum* Flügge) swards has been developed. The model incorporates interrelationships between various morphogenetic and structural components of the sward and uses the inverse of the self-thinning rule as the standard relationship between tiller density and tiller weight (a density-size equilibrium) toward which tiller density progressively changes over time under varying nitrogen (N) rates, air temperature and season. Water and nutrient limitations were not considered except partial consideration of N. The model was calibrated against data from swards subjected to different N rates and cutting intensities, and further validated against data from a grazed sward and swards under different cutting intensities. As the calibration and validation results were satisfactory, the model was used as a tool to investigate the responses of tiller density to various combinations of defoliation frequencies and intensities. Simulations identified defoliation regimes required for stabilizing tiller density at an arbitrary target level, *i.e.*, sustainable use of the sward. For example, the model predicted that tiller density can be maintained at a medium level of about 4000 m^{-2} under conditions ranging from weekly cuttings to an 8 cm height to 8-weekly cuttings to 4 cm. More intense defoliation is needed for higher target tiller density and *vice versa*.

Keywords: model; tiller density; tiller birth; tiller death; self-thinning rule; bahiagrass

1. Introduction

Grasslands are essential to human life. Our challenge is to sustain grasslands and make better use of them for agricultural production, conservation of the environment and wildlife, and other purposes (e.g., recreation and amenity). This requires understanding and predicting sward dynamics in grasslands in response to the environment (e.g., temperature and rainfall) and management (e.g., defoliation and fertilizer application).

Sward dynamics in grasslands can be mechanistically analyzed and understood by breaking down the sward into a set of morphogenetic and structural components [1,2]. This approach was taken initially for temperate forage species and later for tropical species. Based on the morphogenetic and structural mechanisms, persistence of grass (Poaceae) swards is dependent on the ability of the plant to maintain a high tiller density, which in turn depends on the longevity (rate of death) and recruitment (rate of appearance) of the tillers [3].

Bahiagrass (*Paspalum notatum* Flügge), a sod-forming, warm-season perennial, is widespread in the southern USA, and Central and South America [4]. It is also well adapted to the low-altitude regions of south-western Japan, and used for both grazing and hay [5]. This grass forms a highly persistent sward under a wide range of management [6–11]. Among tropical forage species, bahiagrass has been most detailedly studied in terms of the morphogenetic and structural components, providing a good deal of information for modeling sward dynamics of the grass [12–22].

In the present study, a model of tiller density dynamics in bahiagrass swards was built by integrating information selected from the literature. The framework of the model is a combination of interrelationships linking management, environment and morphogenetic and structural components of the sward. The model was calibrated and validated against data from swards subjected to various management conditions, and then used to explore defoliation management for sustainable use of bahiagrass swards. The aims of the study were to examine how the integrated interrelationships work as a whole and to characterize the model in comparison with previous models of tiller density dynamics in grasses. The structure and performance of the model have been partly described in Hirata [22,23].

2. The Model

The model simulates changes in tiller density by calculating tiller appearance and death, which are driven by variables relating to the sward, environment and management (Figure 1). Since the model forms a submodel of an integrated model of sward dynamics in grasslands, it requires daily herbage mass as an input from a submodel of herbage production and utilization. The model also needs mean daily air temperature, annual nitrogen (N) fertilizer rate and month of the year as inputs, and initial tiller density at the commencement of a simulation run. Water and nutrient limitations are not considered except partial consideration of N. Variables used in the model are given in Table 1.

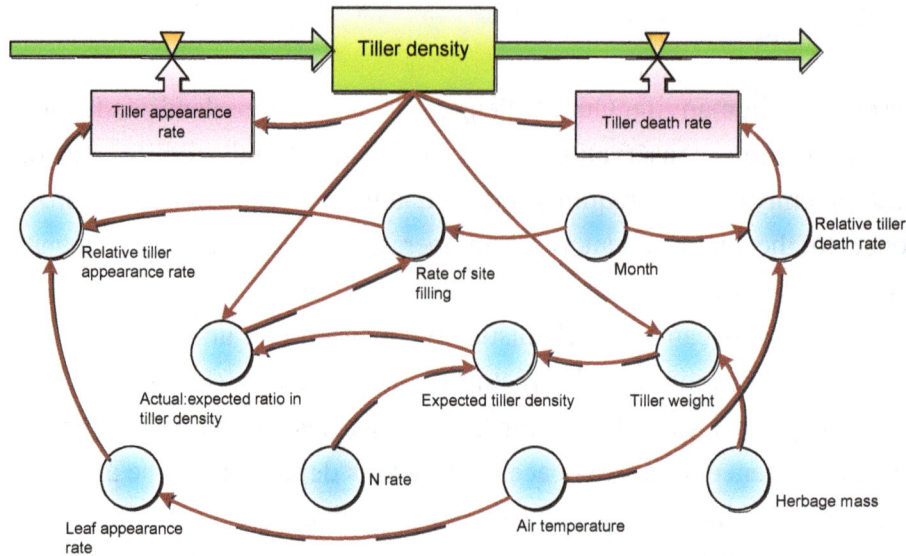

Figure 1. The framework of the model. Reproduced with permission from Hirata [23]; published by Wageningen Academic Publishers.

Table 1. Variables used in the model.

Symbol	Description	Unit
b_0	Intercept of the standard relationship between tiller density and tiller weight	log tillers m^{-2}
b_1	Slope of the standard relationship between tiller density and tiller weight	log tillers m^{-2} (log mg DM tiller^{-1})$^{-1}$
D	Tiller density	tillers m^{-2}
$D_{A:E}$	Actual: expected ratio in tiller density	fraction
D_E	Expected tiller density	tillers m^{-2}
F_N	Annual nitrogen fertilizer rate	g m^{-2} year^{-1}
F_S	Rate of site filling	tillers leaf^{-1}
M	Herbage mass	g DM m^{-2}
$R_{\text{leaf,app}}$	Leaf appearance rate	leaves tiller^{-1} day^{-1}
$R_{\text{tiller,app}}$	Tiller appearance rate	tillers m^{-2} day^{-1}
$R'_{\text{tiller,app}}$	Relative tiller appearance rate	tillers tiller^{-1} day^{-1}
$R_{\text{tiller,death}}$	Tiller death rate	tillers m^{-2} day^{-1}
$R'_{\text{tiller,death}}$	Relative tiller death rate	tillers tiller^{-1} day^{-1}
t	Time	day
T	Mean daily air temperature	°C
W	Tiller weight	mg DM tiller^{-1}

2.1. Rate of Change in Tiller Density

The daily rate of change in tiller density ($\Delta D/\Delta t$, tillers m^{-2} day^{-1}) is expressed as the balance between tiller appearance rate ($R_{\text{tiller,app}}$, tillers m^{-2} day^{-1}) and tiller death rate ($R_{\text{tiller,death}}$, tillers m^{-2} day^{-1}):

$$\frac{\Delta D}{\Delta t} = R_{\text{tiller,app}} - R_{\text{tiller,death}} \qquad (1)$$

The two rates are written as:

$$R_{\text{tiller,app}} = R'_{\text{tiller,app}} \times D \qquad (2)$$

and

$$R_{\text{tiller,death}} = R'_{\text{tiller,death}} \times D \qquad (3)$$

where D is the tiller density (tillers m^{-2}), and $R'_{\text{tiller,app}}$ and $R'_{\text{tiller,death}}$ are the relative rates of tiller appearance and death (tillers tiller^{-1} day^{-1}), respectively.

2.2. Tiller Appearance

Relative tiller appearance rate is known to be the product of leaf appearance rate ($R_{\text{leaf,app}}$, leaves tiller^{-1} day^{-1}) and the rate of site filling (F_S, tillers leaf^{-1}), *i.e.*, the rate at which axillary buds develop into tillers (visible without dissection) in relation to the rate at which leaf axils are formed [24,25]:

$$R'_{\text{tiller,app}} = F_S \times R_{\text{leaf,app}} \qquad (4)$$

Leaf appearance rate is expressed as a threshold response function of the mean daily air temperature (T, °C) [14]:

$$R_{\text{leaf,app}} = 0 \qquad \qquad \text{(when } T \leq 7.6)$$
$$= 0.117 \times \frac{((T - 7.6)/6.4)^{3.6}}{1 + ((T - 7.6)/6.4)^{3.6}} \qquad \text{(when } T > 7.6) \qquad (5)$$

This equation shows that leaves emerge when the temperature exceeds 7.6 °C and that the leaf appearance rate attains its half-maximal response when the temperature is 14.0 °C and approaches the maximal response (asymptote) of 0.117 leaves tiller^{-1} day^{-1} at higher temperatures (Figure 2).

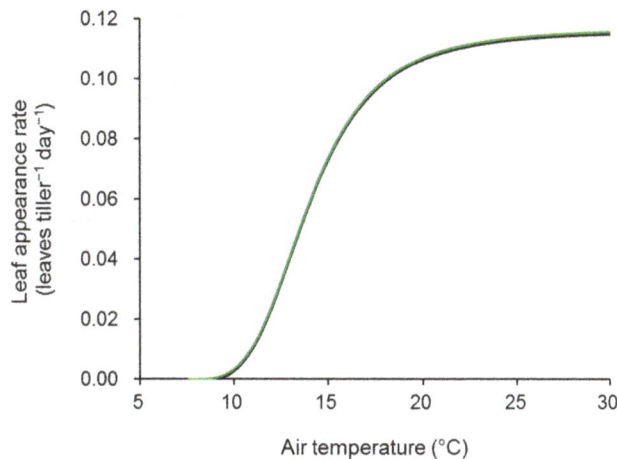

Figure 2. Relationship between leaf appearance rate and mean daily air temperature. Reproduced with permission from Pakiding and Hirata [14]; published by the Tropical Grassland Society of Australia.

The rate of site filling (F_S) is expressed as a function of the actual:expected ratio in tiller density ($D_{A:E}$), *i.e.*, the ratio of actual tiller density (D) to the expected tiller density (D_E, tillers m^{-2}), which was parameterized by revisiting the data given in Hirata and Pakiding [21] and Hirata [22]:

$$F_S = \max[0, -0.120(D_{A:E} - 1.2)] \qquad \text{(for April)}$$
$$= \max[0, -0.110(D_{A:E} - 1.2)] \qquad \text{(for May)}$$
$$= \max[0, -0.100(D_{A:E} - 1.2)] \qquad \text{(for June)} \qquad (6)$$
$$= \max[0, -0.035(D_{A:E} - 1.2)] \qquad \text{(for July–November)}$$
$$= 0 \qquad \text{(for other months)}$$

where

$$D_{A:E} = D/D_E \qquad (7)$$

Equation (6) shows that axillary buds have the potential of developing into tillers only in April–November and only when the actual tiller density is lower than 1.2 times the density expected from the standard relationship, with higher developmental rates at lower actual:expected ratios in tiller density and in April–June (April > May > June) than in July–November (Figure 3).

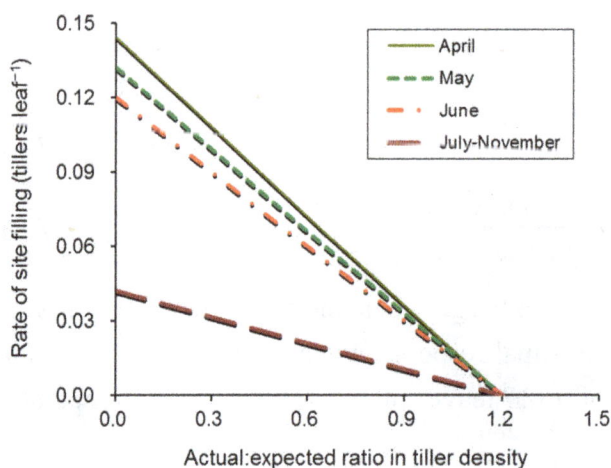

Figure 3. Relationships between rate of site filling and actual: expected ratio in tiller density.

The expected tiller density is calculated from tiller weight (W, mg DM tiller^{-1}), using the standard relationship between tiller density and tiller weight, *i.e.*, relationship in an almost stabilized sward (an equilibrium) under fixed management:

$$D_E = 10^{(b_0 - b_1 \log W)} \qquad (8)$$

This standard density–weight relationship derives from the relationship, $\log D_E = b_0 - b_1 \log W$, a reverse form of the self-thinning rule [26] in terms of x- and y-variables, which makes it possible to estimate tiller density from tiller weight. The two parameters for the standard relationship (b_0 and b_1) are influenced by annual N fertilizer rate (F_N, g m^{-2} year^{-1}) [21,22]:

$$b_0 = 4.355 + 0.0247(F_N - 5) \qquad (9)$$

$$b_1 = -0.376 - 0.0092(F_N - 5) \qquad (10)$$

Equations (8)–(10) show that the number of tillers carried on a unit land area increases as the tiller weight decreases, with a greater rate of increase at a higher N fertilizer rate (Figure 4).

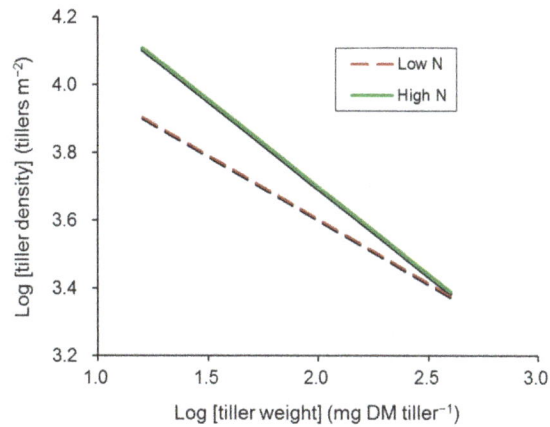

Figure 4. Standard relationships between tiller density and tiller weight. Low nitrogen = 5 g m^{-2} year^{-1}, high nitrogen = 20 g m^{-2} year^{-1}. Reproduced with permission from Hirata and Pakiding [21]; published by the Tropical Grassland Society of Australia.

Tiller weight is calculated as:

$$W = 1000 \times M/D \tag{11}$$

where M is herbage mass (g DM m^{-2}) and 1000 is a unit conversion factor from g to mg.

2.3. Tiller Death

Relative tiller death rate is expressed as a function of the mean daily air temperature [22]:

$$
\begin{aligned}
R'_{\text{tiller,death}} &= 0.00084 \times \exp(0.031 \times T) && \text{(for spring–summer)} \\
&= 0.00020 \times \exp(0.083 \times T) && \text{(for autumn–winter)}
\end{aligned}
\tag{12}
$$

This equation shows that the relative tiller death rate increases exponentially as the mean daily air temperature increases, maintaining higher values in spring–summer (March–August) than in autumn–winter (September–February) until the temperature reaches 27.9 °C (Figure 5).

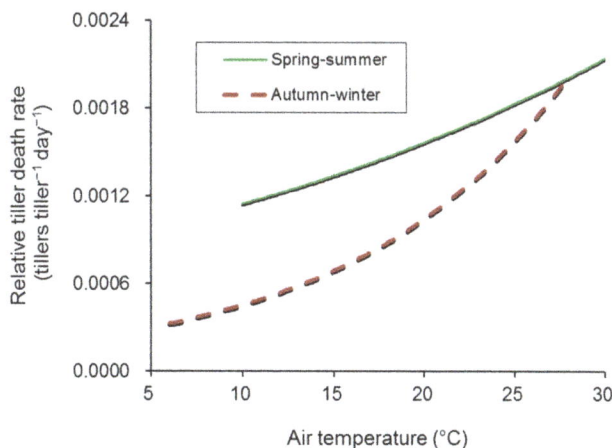

Figure 5. Relationship between relative tiller death rate and mean daily air temperature.

3. Model Performance

3.1. Calibration

The parameters for Equations (5), (6), (9), (10) and (12) were determined using a regression technique to calibrate the model. The performance of the model was first evaluated by simulating tiller density dynamics for the study from which most of the data used for calibrating the model were derived, *i.e.*, bahiagrass tiller density under different N fertilizer rates and cutting heights [18,21]. The simulations for the six experimental treatments (2 N rates × 3 cutting heights) were run for 1341 days ($t = 1, \cdots, 1341$) from 1 June 1996 to 1 February 2000. Daily herbage mass as an input (Figure 1) was estimated by interpolating data from monthly measurements. Mean daily air temperature was derived from daily records. Annual N rate was set at the actual dosage (5 and 20 g m^{-2} year^{-1} for low and high N treatments, respectively). Month of the year was calculated from the day number (t) and the initial date of simulation. Initial tiller density was set at the measured data.

Overall, the simulated tiller densities showed good agreement with the measured data, despite slight to moderate over- or under-prediction in some seasons in some treatments (Figure 6).

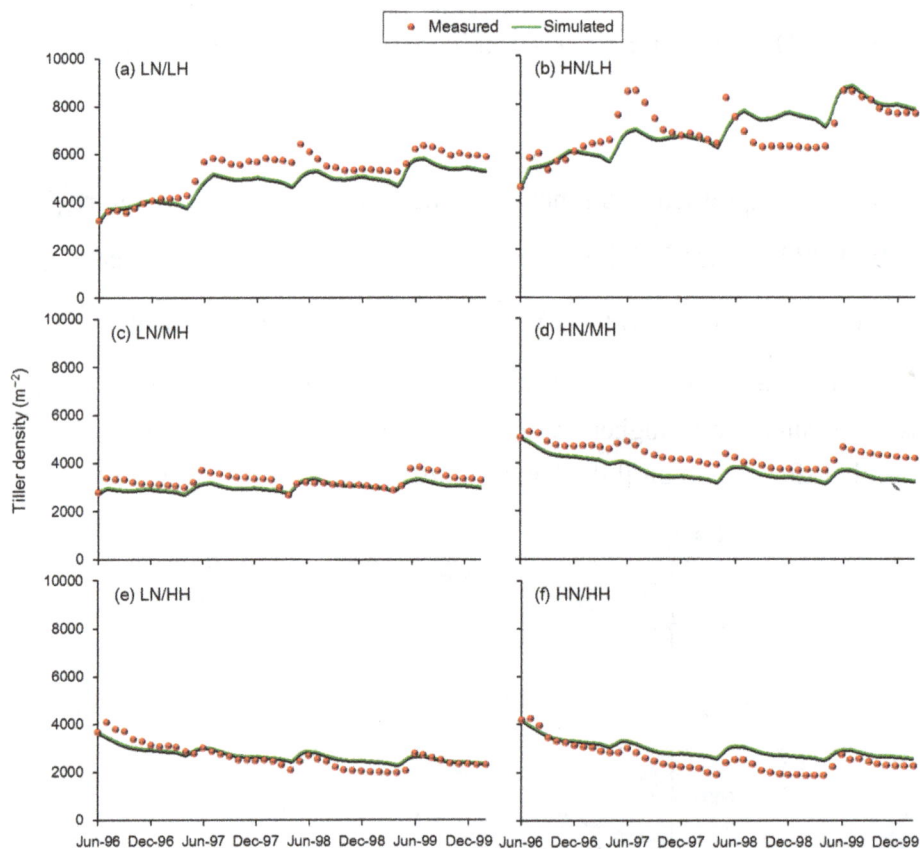

Figure 6. Measured (closed circle) and simulated (line) tiller densities in bahiagrass swards under different nitrogen rates and defoliation intensities (calibration results). Measured data derive from Pakiding and Hirata [18] and Hirata and Pakiding [21]. LN = low nitrogen (5 g m^{-2} year^{-1}), HN = high nitrogen (20 g m^{-2} year^{-1}), LH = low height (2 cm), MH = medium height (12 cm), HH = high height (22 cm) of cutting (heights above ground). Reproduced with permission from Hirata [22]; published by Research Signpost.

3.2. Validation

The model was then validated against independent data sets, *i.e.*, bahiagrass tiller density under cattle grazing [15] and under different cutting heights [10]. The simulations for the former study were run for 1462 days ($t = 1, \cdots, 1462$) from 18 May 1996 to 18 May 2000, and those for the latter study (5 cutting-height treatments) were run for 1092 days ($t = 1, \cdots, 1092$) from 30 May 1986 to 25 May 1989. Daily herbage mass was estimated by interpolating data from 2-weekly to seasonal measurements. Mean daily air temperature was derived from daily records. Annual N rate was set at the actual doses (4.5–9.7 g m^{-2} year^{-1} and 20 g m^{-2} year^{-1} for the former and latter studies, respectively). Month of the year was calculated from the day number (t) and the initial date of simulation. Initial tiller density used the measured data.

Under grazing, the simulated tiller densities showed good agreement with the measured data except some over-prediction in summer (June–August) 1997 (Figure 7). Under cutting, the simulated densities followed the measured values well over the course of 3 years, despite partial inability to follow the seasonal fluctuations within individual years (Figure 8). As a whole, the validation results were acceptable.

4. Use of the Model

As the calibration and validation results were successful, the model was used as a tool to investigate the responses of tiller density to various combinations of defoliation frequencies and intensities (heights above ground). The simulations were run for 3 years ($t = 1, \cdots, 1095$) from 1 May. Daily herbage mass was provided by a model of herbage production and utilization of a bahiagrass sward which was based on the data presented in Hirata [27,28], with no feedback from the tiller dynamics model. The mean daily air temperature (T, °C) was determined using the following equation, which approximates the long-term average of the annual cycle in Miyazaki (31°56′ N, 131°25′ E):

$$T = 17.5 + 10.5 \times \sin(2\pi(t + 9)/365) \tag{13}$$

Annual N rate was set at 10 g m^{-2}. Month of the year was calculated from the day number (t) and the initial date of simulation.

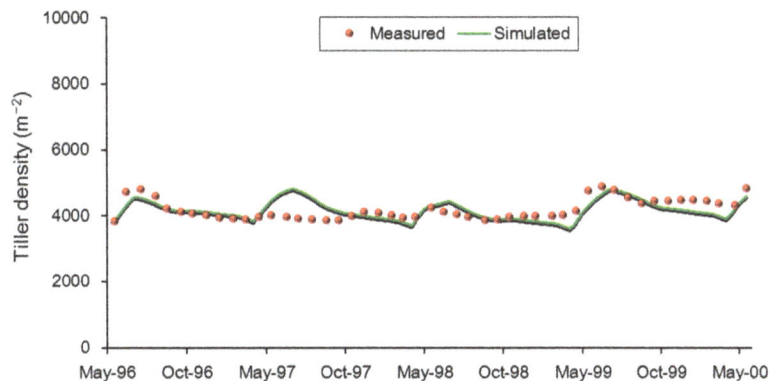

Figure 7. Measured (closed circle) and simulated (line) tiller densities in a bahiagrass pasture under cattle grazing (validation results). Measured data derive from Hirata and Pakiding [15].

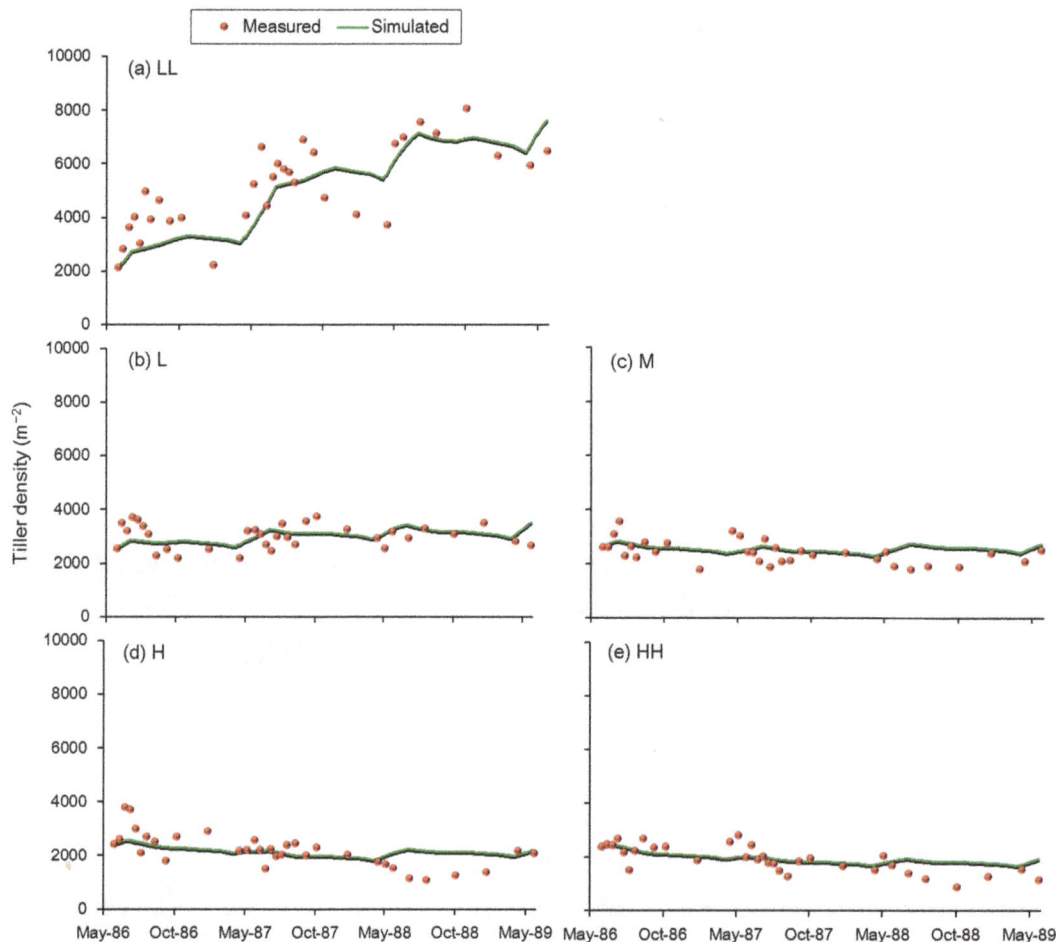

Figure 8. Measured (closed circle) and simulated (line) tiller densities in bahiagrass swards under different defoliation intensities (validation results). Measured data derive from Hirata [10]. LL = 2 cm, L = 7 cm, M = 12 cm, H = 17 cm, HH = 22 cm of cutting height above ground. Reproduced with permission from Hirata [22]; published by Research Signpost.

The simulations predicted gradual decrease in tiller density from the initial value of 4000 m^{-2} under cutting to a medium (12 cm) or high (22 cm) height, irrespective of cutting intervals ranging between 2 and 6 weeks (Figure 9). By contrast, tiller density was predicted to increase with time under a low defoliation height (2 cm), with steeper increases under more frequent defoliation.

Further simulations identified defoliation regimes (frequency and height) required for stabilizing tiller density at a low (2500 m^{-2}), medium (4000 m^{-2}) or high (6000 m^{-2}) level over 3 years, *i.e.*, sustainable use of a sward (Figure 10). The model predicted that the high tiller density can be maintained under conditions ranging from weekly cuttings to a 4 cm height to 8-weekly cuttings to 2 cm, while the low density can be maintained under conditions ranging from weekly cuttings to 19 cm to 8-weekly cuttings to 12 cm. The model also showed that maintaining the medium tiller density requires defoliation with intermediate intensities ranging from 8 cm at weekly cuttings to 4 cm at 8-weekly cuttings. Defoliation height needed decreased as the target tiller density increased and *vice versa*.

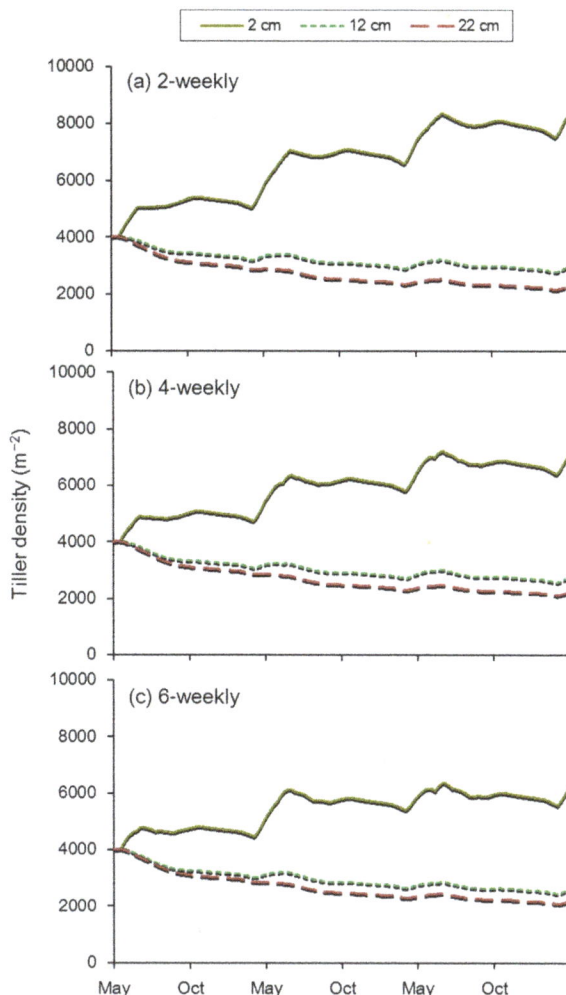

Figure 9. Predicted changes in tiller density in bahiagrass swards under various combinations of defoliation frequencies and intensities (heights above ground) at an annual N rate of 10 g m^{-2}. The initial tiller density (on 1 May) is 4000 m^{-2}.

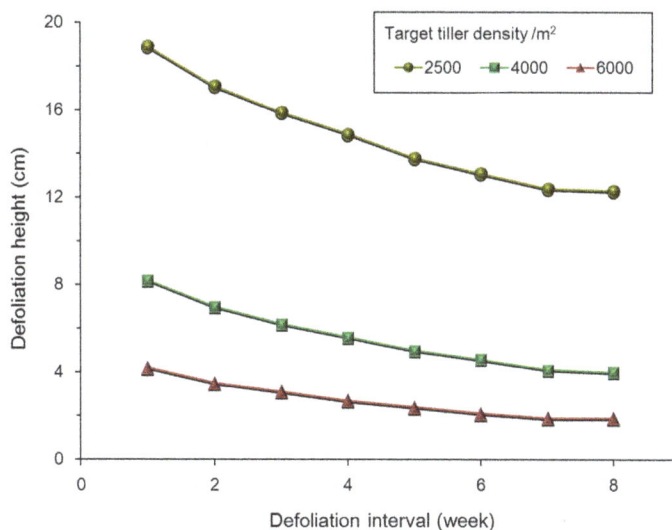

Figure 10. Predicted height (above ground) and interval of defoliation required for maintaining constant tiller density in bahiagrass swards at 2500 (○), 4000 (□) and 6000 (△) m^{-2} at an annual N rate of 10 g m^{-2}.

5. Discussion

Maintaining plant population density is crucial to sustainable use of grasslands for agricultural production, conservation of the environment and wildlife, and other purposes such as recreation and amenity. It has been reported that tiller density in grass swards exhibits considerable response to the environment and management [10,18,21,22,29–31]. It is therefore important to develop a model which can predict tiller density dynamics under varying environmental and management conditions.

Previous models of tiller density dynamics in grass swards described the production of new tillers (tillering) in various ways. Some defined the tiller formation rate (per unit land area or per existing tiller) directly as a function of the nutrient state in the plants (amount and concentration of assimilate, carbon (C) and N) [32,33] and leaf area index (LAI) of the sward canopy (an index of self-shading) [33]. Others described the relative rate of tiller appearance mechanistically, as a product of the leaf appearance rate and the rate of site filling (Equation (4)), and expressed the former as a function of temperature [34,35] and the latter as a function of time after defoliation [34], LAI or light transmission at ground level [34,35] and availability (amount) of C and N within the plants [34,35]. These models described the death of tillers also in various ways, relating tiller mortality (relative rate of tiller death) to the developmental stage of tillers [32], LAI [33,34], thermal time [35], assimilate supply and C and N reserves [35] and self-thinning [35].

The current model differs from the previous models mainly in that it does not include the internal nutrient state of the plants or LAI as factors influencing tiller appearance and death (Figure 1), although the positive effect of plant N and negative effect of LAI (self-shading) on tillering can be achieved indirectly by the N-dependent standard relationships between tiller density and tiller weight (Figure 4 and Equations (8)–(10)). Because of the lack of the internal nutrient state, the present model cannot predict a decrease in tillering or an increase in tiller mortality caused by the nutrient limitations within the plants, which may take place when supply of nutrients via photosynthesis and uptake from the soil are restricted due to the deficiency in soil moisture and nutrients, *i.e.*, the model can be used when the growth of plants is not limited by either water or nutrients. The model is also unable to respond flexibly to N management with varying times and rates of split applications. Incorporating the plant nutrient state into the model and linking the model to a herbage production and utilization model simulating the internal nutrient state as well as herbage mass should thus broaden the management and environmental conditions to which the model can be applied.

Furthermore, the current model uses the standard density–weight relationship (reverse form of the self-thinning rule) only for controlling tiller appearance, leaving it unused for controlling tiller death (Figure 1). This may be a reason why the simulations were not able to follow some of the drastic decreases in tiller density which often followed an increase in mid-to-late spring and/or early-to-mid summer (Figures 6 and 8). Although a previous analysis reported a poor association between the relative tiller death rate and the actual:expected ratio in tiller density [21,22], their relationship may need to be reanalyzed in more detail to find a mechanism which can be incorporated into the model.

Despite the limitations discussed above, the present model can provide important information on how tiller population density changes in response to the environment and management (Figures 6–10) based on a simple and mechanistic structure. It can therefore be concluded that the model is of potential value as a prototype submodel for an integrated model of sward dynamics in grasslands. Refinement of the model needs to trade simplicity for complexity.

Acknowledgments

The author would like to thank Wempie Pakiding for his contribution to data collection in the field and Cory Matthew for his encouragement during manuscript preparation.

Conflicts of Interest

The author declares no conflict of interest.

References

1. Chapman, D.F.; Lemaire, G. Morphogenetic and structural determinants of plant regrowth after defoliation. In Proceedings of the XVII International Grassland Congress, Palmerston North, Hamilton, Lincoln and Rockhampton, New Zealand and Australia, 8–21 February 1993; pp. 95–104.

2. Lemaire, G.; Chapman, D. Tissue flows in grazed plant communities. In *The Ecology and Management of Grazing Systems*; Hodgson, J., Illius, A.W., Eds.; CAB International: Wallingford, UK, 1996; pp. 3–36.

3. Matthew, C.; Agnusdei, M.G.; Assuero, S.G.; Sbrissia, A.F.; Scheneiter, O.; da Silva, S.C. State of knowledge in tiller dynamics. In Proceedings of the 22nd International Grassland Congress, Sydney, Australia, 15–19 September 2013; New South Wales Department of Primary Industry: Orange, New South Wales, Australia, 2013; pp. 1041–1044.

4. Skerman, P.J.; Riveros, F. *Tropical Grasses*; FAO: Rome, Italy, 1989; pp. 571–575.

5. Hirata, M.; Ogawa, Y.; Koyama, N.; Shindo, K.; Sugimoto, Y.; Higashiyama, M.; Ogura, S.; Fukuyama, K. Productivity of bahiagrass pastures in south-western Japan: Synthesis of data from grazing trials. *J. Agron. Crop Sci.* **2006**, *192*, 79–91.

6. Beaty, E.R.; Brown, R.H.; Morris, J.B. Response of Pensacola bahiagrass to intense clipping. In Proceedings of the XI International Grassland Congress, Surfers Paradise, Australia, 13–23 April 1970; University of Queensland Press: St. Lucia, Queensland, Australia, 1970; pp. 538–542.

7. Beaty, E.R.; Engel, J.L.; Powell, J.D. Yield, leaf growth, and tillering in bahiagrass by N rate and season. *Agron. J.* **1977**, *69*, 308–311.

8. Stanley, R.L.; Beaty, E.R.; Powell, J.D. Forage yield and percent cell wall constituents of Pensacola bahiagrass as related to N fertilization and clipping height. *Agron. J.* **1977**, *69*, 501–504.

9. Hirata, M.; Ueno, M. Response of bahiagrass (*Paspalum notatum* Flügge) sward to cutting height. 1. Dry weight of plant and litter. *J. Jpn. Grassl. Sci.* **1993**, *38*, 487–497.

10. Hirata, M. Response of bahiagrass (*Paspalum notatum* Flügge) sward to cutting height. 3. Density of tillers, stolons and primary roots. *J. Jpn. Grassl. Sci.* **1993**, *39*, 196–205.

11. Hirata, M. Response of bahiagrass (*Paspalum notatum* Flügge) sward to nitrogen fertilization rate and cutting interval. 1. Dry weight of plant and litter. *J. Jpn. Grassl. Sci.* **1994**, *40*, 313–324.

12. Pakiding, W.; Hirata, M. Tillering in a bahia grass (*Paspalum notatum*) pasture under cattle grazing: Results from the first two years. *Trop. Grassl.* **1999**, *33*, 170–176.

13. Hirata, M. Effects of nitrogen fertiliser rate and cutting height on leaf appearance and extension in bahia grass (*Paspalum notatum*) swards. *Trop. Grassl.* **2000**, *34*, 7–13.

14. Pakiding, W.; Hirata, M. Leaf appearance, death and detachment in a bahia grass (*Paspalum notatum*) pasture under cattle grazing. *Trop. Grassl.* **2001**, *35*, 114–123.

15. Hirata, M.; Pakiding, W. Tiller dynamics in a bahia grass (*Paspalum notatum*) pasture under cattle grazing. *Trop. Grassl.* **2001**, *35*, 151–160.

16. Hirata, M.; Pakiding, W. Dynamics in tiller weight and its association with herbage mass and tiller density in a bahia grass (*Paspalum notatum*) pasture under cattle grazing. *Trop. Grassl.* **2002**, *36*, 24–32.

17. Hirata, M.; Pakiding, W. Dynamics in lamina size in a bahia grass (*Paspalum notatum*) pasture under cattle grazing. *Trop. Grassl.* **2002**, *36*, 180–192.

18. Pakiding, W.; Hirata, M. Effects of nitrogen fertilizer rate and cutting height on tiller and leaf dynamics in bahiagrass (*Paspalum notatum* Flügge) swards: Tiller appearance and death. *Grassl. Sci.* **2003**, *49*, 193–202.

19. Pakiding, W.; Hirata, M. Effects of nitrogen fertilizer rate and cutting height on tiller and leaf dynamics in bahiagrass (*Paspalum notatum* Flügge) swards: Leaf appearance, death and detachment. *Grassl. Sci.* **2003**, *49*, 203–210.

20. Pakiding, W.; Hirata, M. Effects of nitrogen fertilizer rate and cutting height on tiller and leaf dynamics in bahiagrass (*Paspalum notatum* Flügge) swards: Leaf extension and mature leaf size. *Grassl. Sci.* **2003**, *49*, 211–216.

21. Hirata, M.; Pakiding, W. Tiller dynamics in bahia grass (*Paspalum notatum*): An analysis of responses to nitrogen fertiliser rate, defoliation intensity and season. *Trop. Grassl.* **2004**, *38*, 100–111.

22. Hirata, M. Canopy dynamics in bahia grass (*Paspalum notatum*) swards. In *Recent Research Developments in Crop Science*; Pandalai, S.G., Ed.; Research Signpost: Kerala, India, 2004; Volume 1, pp. 117–145.

23. Hirata, M. Modelling tiller density dynamics in a grass sward. In *XX International Grassland Congress: Offered Papers*; O'Mara, F.P., Wilkins, R.J., t'Mannetje, L., Lovett, D.K., Rogers, P.A.M., Boland, T.M., Eds.; Wageningen Academic Publishers: Wageningen, The Netherlands, 2005; p. 870.

24. Davies, A. Leaf tissue remaining after cutting and regrowth in perennial ryegrass. *J. Agric. Sci. Camb.* **1974**, *82*, 165–172.

25. Thomas, H. Terminology and definitions in studies of grassland plants. *Grass Forage Sci.* **1980**, *35*, 13–23.

26. Yoda, K.; Kira, T.; Ogawa, H.; Hozumi, K. Intraspecific competition among higher plants. XI. Self-thinning in overcrowded pure stands under cultivated and natural conditions. *J. Biol. Osaka City Univ.* **1963**, *14*, 107–129.

27. Hirata, M. Quantifying spatial heterogeneity in herbage mass and consumption in pastures. *J. Range Manag.* **2000**, *53*, 315–321.

28. Hirata, M. Estimating herbage and leaf utilization in bahiagrass (*Paspalum notatum* Flügge) swards from height measurements. *Grassl. Sci.* **2002**, *48*, 105–109.

29. Korte, C.J. Tillering in 'Grasslands Nui' perennial ryegrass swards. 2. Seasonal pattern of tillering and age of flowering tillers with two mowing frequencies. *N. Z. J. Agric. Res.* **1986**, *29*, 629–638.

30. Bullock, J.M.; Hill, B.C.; Silvertown, J. Tiller dynamics of two grasses—Response to grazing, density and weather. *J. Ecol.* **1994**, *82*, 331–340.

31. Sbrissia, A.F.; da Silva, S.C.; Sarmento, D.O.L.; Molan, L.K.; Andrade, F.M.E.; Gonçalves, A.C.; Lupinacci, A.V. Tillering dynamics in palisadegrass swards continuously stocked by cattle. *Plant Ecol.* **2010**, *206*, 349–359.

32. Dayan, E.; van Keulen, H.; Dovrat, A. Tiller dynamics and growth of Rhodes grass after defoliation: A model named TILDYN. *Agro-Ecosystems* **1981**, *7*, 101–112.

33. Coughenour, M.B.; McNaughton, S.J.; Wallace, L.L. Simulation study of East-African perennial graminoid responses to defoliation. *Ecol. Model.* **1984**, *26*, 177–201.

34. Schapendonk, A.H.C.M.; Stol, W.; van Kraalingen, D.W.G.; Bouman, B.A.M. LINGRA, a sink/source model to simulate grassland productivity in Europe. *Eur. J. Agron.* **1998**, *9*, 87–100.

35. Soussana, J.F.; Oliveira Machado, A. Modelling the dynamics of temperate grasses and legumes in cut mixtures. In *Grassland Ecophysiology and Grazing Ecology*; Lemaire, G., Hodgson, J., de Moraes, A., de F. Carvalho, P.C., Nabinger, C., Eds.; CABI Publishing: Wallingford, UK, 2000; pp. 169–190.

15

Effect of Date Palm Cultivar, Particle Size, Panel Density and Hot Water Extraction on Particleboards Manufactured from Date Palm Fronds

Said S. Hegazy [1,2,*] and Khaled Ahmed [1,3]

[1] Chair of Dates Industry and Technology, College of Food and Agricultural Sciences, King Saud University, Riyadh 11451, Saudi Arabia; E-Mail: kehmed@ksu.edu.sa
[2] Timber Trees Department, Horticulture Research institute, Agriculture Research Center, Giza, Cairo 12619, Egypt
[3] Agricultural Engineering Institution, Giza, Cairo2450, Egypt

* Author to whom correspondence should be addressed; E-Mail: sashegazy@ksu.edu.sa

Academic Editor: Stephen R. Smith

Abstract: The objective of this work was to evaluate some of the important physical and mechanical properties of particleboard panels manufactured from three different cultivars of date palm (*Phoenix dactylifera*) fronds, namely Saqui, Barhi and Sukkari. Experimental panels were manufactured from hot water extracted and non-extracted, and fine and coarse particles of the raw material under two target panel densities of 650 and 750 kg/m³. Bending properties and internal bond strength, along with dimensional stability in the form of thickness swelling, water absorption, and linear expansion of the samples was tested. Based on the findings of this work, panels manufactured from high density level and Saqie cultivar, as well as fine particles, had better performance for their mechanical properties. The effect of hot water-treatment had less robust mechanical and physical properties. It appears that date palm fronds are underutilized resources that have the potential to be used in the manufacture of value-added panel products.

Keywords: date palm fronds; particleboard; particle size; strength properties; hot water treatment; dimensional stability

1. Introduction

Date palm (*Phoenix dactylifera*) is a significant agricultural crop thought to have originated from the lands around the Arabian Gulf in Saudi Arabia [1]. The arid climate of Middle Eastern countries is ideal for date plantations and Saudi Arabia is one of these countries, possessing a major share of date production with seven million tons in the region [2]. There are more than 120 million date palm trees in different countries worldwide. Over two-thirds of such palms are in Arab countries and it is estimated that there are 62 million trees in the Middle East and North Africa [3].

In general, the date palm tree has an average production life of 150 years and the trees are pruned annually to eliminate broken leaves to enhance the quality of the dates. Once the date palms' fruit are harvested, large quantities of date palm rachis and leaf waste accumulates every year in agricultural lands of different countries. It is estimated that 100,000 tons/year of date palm fronds and 15,000 tons of leaves are created as a result of the pruning process in Saudi Arabia [4], and these estimations might be doubled in recent years. Both pruning and cutting old trees produces a substantial amount of biomass that is currently not efficiently and effectively used in Saudi Arabia. Burning and land filling are some of the current practices, creating significant environmental problems. Bashah [5] reported that the raw material from palm waste and residues is likely to be highly flammable if left on the ground for a long time. Thus, innovative ways of using this abundant renewable resource should be found [6]. One of these ideas is to use such natural fibers in natural fiber composites suitable for different industrial applications to meet the increasing demand in renewable and biodegradable materials [7].

The limited availability of wood resources due to the depletion of natural and plantation forestland has increased the cost of the raw material. As a result of limited raw material, supply waste from lumber manufacture using non-wood based agricultural products is becoming a substitute raw material supply in particleboard production. However, having very few domestic wood resources in Saudi Arabia and the surrounding countries, utilization of non-wood lignocellulosic fiber resources such as date palm could become ideal as a raw material for the manufacture of value-added panel products.

Date palm fronds, having rich fiber content, were investigated to be used as raw material for experimental particleboard manufacture in several past works [8–11]. The four investigations above revealed that particleboard panels made from date palm fronds resulted in satisfactory mechanical and physical characteristics. Basic physical and mechanical properties of experimental particleboard panels from date palm fronds have also been evaluated and studied [10]. Although there are many cultivars of date palm, three of them, namely Barhi, Saqie, and Sukkari, are the most common cultivars in Saudi Arabia. However, there is no or very little information on the properties of particleboard panels manufactured from these three different cultivars of date fronds regarding the function of their particle size and treatment with water.

Therefore, the objective of this work was to determine both physical and mechanical properties of particleboard samples made from fine and coarse frond particles of the Barhi, Saqie, and Sukkari cultivars of date palms and the effect of using hot water extraction on the panel properties. It is expected that the initial data from this work will aid in the consideration of using such underutilized species to manufacture value-added panel products so that a major environmental problem can possibly be solved to a certain extent in Saudi Arabia.

2. Methods

2.1. Frond Materials

Fronds of three date palm cultivars, namely: Barhi, Saqie, and Sukkari were collected from date palm farms in Al-Kharj located 100 km east of Riyadh, Saudi Arabia. First, leaflets were mechanically stripped from the frond stalks using a commercially manufactured stripper machine. Later, these stalks were cut into sections with 150 mm length before they were converted into particles in a laboratory ring flaker machine (Model BX-466, from Changzhou Jinmu Forestry Machinary Co. Ltd, Changzhou, China). Figure 1 illustrates date palm fronds after the leaflets were stripped and the ring flaker machine used for flaking the fronds.

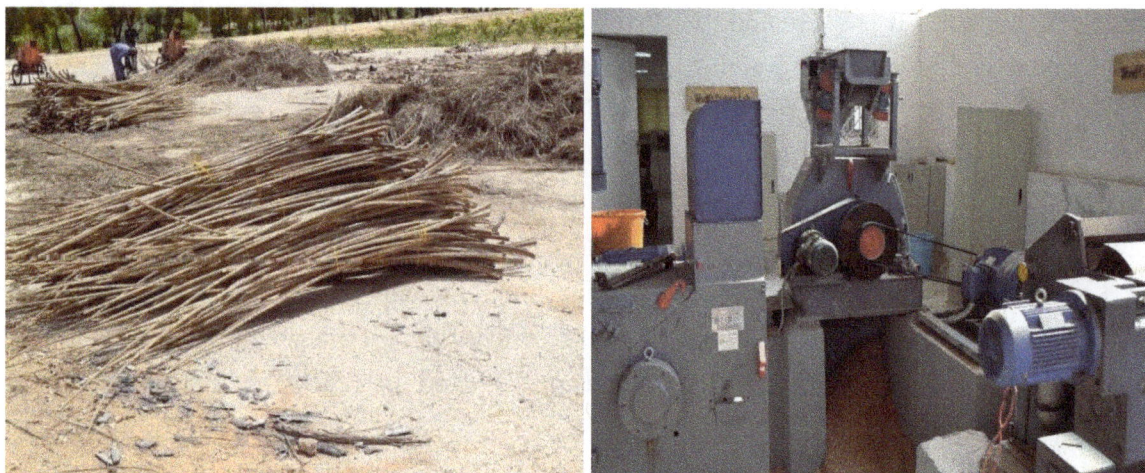

Figure 1. Date palm fronds without leaflets and the ring flaker machine (BX-466).

A shaker-type screen (domestically manufactured) was employed to classify the obtained raw material into different flake and particle size classes through the following square screen openings: class A (<2.54 mm); class B (2.54 mm < B < 1.27 mm); class C (1.27 mm < C < 0.64 mm); class D (0.64 mm < D < 0.25 mm); class E (0.25 mm < E < 0.12 mm); and class F ($\geq$ 0.12 mm). The average percentage of the obtained particle size was 1.7, 11.9, 26.8, 32.4, 24.4 and 1.8% for abovementioned size class, respectively. In our experiment, only the two class categories of D and E were used as they considered particleboard particles, which would be mentioned later in the text as coarse particles (class D) and fine particles (class E). Classes A, B, and C were considered as flakes and were used in another study for manufacturing oriented strand board [11].

To study the effect of hot water extraction on the panel's performance, half of the obtained particles (both sizes) were soaked in a hot water container at a fixed temperature of 80 °C for 16 h to eliminate the extractive and sugar contents from the raw material. In the next step, particles exposed to the hot water treatment were extracted with distilled water at room temperature and dried in a laboratory oven dryer to 3% moisture content along with un-extracted particles used in controlled panels. The particles were stored inside plastic bags in an ordinary room to maintain constant 3 percent moisture content at 22 percent relative humidity and 21 °C ± 1 °C until they were used.

2.2. Fiber Length Determination

Thin chips obtained from the fronds samples assigned for fiber-length determination were macerated in a 1:1 (by volume) solution of glacial acetic acid and 30 percent hydrogen peroxide at 60 °C for 48 h. After delignification was completed, the macerated fibers were washed several times, with mild shaking in distilled water, and then stained with Safranin. Lengths of 50 randomly selected fibers from each sampling specimen were measured in a wet condition to the nearest 0.01 mm, using a projection microscope connected to a TV screen.

2.3. Specific Gravity Determination

Specific gravity of fronds' lignocellulosic materials was determined according to American Standard Testing Methods [12]. This standard is used to evaluate the engineering performance of wood-based panels, such as particleboard, medium density fiberboard, and hardboard. It was based on oven-dry weight and green volume measured using displacement method.

2.4. Chemical Determinations of Date Palm Cultivars

2.4.1. Extractives Content Determination

The extractives content of each material was determined according to the American Standard Testing Methods [13] in three steps of 4 h each, using a Soxhlet apparatus. The percentage of extractives was calculated based on the oven-dry weight of sawdust samples.

2.4.2. Cellulose Content Determination

Cellulose was determined by the treatment of extractive-free sawdust meal with nitric acid and sodium hydroxide: one gram of extractive-free sawdust meal was treated with 20 mL of a solution of nitric acid 3% in a flask and was boiled for 30min. The solution was filtered in crucible G3. The residue was treated with 25 mL of a solution of sodium hydroxide 3% and was boiled for 30 min. The residue was filtered, washed with warm water to neutral filtrate, oven dried, and weighed [14].

2.4.3. Hemicelluloses Content Determination

Hemicellulose content was determined by the treatment of extractive free wood meal (1–2 g) with 50–100 mL sulfuric acid 2% and boiled for 1 hr. under a reflex condenser and filtrated in crucible G2. After that the residue was washed with 500 mL of hot distilled water to free the acid, and the contents were dried in an oven at 105 ± 2 °C, cooled in a desecrator, and weighed [15].

2.5. Anatomical Study of the Frond Midrib Segments

Anatomical study of midrib segments in respect of vascular bundles (VB) density, vessel dimension, vessel wall thickness, type of end wall, and vessel density per vascular bundle was performed on transverse (TS), tangential (TLS) sections. Data on VB density, vessel dimension, vessel wall thickness, vessel density/VB were collected from the sections of abaxial and adaxial surface area covering from the periphery up to the center of the basal and middle segments of the midrib. All the vessels present in the VB from the periphery up to the center of the midrib segments were measured on an Olympus CX41 (Olympus Corporation, Shinjuku-ku, Tokyo, Japan), Japan microscope, in respect of radial diameter, tangential diameter, and vessel wall thickness. Analysis of area fraction of vascular bundles, vessels lumen, fiber transverse wall, and ground parenchymatous tissue in the basal and middle segments of each replicate (frond) in each cultivar was done on an Olympus BX 51 microscope (Olympus Corporation, Shinjuku-ku, Tokyo, Japan).

2.6. Panel Manufacturing

Particles were oven dried at 90 ± 5 °C for 40–48 h until the moisture content (MC) of particles reached and equilibrated to 3% MC (by taking MC samples, until having two constant weights). Particles were then blended with urea-formaldehyde resin (UF, TIONES 5100C, from BOSSN Chemicals Co., Beijing, China) using a pneumatic spray gun and mixing the shaker for 10 minutes at room temperature. Based on the oven dry particle weight, a 10% UF resin (50% solid content) and 1% liquid paraffin as wax were applied for all boards, to enhance the dimensional stability of the panels and for an efficient press cycle. As a hardener, ammonium chloride (NH4Cl, 2%, based on the resin weight) was applied to the UF solution. The resinated particles were placed in a molding box. Furnishes were manually formed, and prepressed into 50 by 50-cm mats inside the box. The adhesive-coated mats were then compressed on steel cauls in a computer controlled press (Carver Laboratory Press), using a pressure of 5 MPa and a temperature of 140 °C for 10 min to a target thickness of 12.7 mm and two target densities of 650 and 750 kg/m^3.

A total of 72 single-layer panels, three panels for each treatment combination, from three cultivars (Barhi, Saqie, and Sukkari) of date palm, two particle sizes (sizes D and E), with or without hot water treatment, and at two density levels (650 and 750 kg/m^3) were manufactured. After pressing, the panels were trimmed to a final size of 48 by 48 cm to avoid edge effects. Particleboards were conditioned for 3 weeks in a special chamber cabinet to maintain an RH of 65% $\pm$ 3% and temperature of 20 °C $\pm$ 1 °C. The conditioned panels were cut later into various sizes for property evaluations. Mechanical properties evaluation, including modulus of rupture (MOR), modulus of elasticity (MOE), and internal bond strength (IB), as well as the physical properties, including linear expansion (LE), thickness swelling (TS), and water absorption (WA), are the most important specifications required for particleboard evaluation. They were measured for each finished panel.

2.7. Mechanical Testing

Finished particleboards were cut into various specimens following the American Standard Testing Methods [12]. Figure 2 represents the cutting diagram for mechanical and physical samples taken from the manufactured panels. For the bending test, four rectangular (7.5 by 32-cm) pieces were used for three-point flex measurement of MOR and MOE. The mechanical properties were determined using a Universal Testing Machine (Model MTI-20K, Measurements Technology Inc., Roswell, GA, USA, equipped with 5000-kg load cell). The span of bending test samples was 28 cm with rounded supports. Samples were loaded at the center of span, and load was applied to the top surface of samples, with a uniform loading rate of 6 mm/min (as the thickness of panels was 12.7 mm). The load-deflection data were obtained until the maximum load was achieved. Both MOR and MOE were calculated from data obtained from computer software attached to the testing machine. Each reported value is an average of 12 measurements.

For internal bond test (IB), four 5 cm square pieces were used to determine the cohesion of panels. The square faces of the samples were effectively bonded with high quality adhesive to two loading blocks of steel alloy 5 cm square and 2.5 cm in thickness. The blocks were then attached to the same testing machine mentioned previously and a uniform rate of tension motion of 1 mm/min was applied. The maximum load was recorded and divided by the sample surface area (25 cm^2) to calculate the internal bond for each sample. Each reported value is an average of 12 measurements.

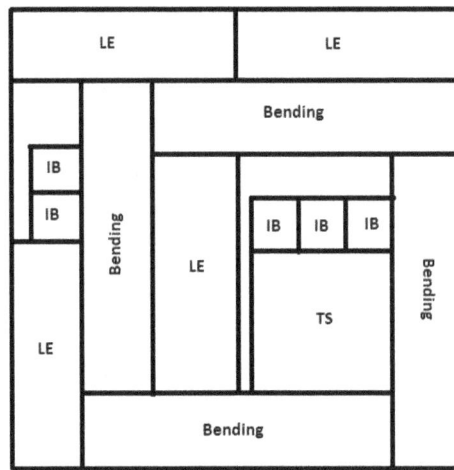

Figure 2. Cutting diagram for mechanical and physical samples taken from the manufactured panels (48 × 48 cm). Bending samples = 32 × 7.5 cm, internal bond (IB) = 5 × 5 cm, linear expansion (LE) = 24 × 7.5 cm, thickness swelling and water absorption (TS) = 15 × 15 cm.

2.8. Physical Testing

For the LE test, four rectangular 7.5 by 24-cm pieces from each panel were used for determining LE according to American Standard Testing Methods [12]. All the samples assigned for the test were conditioned for 2 weeks at an RH of 50 percent and a temperature of 20 °C ± 2 °C. Measurements of the samples' length were recorded to the nearest 0.02 mm with a digital caliper. Samples were conditioned again for 2 weeks at an RH of 90% ± 5% and a temperature of 20 °C ± 2 °C and measured again at the same previous position. The difference between the two measurements was used to calculate LE as percentages of the first conditioning values. Each reported value for LE is an average of six measurements. For the TS and WA test, one square 15 by 15-cm piece from each panel was used for determining TS and WA according to American Standard Testing Methods [12].TM Standard (ASTM D1037-2006). Samples were soaked in water at room temperature (20 °C–22 °C) for 2 and 24 h to determine the short- and long-term properties. The weight and thickness of the samples was measured before and immediately after soaking and used to calculate WA and TS, which are reported as percentages of the values before soaking. Each reported value for WA is an average of three measurements.

2.9. Statistical Analysis

Analysis of variance (ANOVA) using a four factorial experiment, with a complete randomized design (CRD), was performed by SAS software package [16]. The significance of different treatments was determined with analysis of variance and a least significant difference test ($\alpha = 0.05$). The specific methods used for evaluation of various properties are described below.

3. Results and Discussion

The aim of this study was collecting the basic data for using date palm frond residues to be used in particleboard manufacture in Saudi Arabia. For this reason, all the data required about the raw material, such as chemical composition, anatomy structure, fiber length, and particle dimensions were determined. The study is also focused on evaluating the strength properties of modulus of rupture (MOR), modulus of elasticity (MOE), and internal bond strength (IB), and responses of these boards to the linear expansion (LE) and thickness swelling (TS). This information is required prior to the commercialization of these residues as value-added products.

3.1. Particle size and Geometry

Coarse particles (0.64 mm < D < 0.25 mm) and fine particles (0.25 mm < E < 0.12 mm) used for manufacturing particleboard in this study are shown in Figure 3, and the measurements of length, width, thickness, aspect ratio (length/width), and slenderness ratio (length/thickness) are listed in Table 1.

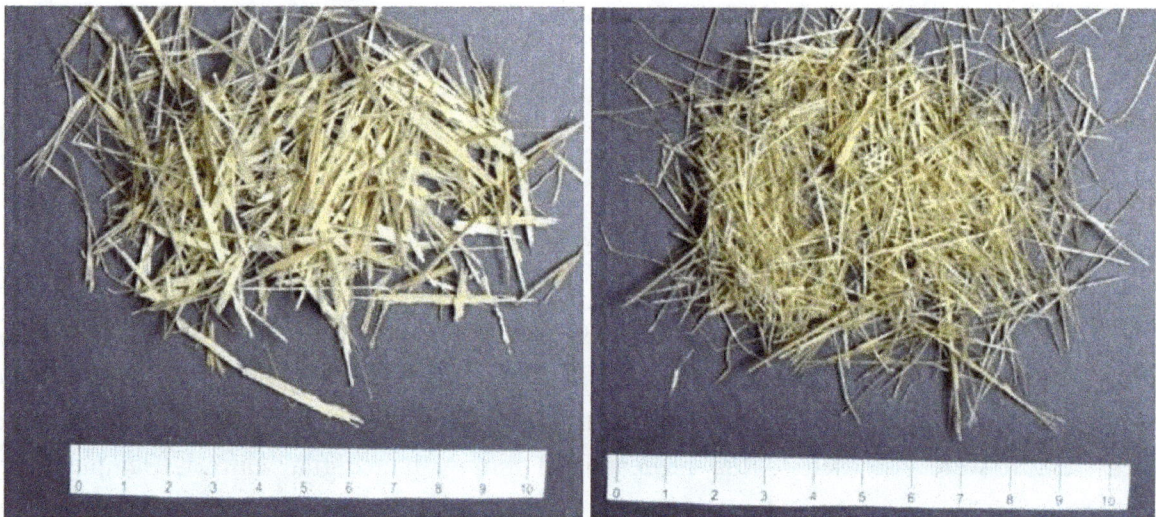

Figure 3. Coarse and fine date palm frond particles used in this study.

It is clear from Table 1 that coarse and fine particles have average length values between 42.3 mm–48.6 mm and 42.3 mm–48.6 mm, respectively. The range value of thickness for coarse and fine particles was 0.53 mm–0.83 mm, respectively. According to these measurements, the slenderness ratio values for coarse and fine particles ranged 33.5–58.3 and 48.1–78.9, respectively. However, both particle sizes of Saqie cultivar have attained the highest slenderness ratio, recording 58.3 and 78.9 for coarse and fine particles, respectively.

Table 1. Mean values measurements of coarse and fine particles for the three date palm cultivars used in this study.

Cultivar	Size	Length (mm)	Width (mm)	Thickness (mm)	Aspect ratio	Slenderness ratio
Saqui	Coarse	48.6 (9.4)	1.9 (0.73)	0.62 (0.18)	15.6 (3.1)	58.3 (12.1)
	Fine	16.2 (6.6)	1.3 (0.72)	0.21 (0.10)	19.8 (7.1)	78.9 (19.7)
Barhi	Coarse	45.9 (11.4)	2.3 (0.82)	0.83 (0.25)	20.9 (4.8)	36.9 (15.3)
	Fine	12.8 (4.7)	1.2 (0.45)	0.28 (0.11)	12.4 (6.2)	54.6 (20.3)
Sukkari	Coarse	42.3 (11.4)	2.2 (0.82)	0.53 (0.25)	20.9 (4.8)	33.5 (12.3)
	Fine	14.8 (4.7)	1.3 (0.45)	0.22 (0.11)	13.4 (6.2)	48.1 (20.3)

Each value is an average of 150 measured particles. Values in parentheses are standard deviations.

3.2. Frond Density, Fiber Length, and Chemical Composition

It is clear from Table 2 that Saquie cultivar has recorded the highest mean values for frond density and fiber length, recording 0.73 g/cm^3 and 1.14 mm, respectively. Saquie cultivar has also recorded the highest mean values for cellulose, lignin, and extractive content, recording 48.86, 31.28 and 23.51%, respectively, while this cultivar has recorded the lowest mean value of 19.86% for hemicellullose content. More fiber length would increase the density of wood. Increasing wood density improves the mechanical properties of wood [17]. The cellulose is the main component of wood. It gives strength to wood. Low hemicellulose content decreases the water diffusion and thickness swelling. Lignin is a hydrophilic component and it shows water repellent effectiveness [17]. Larger amounts of extractives in wood cause poorer mechanical properties due to breaking down of the adhesive to fiber linkage [18]. Extractives negatively affect adhesive bonding and adhesion.

Table 2. The mean values for frond density, fiber length, and the chemical composition for the three date palm cultivars used in this study.

Frond cultivar	Biomass density (gm/cm³)	Fiber length (mm)	Extractives content (%)	Cellulose content (%)	Hemicellulose content (%)	Lignin content (%)
Barhi	0.628 C (0.08)	1.09 AB (0.01)	19.13 B (1.8)	47.84 B (2.4)	23.10 A (1.8)	29.6
Saqie	0.731 A (0.05)	1.14 A (0.02)	23.51 A (1.7)	48.86 A (2.1)	19.86 C (1.3)	31.28
Sukkari	0.683 B (0.05)	1.06 B (0.01)	22.64 A (1.6)	47.17 A (1.3)	22.30 B (1.9)	30.19
L.S.D$_{0.05}$	0.040	0.06	0.065	0.77	2.26	----

Values in parentheses are standard deviations.

3.3. Date Palm Anatomy

Regarding the frond anatomy, sections were prepared to evaluate the anatomical structure of the different cultivars using a light microscope. Figure 4 illustrates the typical cross section taken from the basal frond segments of A: Barhi, B:Sukkari, and C: Saquie cultivars, respectively, showing the variation in size of vascular bundles rich in fibers embedded in the parenchymatous tissues. However, it can be noticed that there are three zones of vascular bundles which can be distinguished across the midrib, the peripheral, transitional zone, and inner zone. In the first and second zones, the fiber sheath is thick and the vascular bundles are numerous, with small parenchyma cells between them. The third zone is the broadest, where the bundles reach their highest diameter. Fiber tissue percent is higher in the periphery and transition zones (38%) than in the inner or central zone (10%). The size and shape of fiber strands in the vascular bundle are the most important structural factor that determines the date palm leaves' behavior. The periphery and transition zone across the midrib, which is characterized by the higher percentage of fiber tissue and larger number of bundles, affects the density and strength properties of the frond. Generally, as shown in Figure 3 Saqie cultivar was found to be superior on account of having the highest number of vascular bundles per cm^2 cross-sectional area (205), and a higher fiber transverse wall area fraction (22%) and a narrow vessel diameter (55.42 μm).

More fiber length and a higher number of cells increase the density of wood. Increasing wood density improves the mechanical properties of wood [10,19]. Bhat *et al.* [20] mentioned that the average fiber length of the date palm midrib is within the average of dicotyledons and hardwood species and shorter than the fiber length of stems of some other palm species.

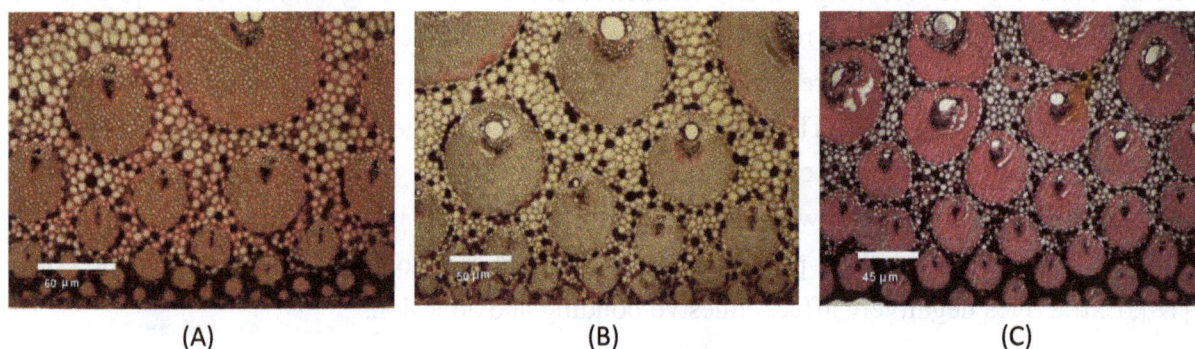

(A)   (B)   (C)

Figure 4. Cross section of fronds of the basal segment of (**A**) Barhi, (**B**) Sukkari and (**C**) Saquie cultivars passing from the peripheral region, showing varying sizes of vascular bundles rich in fibers embedded in the parenchymatous tissues.

3.4. Effects of Individual Parameters

Table 3 summarizes the F-values (as an indicator for parameter significance) obtained from the statistical analysis and the ANOVA results for the effects of date palm cultivar, particle size, water treatment, and panel density on both mechanical properties (MOR, MOE and IB) and dimensional stability properties (LE, TS and WA) for all panel combinations under investigation. It is clear that **date palm cultivar** has a significant effect on all the mechanical and dimensional stability properties under this study, while **particle size** has a significant effect only on MOR, IB and LE. **Hot water extraction** treatment has also a significant effect on all dimensional stability properties and MOE, while **panel density** has a significant effect on all mechanical properties as well as LE (Table 3). The mean values for each individual parameter of date palm cultivars, particle size, water extraction treatments, and panel densities on both the mechanical and dimensional stability properties are represented through Figures 5–9.

Table 3. F-values obtained from the statistical analysis and the ANOVA results for the main effects as well as the interactions of date palm cultivar, particle size, water treatment, and panel density on both mechanical properties and dimensional stability properties.

Parameters	MOR	MOE	IB	LE	2 h-TS	2 h-WA	24 h-TS	24 h-WA
Cult	$P < 0.0001$	$P < 0.0001$	$P < 0.0001$	$P = 0.0004$	$P = 0.0012$	$P < 0.0001$	$P = 0.0003$	$P < 0.0001$
Size	$P < 0.0001$	$P = 0.1596$	$P < 0.0001$	$P < 0.0001$	$P = 0.0294$	$P = 0.8766$	$P = 0.1325$	$P = 0.1686$
Cult × Size	$P = 0.3286$	$P = 0.0907$	$P = 0.0381$	$P < 0.0001$	$P < 0.0001$	$P < 0.0001$	$P < 0.0001$	$P < 0.0001$
Wt	$P = 0.7550$	$P = 0.0013$	$P = 0.8087$	$P < 0.0001$	$P < 0.0001$	$P < 0.0001$	$P < 0.0001$	$P < 0.0001$
Cult × Wt	$P < 0.0001$	$P < 0.0001$	$P < 0.0001$	$P < 0.0001$	$P < 0.0001$	$P < 0.0001$	$P < 0.0001$	$P < 0.0001$
Size × Wt	$P = 0.0016$	$P = 0.0781$	$P < 0.0001$	$P < 0.0001$	$P = 0.4749$	$P < 0.0001$	$P = 0.0059$	$P = 0.0061$
Cult × Size × Wt	$P = 0.0572$	$P < 0.0001$	$P < 0.0001$	$P < 0.0001$	$P < 0.0001$	$P < 0.0001$	$P < 0.0001$	$P < 0.0001$
D	$P < 0.0001$	$P < 0.0001$	$P < 0.0001$	$P < 0.0001$	$P = 0.0898$	$P = 0.4883$	$P < 0.0005$	$P = 0.4142$
Cult × D	$P < 0.0001$	$P = 0.1669$	$P < 0.0001$	$P = 0.0001$	$P < 0.0001$	$P < 0.0001$	$P < 0.0001$	$P < 0.0001$
Size × D	$P = 0.9588$	$P = 0.8872$	$P = 0.0110$	$P = 0.0006$	$P = 0.0487$	$P = 0.8469$	$P < 0.0001$	$P = 0.3242$
Cult × Size × D	$P = 0.0122$	$P = 0.2434$	$P < 0.0001$	$P < 0.0001$	$P < 0.0001$	$P < 0.0001$	$P < 0.0001$	$P < 0.0001$
Wt × D	$P = 0.7867$	$P = 0.1432$	$P = 0.0007$	$P < 0.0001$	$P = 0.0058$	$P = 0.1165$	$P = 0.0001$	$P = 0.1034$
Cult × Wt × D	$P = 0.6051$	$P = 0.2311$	$P < 0.0001$	$P = 0.0329$	$P < 0.0001$	$P < 0.0001$	$P < 0.0001$	$P = 0.0156$
Size × Wt × D	$P = 0.4103$	$P = 0.5799$	$P = 0.0003$	$P = 0.5833$	$P < 0.0001$	$P = 0.0426$	$P < 0.0001$	$P = 0.0019$
Cult × Size × Wt × D	$P = 0.0422$	$P = 0.3177$	$P < 0.0001$	$P = 0.0035$	$P < 0.0001$	$P < 0.0001$	$P < 0.0001$	$P < 0.0001$

Abbreviations: Cult = date cultivar, Size = particle size, Wt = water treatment, and D = panel density.

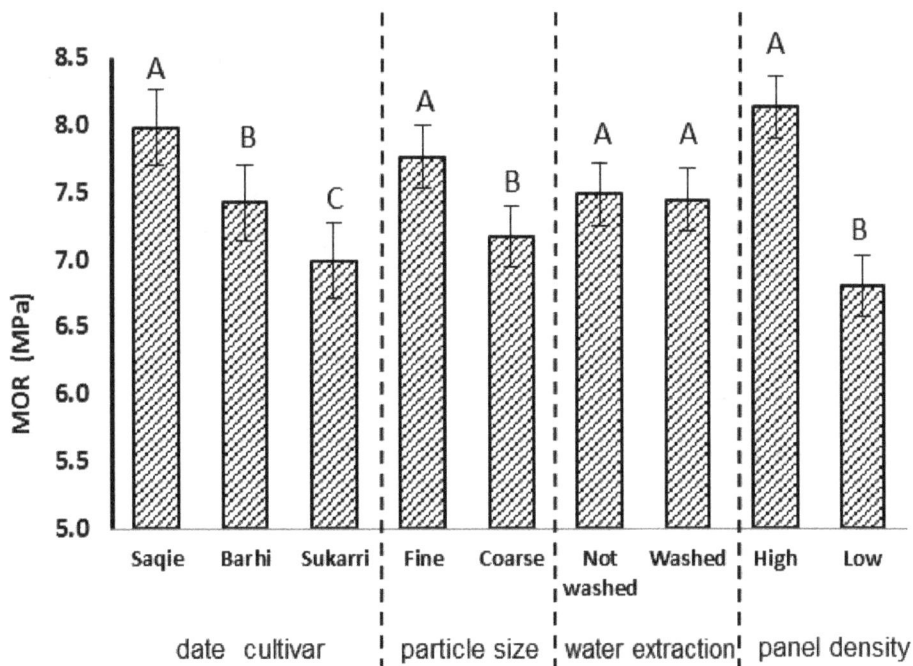

Figure 5. Effect of date palm cultivar, particle size, hot water extraction, and panel density on Modulus of Rupture (MOR) property. (Means with same letter do not significantly different at L.S.D.$_{0.05}$).

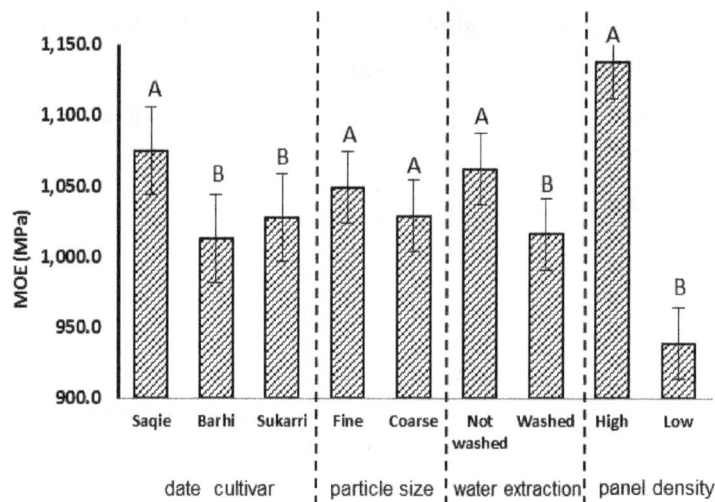

Figure 6. Effect of date palm cultivar, particle size, hot water extraction and panel density on Modulus of Elasticity (MOE) property. (Means with same letter do not significantly different at L.S.D.$_{0.05}$).

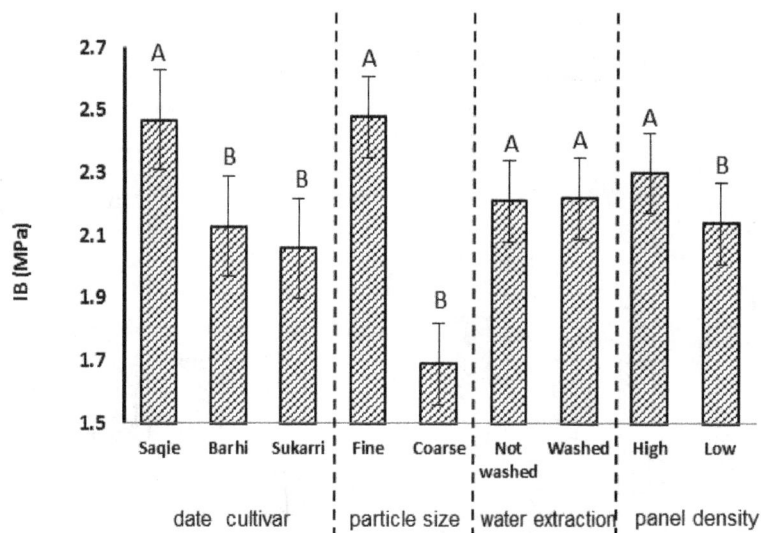

Figure 7. Effect of date palm cultivar, particle size, hot water extraction and panel density on internal bond (IB) property. (Means with same letter do not significantly different at L.S.D.$_{0.05}$).

3.5. Effect of Panel Density

As the board density increased, the compaction ratio increased providing a higher contact surface between the particles, and more efficient glue bonds were improved compared to lower compaction ratio. This caused higher flexural properties and internal bond [10,19,21]. This conclusion obviously applies to our experiment (Figures 5–9), where increasing board density resulted in an increase in all properties except 24h-TS. High density panels of 750 kg/m^3 have recorded higher mean values for MOR, MOE, and IB, recording mean values of 8.13, 1138, and 2.3 MPa, respectively. Low density panels (650 kg/m^3) have recorded lower mean values (better performance) for LE property, recording 0.147%.

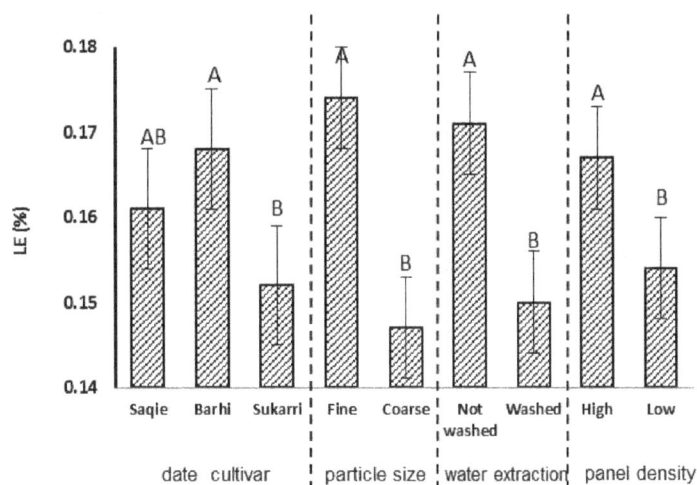

Figure 8. Effect of date palm cultivar, particle size, hot water extraction and panel density on linear expansion (LE) property. (Means with same letter do not significantly different at L.S.D.$_{0.05}$).

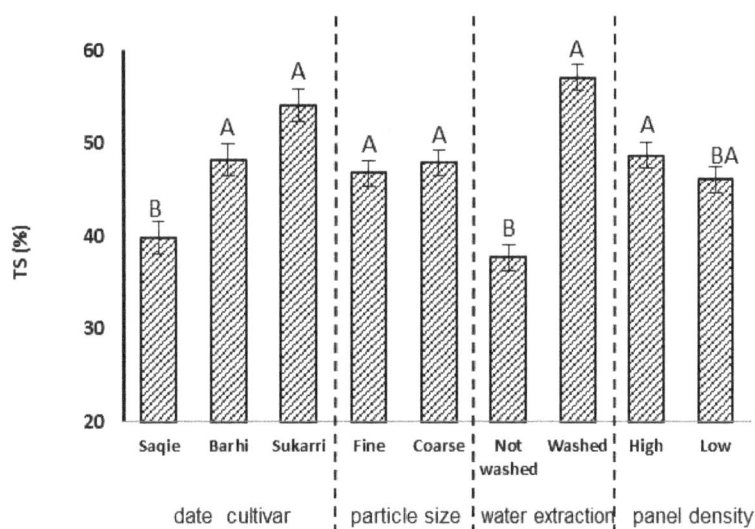

Figure 9. Effect of date palm cultivar, particle size, hot water extraction and panel density on thickness swelling (24-h TS) property. (Means followed by with same letter do not significantly different at L.S.D.$_{0.05}$).

3.6. Effect of Particle Size and Shape

The fine particles resulted in higher mean values than coarse ones for MOR, MOE, IB, and LE, recording 7.76, 1049, 2.48 MPa, and 0.174%; respectively, while particle size had no effect on 24 h-TS property (Figures 5–9). The higher strength values for the fine particles may be attributed to the higher slenderness ratio recorded by the fine particles compared to the coarse ones, as shown in Table 1. The fine particles had slenderness mean values of 78.9, 54.6, and 48.1 for Saqie, Barhi, and Sukkari cultivars, respectively, while for coarse particles these values were 58.3, 36.9, and 33.5, respectively, at the same order. Our results were supported by Biswas *et al.* [22], as they mentioned that properties of particleboards could be affected significantly by particle geometry, which includes the shape and particle size. Particle size is one of the factors that can be manipulated to improve the physical and mechanical properties of particleboard. Increasing the length of particles and the slenderness ratio (length/thickness) also increases both MOR and MOE, but decreases the IB [23]. Osarenmwinda and Nwachukwu [24] also reported that the longer particle size gave better mechanical properties of the particleboard, while Ngueho Yemele *et al.* [25] mentioned that IB strength mostly increased with decreasing particle size. However, thicker and shorter particles have a higher specific surface area, and they receive more resin and provide better inter-particle contacts, which may improve IB property.

Regarding the effect of particle size on dimensional stability properties, Viswanathan and Kailappan [26] reported that WA and TS were least for the board made from the largest particles. They explained that the higher WA values attained by the boards made from smaller particles would be because of the larger surface area, which absorbs more water. However, the high values for both WA and TS recorded in our study may be due to the fact that date palms are monocotyledons, and there are huge amounts of parenchymatous tissues around the vascular bundles, which might absorb more water than the wood inside the vascular bundles as shown in Figure 4.

3.7. Effect of Hot Water Extraction

The extracted particles with hot water have significantly lower mean value for LE property than the un-extracted ones, recording 0.15%, while the un-extracted ones have recorded significantly higher mean values for MOE and lower mean value for 24h-TS compared with the extracted particles, recording 1062 MPa and 37.7%, respectively. On the other hand, water treatments have no effect on both MOR and IB properties (Figures 5–9). However, the affinity of particleboard to water is one of the main limitations for using these panels in moisture-rich environments. Dimensional stability and durability can be improved by hot water extraction, which increases panels resistance to moisture uptake and it does not need additional chemicals [27]. Hot water extraction is an autocatalytic thermo-chemical process for fractionation of easily accessible sugars in lignocellulosic biomass [28]. Sweet and Winandy [29] found a negative effect of hemicellulose reduction on the wood's mechanical properties after such treatment. It is a fact that some of the extractives and starch content are eliminated during the washing process of the particles, resulting in a negative impact on the mechanical properties. The amount of extractives content in date palm cultivars as shown in Table 2 was about 19.1%–23.5%, which is considered a relatively high percentage compared to wood. Similar findings were also determined in a past study [30]. Particleboard made from date palm treated with hot water extraction had lower bending values than those made from unwashed control panels [31].

3.8. Effect of Date Palm Cultivar

Saqie cultivar has significantly recorded the highest mean values for MOR, MOE, and IB, recording 7.98, 1075, and 2.47 MPa, respectively, and has recorded the lowest mean value (more dimensionally stable) for 24h-TS recording 39.8%. On the other hand, Sukkari cultivar has recorded the lowest mean value of 0.15% for LE property.

3.9. The Combined Effect of All Parameters on Mechanical Properties

Table 4 displays the average values of the mechanical properties of the panel samples as they were affected by the above-mentioned four parameters. The highest MOR value of 9.82 MPa was recorded by the panels made from the un-extracted fine particles from Barhi cultivar at the 750 kg/m^3 density level, while the highest MOE value of 1245.4 MPa was recorded by the panels made from the un-extracted fine particles from Saqie cultivar at the 750 kg/m^3 density level. The highest IB value of 3.62 MPa was recorded by the panels made from the extracted fine particles from Saqie cultivar at the 750 kg/m^3 density level.

However, it seems that high density panels with fine particles of Saqie cultivar for both water treatments have obtained the best parameters combination for achieving higher mechanical properties in our study, recording MOR mean values of 9.11 and 9.03 MPa, MOE values of 1130 and 1245 MPa, and IB values of 3.62 and 2.43 MPa for extracted and un-extracted particles, respectively. Conversely, the lowest mean values for MOR, MOE, and IB were 5.06, 800 MPa, and 0.67 MPa, respectively for panels made from coarse un-extracted particles of Barhi cultivar at low density level.

Generally, Table 4 made it clear that manufacturing high density panels at 750 kg/m^3 using fine date palm particles have obtained better values for MOR, MOE, and IB properties compared to the low density panels using coarse particles manufactured under our investigation, while Saqie frond cultivar has achieved better performance for the same previous properties compared to the other two cultivars of Barhi and Sukkari. There are many reasons for the better results of Suqie cultivar such as higher values for frond density, fiber length, and particle slenderness ratio. Saquie cultivar has also had the highest values for cellulose, lignin, and extractive content, and the lowest value for hemicellullose content, as well as having the highest number of vascular bundles per cm^2 cross sectional area.

The American National Standards (ANSI A208.1-2009) [32] stated that the minimum requirements for MOR, MOE, and IB were 10, 1550, and 0.36 MPa, respectively, for commercial particleboard (Grade M1), and were13, 2000, and 0.4 MPa, respectively, at the same order, for the industrial particleboard (Grade M2). However, all the MOR and MOE values found in this study were lower than those stated in ANSI standards for both M1 and M2 grades, but at least 25% of the samples satisfied the minimum MOR limit of 8.0 MPa for particleboard (Type 8) for Japanese industrial standards [33]. On the other hand, all of IB samples have satisfied both the American and Japanese standards [32,33].

3.10. The Combined Effect of all Parameters on Dimensional Stability Properties

Table 4 displays the average values of dimensional stability properties of the panel samples (LE, TS and WA) for the panels under investigation. The lowest LE value of 0.09% was recorded by the panels made from the extracted coarse particles of Saqie cultivar at the 0.65 g/cm^3 density level, while the LE range values was 11%–23% for all other panel combinations. However, all the LE values tabulated in this study were lower than those stated in ANSI requirements [32] for both M1 and M2 grades (<0.40%). Generally, within each cultivar, we could say that low density panels at 650 kg/m^3 with hot water treated coarse particles have obtained lower values for LE property compared to the high density panels with fine particles manufactured under our investigation, In addition, Miyamoto et al. [23] mentioned that LE of particleboard decreased with increasing particle length and size.

Table 4. Average values of mechanical and dimensional stability properties of manufactured panels.

Cultivar	Particle Size	Water Extraction	Target Density (g/cm³)	Actual Density (g/cm³)	Mechanical Properties				Dimensional Stability Properties			
					MOR (MPa)	MOE (MPa)	IB (MPa)	2-h. TS (%)	2-h. WA (%)	24-h. TS (%)	24-h. WA (%)	LE (%)
Barhi	Fine	Extracted	0.65	0.67	6.2 (1.1)	874 (78)	2.61 (0.27)	41.6 (2.4)	83.2 (4.5)	46.0 (2.3)	104.6 (2.8)	0.12 (0.02)
			0.75	0.72	7.8 (0.5)	1071 (98)	2.56 (0.38)	31.5 (13.8)	79.0 (4.7)	46.4 (2.2)	104.5 (5.1)	0.14 (0.02)
		Un-extracted	0.65	0.66	7.5 (0.8)	949 (79)	2.12 (0.22)	29.3 (1.2)	61.4 (5.2)	34.9 (0.8)	86.9 (4.6)	0.18 (0.03)
			0.75	0.74	9.8 (0.7)	1178 (66)	2.52 (0.39)	27.5 (1.6)	51.2 (2.2)	36.4 (1.9)	79.1 (2.9)	0.22 (0.03)
	Coarse	Extracted	0.65	0.66	5.1 (0.7)	800 (63)	0.67 (0.17)	80.0 (8.6)	102.4 (12)	98.6 (8.8)	133.5 (11.9)	0.23 (0.03)
			0.75	0.72	7.9 (0.9)	1018 (120)	1.78 (0.32)	43.7 (4.7)	77.3 (4.9)	51.8 (5.8)	103.3 (5.3)	0.19 (0.02)
		Un-extracted	0.65	0.67	6.2 (0.8)	958 (101)	2.46 (0.38)	23.8 (1.5)	56.7 (2.9)	32.1 (1.0)	82.4 (2.7)	0.12 (0.03)
			0.75	0.74	8.8 (1.3)	1218 (154)	2.31 (0.29)	29.7 (2.8)	52.5 (2.4)	40.0 (1.5)	78.5 (2.5)	0.15 (0.03)
Saqie	Fine	Extracted	0.65	0.69	7.5 (1)	986 (84)	2.40 (0.32)	38.8 (1.6)	82.4 (3.7)	44.2 (1.6)	107.1 (2.5)	0.19 (0.03)
			0.75	0.73	9.2 (0.9)	1130 (75)	3.62 (0.38)	31.7 (1.8)	49.7 (11)	42.0 (3.7)	81.3 (16.9)	0.22 (0.02)
		Un-washed	0.65	0.67	7.4 (1.1)	1055 (120)	2.38 (0.36)	24.7 (5.0)	46.8 (9.5)	33.5 (2.9)	82.8 (2.2)	0.13 (0.03)
			0.75	0.74	9.0 (1.23)	1245 (109)	2.43 (0.30)	27.7 (1.1)	55.4 (2.9)	36.1 (1.0)	82.6 (0.9)	0.16 (0.02)
	Coarse	Extracted	0.65	0.68	8.2 (0.7)	975 (111)	2.41 (0.26)	33.1 (1.9)	70.8 (8.2)	40.9 (1.6)	94.9 (6.3)	0.09 (0.02)
			0.75	0.72	8.9 (1.5)	1213 (151)	2.33 (0.24)	42.1 (1.5)	77.9 (4.9)	49.1 (1.7)	102.1 (4.5)	0.11 (0.02)
		Un-extracted	0.65	0.68	6.5 (0.5)	924 (78)	2.16 (0.30)	26.8 (0.6)	62.5 (3.4)	33.3 (1.5)	83.6 (4.5)	0.14 (0.03)
			0.75	0.71	7.2 (0.8)	1071 (84)	2.01 (0.21)	31.4 (2.9)	64.9 (7.4)	39.0 (3.4)	87.0 (7.2)	0.18 (0.03)
Sukkari	Fine	Extracted	0.65	0.66	6.8 (0.6)	925 (51)	2.32 (0.28)	51.3 (0.7)	96.9 (3.2)	66.7 (3.7)	125.0 (2.0)	0.16 (0.02)
			0.75	0.72	7.8 (1.1)	1098 (155)	1.89 (0.17)	84.8 (0.5)	133.7 (3.1)	100.2 (3.5)	158.1 (3.6)	0.11 (0.02)
		Un-extracted	0.65	0.68	7.2 (0.9)	909 (98)	2.25 (0.33)	29.2 (1.2)	70.7 (0.2)	35.9 (0.3)	93.1 (0.3)	0.23 (0.01)
			0.75	0.71	6.9 (1.3)	1160 (94)	2.60 (0.24)	23.5 (1.8)	68.4 (2.6)	39.6 (1.5)	95.3 (3.3)	0.22 (0.03)
	Coarse	Extracted	0.65	0.67	7.0 (0.8)	993 (121)	2.06 (0.13)	42.3 (3.09)	81.6 (4.9)	47.0 (3.3)	108.4 (5.5)	0.11 (0.02)
			0.75	0.71	7.1 (0.8)	1103 (15)	2.03 (0.18)	46.9 (0.9)	81.8 (4.5)	52.3 (0.5)	104.0 (4.6)	0.14 (0.02)
		Un-extracted	0.65	0.61	6.1 (1.3)	919 (150)	1.82 (0.222)	32.4 (0.8)	71.8 (0.9)	39.3 (1.3)	93.4 (1.0)	0.15 (0.02)
			0.75	0.72	7.0 (1.0)	1117 (87)	1.48 (0.21)	43.1 (0.7)	84.5 (2.4)	51.6 (1.7)	106.6 (2.0)	0.17 (0.02)
L.S.D.$_{0.05}$					0.79	8.2	0.26	6.43	9.01	4.96	9.27	0.02
ANSI standard	M1 commercial panels				10	1550	0.36	....	....	....	....	0.40%
	M2 industrial panels				13	2000	0.40	....	....	....	....	0.40%
CEN standard					....	....	....	8%	....	15%	....	....
JIS Type 8					8	2000	....	....	....	....	....	....

Numbers in parentheses are standard deviation values. ANSI: American National Standards Institute, CEN: European Committee for Standardization, JIS: Japanese industrial standards.

Regarding the thickness swelling test (TS), the 2h-TS samples have recorded range values of 23.8%–84.0% for all panels, while the range values for 24h-WA was 32.1%–100.3%. On the other hand, for water absorption test (WA), the 2h-WA samples have recorded range values of 46.8%–102.4% for all panels, while the range values for 24h-WA was 78.5%–158.2%. Based on European Committee for Standardization (CEN) [34], particleboard should have a maximum thickness swelling value of 8% and 15%, for 2-h and 24-h immersion,; respectively.

In general, the observed TS and water WA values for particleboards in our study were too much higher than 15% (as a maximum requirement). Similar high TS values have been reported for the particleboards that were produced using agricultural residues, such as 60.7% for tobacco and tea leaves [35], 35% for cotton stalks [36], and 19.6% for hazelnut hulls [37], after 24h water soaking. Many treatments could be utilized in the particleboard production to improve these properties, such as the use of phenolic resins, coating the particleboard surfaces, and acetylating of particles to improve the water repellency of the panels [38–40].

3.11. Mechanical Properties of Date Palm in the Literature

Regarding to the mechanical properties published in the past few years for date palm frond, El-Mously et al. [8] obtained MOR, MOE, and IB values of 10.5, 18,512, and 0.43 MPa, respectively, for board density of 650 kg/m^3, while Nemli et al. [41] obtained MOR values in the range of 15.3–18.9 MPa and IB values in the range of 0.35–0.83 MPa for the same density. On the other hand, Ashori and Nourbakhsh [42], using a board density of 750 kg/m^3 and resin content between 9 and 11 percent, attained MOR, MOE, and IB range values of 10–16.6 MPa, 1333–1861 MPa, and 0.38–0.63 MPa, respectively. Hegazy and Aref [10], using a laboratory hammer mill, mentioned MOR, MOE, and IB values of 13.3, 2018, and 0.53 MPa, respectively, for a panel density of 790 kg/m^3, while they got 9.04, 1443, and 0.43 MPa, respectively, for a panel density of 670 kg/m^3. Lower bending properties of the samples made in this work could be related to the particles configuration as a result of the flaking machine and the substantial amount of parenchyma cells and non-fibrous structure of the frond particles, which was observed during microscopic evaluation of the sections taken from the samples, where walls of the parenchyma cells is thin, in contrast to the thick cell wall of fibers. Hashim et al. [43] and Wazzan [44] revealed similar findings in a past investigation related to particleboard panels manufactured from oil palm fronds. Usually, boards having the lower mechanical properties can be used as insulating material in buildings because such boards would not be subjected to any mechanical stress or mechanical properties.

4. Conclusions

1. This study showed that raw material from date palm fronds has the potential to be used in the manufacture of experimental particleboard panels.
2. Increasing board density would increase all mechanical properties by providing a higher contact surface between the particles and more efficient glue bonds.
3. The internal bond strength of the samples was found to be satisfactory, but the bending properties of the samples need to be improved using different approaches, including higher resin distribution or modifying the particle size.

4. Better results for all mechanical properties could be obtained when both parameters of Saqie cultivar and fine particles were used.

5. Hot water treatments have no effect on both MOR and IB properties, but improved the LE performance.

Acknowledgments

This Project was funded by the National Plan for Science, Technology and Innovation (MAARIFAH), King Abdul-Aziz City for Science and Technology, Kingdom of Saudi Arabia, Award Number (11-AGR 1745-02.)

Author Contributions

Said S. Hegazy conducted the panel manufacturing, physical testing, collection of data, data analysis, writing of the final paper, and preparation of figures and graphs. Khaled Ahmed conducted the mechanical testing. All authors read and approved the manuscript.

Conflicts of Interest

The authors declare no conflict of interest.

References

1. Agoudjila, B.; Benchabaneb, A.; Boudennec, A.; Ibosc, L.; Foisc, M. Renewable materials to reduce building heat loss: Characterization of date palm wood. *Energy Build.* **2011**, *43*, 491–497.

2. Barreveld, W.H. Date palm products. In *Food and Agriculture Organization of the United Nations*; FAO Agricultural Services: Rome, Italy, 1993.

3. Al-Sulaiman, F.A. Date palm fiber reinforced composite as a new insulating material. *Int. J. Energy Res.* **2003**, *27*, 1293–1297.

4. Al-Jurf, R.S.; Ahmed, F.A.; Allam, I.A.; Abdel-Rehman, H.H. Development of new building material using date palm fronds; Final Report ARP-6-141; KACST: Riyadh, Saudi Arabia, October 1998.

5. Bashah, M.A. Date varieties in the Kingdom of Saudi Arabia. In *Guidance Booklet for Palms and Dates*; King Abdulaziz University Press: Jeddah, Kingdom of Saudi Arabia, 1996; pp. 51–62.

6. Chandrasekaran, M.; Bahka, A.H. Valorization of date palm (*Phoenix dactylifera*) fruit processing by-products and wastes using bioprocess technology – Review. *Saudi J. Biol. Sci.* **2013**, *20*, 105–120.

7. Al-Oqla, F.M.; Sapuan, S.M. Natural fiber reinforced polymer composites in industrial applications: Feasibility of date palm fibers for sustainable automotive industry. *J. Clean Prod.* **2014**, *66*, 347–354.

8. El-Mously, H.; El-Morshedy, M.M.; Megahed, M.M.; Abd El-Hai, Y. Evaluation of particleboard made of palm leaves midribs as compared with flaxboard. In Proceedings of the 4th International Conference on Production Engineering and Design for Development, Cairo, Egypt, 27–29 December 1993.

9. Iskanderani, F.I. Influence of process variables on the bending strength of particleboard produced from Arabian date palm mid-rib chips. *Int. J. Polym. Mat.* **2009**, *58*, 49–60.

10. Hegazy, S.; Aref, I. Suitability of some fast growing trees and date palm fronds for particleboard production. *For. Prod. J.* **2010**, *60*, 599–604.

11. Hegazy, S.; Ahmed, K.; Hiziroglu. S. Oriented strand board production from water -treated date palm fronds. *Bioresources* **2015**, *10*, 448–456.

12. American Society for Testing and Materials. ASTM D 2395–07a. Standard test method for specific gravity of wood and wood-based materials. In *Annual Book of ASTM Standards*; Volume 04.10—Wood. American Society for Testing and Materials: West Conshohocken, PA, USA, 2010.

13. American Society for Testing and Materials. *Standard Test Method for Preparation of Extractive-Free Wood, ASTM D-1105-84*; American Society for Testing and Materials: Philadelphia, PA, USA, 1989.

14. Nikitin, V.M. Himia drevesini i telliulozi. *Chimia Lemnului SI A Celuloze I Vol I si II* **1973**, *1–2*, 233.

15. Rozmarin, G.; Simionescu, C. Determining cellulose content. *Wood Chem. Cell. (Romanian)* **1960**, *2*, 392–396.

16. SAS Institute. *SAS User's Guide: Statistics*; SAS Institute: Cary, NC, USA, 2000.

17. Baharoğlu, M.; Nemli, G.; Sarı, B.; Birtürk, T.; Bardak, S. Effects of anatomical and chemical l properties of wood on the quality of particleboard. *Comp. Part B: Eng.* **2013**, *52*, 282–285.

18. Cameron, F.A.; Pizzi, A. *American Chemical Society Symposium Series: 316*; ACS: Washington, DC, USA, 1985; Chapter 15, p. 205.

19. Sackey, E.K.; Semple, K.E.; Oh, S.W.; Smith G.D. Improving core bond strength of particleboard through particle size redistribution. *Wood Fiber Sci.* **2008**, *40*, 214–224.

20. Bhat, K.M.; Mahmamed Nasser, K.M.; Thulasidas, P.K. Anatomy and identification of South Indian rattans (*Calamus* species). *IAWA J.* **1993**, *14*, 63–76.

21. Dias, F.M.; Nascimento, M.F.; Martinez-Espinosa, M.; Lahr, F.A.R.; Valarelli, I.D. Relation between the compaction rate and physical and mechanical properties of particleboards. *Mater. Res.* **2005**, *8*, 329–333.

22. Biswas, D.; Kanti Bose, S.; Mozaffar Hossain, M. Physical and mechanical properties of urea formaldehyde-bonded particleboard made from bamboo waste. *Inter. J. Adh. Adhes.* **2010**, *31*, 84–87.

23. Miyamoto, K.; Nakahara, S.; Suzuki, S. Effect of particle shape on linear expansion of particleboard. *J. Wood Sci.* **2002**, *48*, 185–190.

24. Osarenmwinda, J.; Nwachukwu, J. Effect of particle size on some properties of Rice Husk Particleboard. *Adv. Mat. Res.* **2007**, *18*, 43–48.

25. Ngueho Yemele, M.C.; Blanchet, P.; Cloutier, A.; Koubaa, A. Effects of bark content and particle geometry on the physical and mechanical properties of particleboard made from black spruce and trembling aspen bark. *For. Prod. J.* **2008**, *58*, 48–56.

26. Viswanathan, R.L.; Kailappan, G.R. Water absorption and swelling characteristics of coir pith particle board. *Biol. Technol.* **2000**, *71*, 93–94.

27. Garrote, G.; Dominguez, H.; Parajo, J.C. Hydrothermal processing of lignocellulosic materials. *Holz Roh- Werkst.* **1999**, *57*, 191–202.

28. Boonstra, M.J.; Blomberg, J. Semi-isostatic densification of heat-treated radiata pine. *Wood Sci. Technol.* **2007**, *41*, 607–617.

29. Sweet, M.S.; Winandy, J.E. Influence of degree of polymerization of cellulose and hemicellulose on strength loss in fire-retardant-treated southern pine. *Holzforschung* **1999**, *53*, 311–317.

30. Saadaou, N.; Rouilly, A.; Fares, K.; Rigal, L. Characterization of date palm lignocellulosics by products and self-bond composite material obtained thereof. *Mater. Des.* **2013**, *50*, 302–308.

31. Kriker, A.; Bali, B.; Debicki, G.; Bouziane, M.; Chabannet, M. Durability of date palm fibers and their use as reinforcement in hot dry climates. *Cement Concr. Compos.* **2008**, *30*, 639–648.

32. ANSI. *A 208-1. American National Standards Institute. 1899 L Street NW*; ANSI: Washington, DC, USA, 2009.

33. Japanese Industrial Standards (JIS)-A5908. *Particleboards*; Japanese Standards Association: Tokyo, Japan, 2003.

34. European Committee for Standardization (CEN). *EN 317. Particleboards and Fiberboards, Determination of Swelling in Thickness After Immersion*; CEN: Brussels, Belgium, 1993.

35. Kalaycioglu, H. Utilization of annual plant residues in the production of particleboard. In Proceedings of the ORENCO 92, 1st Forest Product Symposium, Trabzon, Turkey, 1–4 November 1992; pp. 288–292.

36. Guler, C.; Ozen, R. Some properties of particleboards made from cotton stalks (*Gossypium hirsitum* L.). *Holz Roh Werkst.* **2004**, *62*, 40–43.

37. Copur, Y.; Guler, C.; Akgul, M.; Tascioglu, C. Some chemical properties of hazelnut husk and its suitability for particleboard production. *Build. Environ.* **2007**, *42*, 2568–2572.

38. Nemli, G.; Ors, Y.; Kalaycioglu, H. The choosing of suitable decorative surface coating material types for interior end use applications of particleboard. *Constr. Buil. Mater.* **2005**, *19*, 307–312.

39. Guntekin, E.; Uner, B.; Sahin, H.T.; Karakus, B. Pepper stalks (*Capsicum annuum*) as raw material for particleboard manufacturing. *J. Appl. Sci.* **2008**, *8*, 2333–2336.

40. Ayrilmis, N.; Buyuksari, U.; Avci, E.; Koc, E. Utilization of pine (*Pinus pinea* L.) cone in manufacture of wood based composite. *For. Ecol. Manag.* **2009**, *259*, 65–70.

41. Nemli, G.; Kalaycıoğlu, H.; Alp, T. Suitability of date (*Phoenix dactyiferis*) branches for particleboard production. *Holz als Roh- und Werkstoff.* **2001**, *59*, 411–412.

42. Ashori, A.; Nourbakhsh, A. Effect of press cycle time and resin content on physical and mechanical properties of particleboard panels made from the underutilized low-quality raw materials. *Ind. Crops Prod.* **2008**, *28*, 225–230.

43. Hashim, R.; Saari, N.; Sulaiman, O.; Sugimoto, T.; Hiziroglu, S.; Sato, M.; Tanaka, R. Effect of particle geometry on the properties of binderless particleboard manufactured from oil palm trunk. *Mater. Des.* **2008**, *31*, 4251–4257.

44. Wazzan, A.A. Effect of fiber orientation on the mechanical properties and fracture characteristics of date palm fiber reinforced composites. *Int. J. Polym. Mater.* **2005**, *54*, 213–225.

Permissions

All chapters in this book were first published in Agriculture, by MDPI; hereby published with permission under the Creative Commons Attribution License or equivalent. Every chapter published in this book has been scrutinized by our experts. Their significance has been extensively debated. The topics covered herein carry significant findings which will fuel the growth of the discipline. They may even be implemented as practical applications or may be referred to as a beginning point for another development.

The contributors of this book come from diverse backgrounds, making this book a truly international effort. This book will bring forth new frontiers with its revolutionizing research information and detailed analysis of the nascent developments around the world.

We would like to thank all the contributing authors for lending their expertise to make the book truly unique. They have played a crucial role in the development of this book. Without their invaluable contributions this book wouldn't have been possible. They have made vital efforts to compile up to date information on the varied aspects of this subject to make this book a valuable addition to the collection of many professionals and students.

This book was conceptualized with the vision of imparting up-to-date information and advanced data in this field. To ensure the same, a matchless editorial board was set up. Every individual on the board went through rigorous rounds of assessment to prove their worth. After which they invested a large part of their time researching and compiling the most relevant data for our readers.

The editorial board has been involved in producing this book since its inception. They have spent rigorous hours researching and exploring the diverse topics which have resulted in the successful publishing of this book. They have passed on their knowledge of decades through this book. To expedite this challenging task, the publisher supported the team at every step. A small team of assistant editors was also appointed to further simplify the editing procedure and attain best results for the readers.

Apart from the editorial board, the designing team has also invested a significant amount of their time in understanding the subject and creating the most relevant covers. They scrutinized every image to scout for the most suitable representation of the subject and create an appropriate cover for the book.

The publishing team has been an ardent support to the editorial, designing and production team. Their endless efforts to recruit the best for this project, has resulted in the accomplishment of this book. They are a veteran in the field of academics and their pool of knowledge is as vast as their experience in printing. Their expertise and guidance has proved useful at every step. Their uncompromising quality standards have made this book an exceptional effort. Their encouragement from time to time has been an inspiration for everyone.

The publisher and the editorial board hope that this book will prove to be a valuable piece of knowledge for researchers, students, practitioners and scholars across the globe.

List of Contributors

Kristin M. Trippe, Stephen M. Griffith, Gary M. Banowetz and Gerald W. Whitaker
US Department of Agriculture Agricultural Research Service National Forage Seed and Production Research Center, Corvallis, OR 97331, USA

Jean C. Buzby and Jeanine T. Bentley
Economic Research Service, U.S. Department of Agriculture, 1400 Independence Ave., Mail Stop 1800, SW, Washington, DC 20250-1800, USA

Beth Padera
MobiSave, 712 5th Avenue, 14th Floor, New York, NY 10019, USA

Cara Ammon
Beacon Research Solutions, 4556 N. Beacon St. No. 3, Chicago, IL 60640, USA

Jennifer Campuzano
Nielsen Perishables Group Inc., 1700 West Irving Park Road, Suite 310, Chicago, IL 60613, USA

Sébastien Fournel
Department of Chemical and Biotechnological Engineering, Université de Sherbrooke, 2500 Université Boulevard, Sherbrooke QC J1K 2R1, Canada
Research and Development Institute for the Agri-Environment (IRDA), 2700 Einstein Street, Quebec City QC G1P 3W8, Canada

Joahnn H. Palacios and Stéphane Godbout
Research and Development Institute for the Agri-Environment (IRDA), 2700 Einstein Street, Quebec City QC G1P 3W8, Canada

Michèle Heitz
Department of Chemical and Biotechnological Engineering, Université de Sherbrooke, 2500 Université Boulevard, Sherbrooke QC J1K 2R1, Canada

Máximo Lorenzo
INTA, Estación Experimental Balcarce, C.C. 276, Balcarce 7620, Argentina

Silvia G. Assuero
Laboratorio de Fisiología Vegetal, Facultad de Ciencias Agrarias, Universidad Nacional de Mar del Plata, C.C. 276, Balcarce 7620, Argentina

Jorge A. Tognetti
Laboratorio de Fisiología Vegetal, Facultad de Ciencias Agrarias, Universidad Nacional de Mar del Plata, C.C. 276, Balcarce 7620, Argentina
Comisión de Investigaciones Científicas de la Provincia de Buenos Aires, La Plata 1900, Argentina

Athole H. Marshall, Matthew Lowe and Rosemary P. Collins
Institute of Biological, Environmental and Rural Sciences, Aberystwyth University, Gogerddan, Aberystwyth, Ceredigion SY233EE, UK

Hongxiang Zhang
Northeast Institute of Geography and Agroecology, Chinese Academy of Sciences, Changchun 130012, China

Yu Tian
Animal Science and Technology College, Jilin Agricultural University, Changchun 130118, China

Daowei Zhou
Northeast Institute of Geography and Agroecology, Chinese Academy of Sciences, Changchun 130012, China

Lisa Kitinoja
World Food Logistics Organization, 1500 King Street, Alexandria, VA 22314, USA

Diane M. Barrett
Department of Food Science and Technology, University of California, Davis, One Shields Ave, Davis, CA 95616, USA

Nora Tilly, Victoria Lenz-Wiedemann, Dirk Hoffmeister and Georg Bareth
ICASD-International Center for Agro-Informatics and Sustainable Development, Institute of Geography (GIS & Remote Sensing Group), University of Cologne, 50923 Cologne, Germany

Qiang Cao and Yuxin Miao
ICASD-International Center for Agro-Informatics and Sustainable Development, Department of Plant Nutrition, China Agricultural University, 100193 Beijing, China

W. Paul Williams and Gary L. Windham
United States Department of Agriculture, Agricultural Research Service, Corn Host Plant Resistance Research Unit, P.O. Box 9555, MS 39762, USA

Burtram C. Fielding
Molecular Biology and Virology Laboratory, Department of Medical BioSciences, Faculty of Natural Sciences, University of the Western Cape, Bellville 7535, South Africa

Cindy-Lee Knowles
Plant Pathology Laboratory, Department of Medical BioSciences, Faculty of Natural Sciences, University of the Western Cape, Bellville 7535, South Africa

Filicity A. Vries
Fruit, Vine and Wine Institute of the Agricultural Research Council, ARC Infruitec-Nietvoorbij, Private Bag X5026, Stellenbosch 7599, South Africa

Jeremy A. Klaasen
Plant Pathology Laboratory, Department of Medical BioSciences, Faculty of Natural Sciences, University of the Western Cape, Bellville 7535, South Africa

Jorge F. S. Ferreira, Monica V. Cornacchione, Xuan Liu and Donald L. Suarez
US Salinity Laboratory, 450 W. Big Springs Rd., Riverside, CA 92507, USA

Quentin Farmar-Bowers
17 The Grange, East Malvern, Victoria 3145, Australia

Chandran M. Ajila, Saurabh J. Sarma, Satinder K. Brar and Jose R. Valéro
INRS-ETE, Université du Québec, 490, Rue de la Couronne, QC G1K 9A9, Canada

Stephane Godbout and Michel Cote
Institut de recherche et de développement en agroenvironnement inc (IRDA), 2700 rue Einstein, QC G1P 3W8, Canada

Frederic Guay
Department of Animal Science and Center de Recherche en Biologie de la Reproduction, Laval University, Sainte-Foy, QC G1K 7P4, Canada

Mausam Verma
Institut de recherche et de développement en agroenvironnement inc (IRDA), 2700 rue Einstein, QC G1P 3W8, Canada
CO2 Solutions Inc., 2300, rue Jean-Perrin, QC G2C 1T9, Canada

Masahiko Hirata
Department of Animal and Grassland Sciences, Faculty of Agriculture, University of Miyazaki, Miyazaki 889-2192, Japan

Said S. Hegazy
Chair of Dates Industry and Technology, College of Food and Agricultural Sciences, King Saud University, Riyadh 11451, Saudi Arabia
Timber Trees Department, Horticulture Research institute, Agriculture Research Center, Giza, Cairo 12619, Egypt

Khaled Ahmed
Chair of Dates Industry and Technology, College of Food and Agricultural Sciences, King Saud University, Riyadh 11451, Saudi Arabia
Agricultural Engineering Institution, Giza, Cairo2450, Egypt